ECONOMIC STUDIES ON FOOD, AGRICULTURE, AND THE ENVIRONMENT

ECONOMIC STUDIES ON FOOD, AGRICULTURE, AND THE ENVIRONMENT

Edited by

Maurizio Canavari

Università di Bologna
Bólogna, Italy

Paolo Caggiati

Centro di Studio sulla Gestione dei Sistemi Agricoli e Territoriali
CNR
Bologna, Italy

and

K. William Easter

University of Minnesota
St. Paul, Minnesota

Kluwer Academic / Plenum Publishers
New York, Boston, Dordrecht, London, Moscow

Library of Congress Cataloging-in-Publication Data

Joint Minnesota-Padova Conference on "Food, Agriculture, and the Environment" (7th: 2000: Bologna, Italy)
Economic studies on food, agriculture, and the environment/edited by Maurizio Canavari, Paolo Caggiati, and K. William Easter.
p.   cm.
"Proceedings of the 7th Joint Minnesota–Padova Conference on "Food, Agriculture, and the Environment," held June 12–14, 2000, in Bologna, Italy—T.p. verso.
Includes bibliographical references and index.
ISBN 0-306-47242-2
1. Agriculture—Economic aspects—Congresses.   2. Agriculture—Economic aspects—Italy—Congresses.   2. Agriculture—Economic aspects—United States—Congresses.   4. Agriculture—Environmental aspects—Congresses.   5. Agriculture—Environmental aspects—Italy—Congresses.   6. Agriculture—Environmental aspects—United States—Congresses.   I. Canavari, Maurizio, 1964–   II. Caggiati, Paolo, 1947–   III. Easter, K. William.   IV. Title.

HD1405 .J64 2000
338.1—dc21

2002067804

Proceedings of the 7th Joint Minnesota–Padova Conference on "Food, Agriculture, and the Environment," held June 12–14, 2000, in Bologna, Italy

ISBN: 0-306-47242-2

© 2002 Kluwer Academic/Plenum Publishers, New York
233 Spring Street, New York, N.Y. 10013

10   9   8   7   6   5   4   3   2   1

A C.I.P. record for this book is available from the Library of Congress

DEDICATION: IN MEMORY OF
MAURIZIO GRILLENZONI AND FRANCO ALVISI

Memento of Maurizio Grillenzoni

On May 9th 1997 Professor Maurizio Grillenzoni, who held the Chair in Rural Appraisal in the Faculty of Agriculture in the University of Bologna and who also was Director of the Agricultural and Territorial Systems Management Research Center of the Italian National Research Council, passed away.

He became Assistant Professor at this Faculty while still very young, and in the 1970s was appointed Full Professor.

From among national and international scientific societies he belonged to, I would like to mention the National Advisory Committee on Agricultural Sciences of the Italian National Research Council; the Central Census Commission—where he presided over the Land Bureau; and the Territorial Appraisal and Territorial Economics Study Center, where his efforts to develop an interdisciplinary approach in the field began an innovative trend. His more than 250 publications touched upon a vast range of subjects. Among these were rural appraisal, economic planning, territorial economics, farm management, and farm accounting. He was often the coordinator of national research projects. His intellectual rigor and vast knowledge was an inspiration for the young researchers working under him.

He was particularly aware of environmental issues and encouraged his teaching assistants and researchers to further look into them.

He was among the promoters of the agreement of collaboration between the University of Minnesota and universities in Italy.

This conference is a outcome of his effort.

Professor Grillenzoni was an acute observer of the problems characterizing the changes in and the growth of our discipline. His scholarly status has contributed to promote both. Besides his unquestionable ability as a teacher and the dedication he put into various institutional roles he held, Professor Grillenzoni remains for all those who knew him a model worthy of imitation.

Paolo Caggiati

Memento of Franco Alvisi

Friends, family, and colleagues are still mourning Professor Franco Alvisi's death. He passed away two years ago and since then the Faculty of Agriculture of the University of Bologna has undergone profound changes. A lifetime of work for this institution on his part set the groundwork for these changes. We are indebted to him for this.

In 1955, after having completed a degree in Agricultural Sciences, Professor Alvisi joined the Faculty's teaching staff first as a volunteer Teaching Assistant and two years later as an Associate Professor. In 1963 he was appointed to a Professorship in Rural Appraisal and Farm Accounting, a chair with which his previous appointments had been affiliated.

His 350 publications cover areas such as the production and marketing of agricultural products, the internalization process of the agricultural market, farm accounting, farm management, territorial economics, and rural appraisal.

His style was concrete; his method, interdisciplinarian. The prestige derived from his scientific work earned him an important place in the academic community. Among the numerous prestigious positions he held, some were: from 1971 to 1990, Director of the Agricultural and Territorial Systems Management Research Center of the Italian National Research Council; from 1977 to 1994, Director of this Faculty's Institute of Rural Appraisal and Farm Accounting; from 1982 to 1988, President of the Course Degree in Agricultural Sciences Didactic Council; from 1982 to 1999, Member of the Administration Council of the University of Bologna. Among many scientific societies he belonged to were: the Italian Society of Agricultural Economists—first as member of the Presidential Council, then as President; the Italian Society of Agri-food Economics—first as founding member and then as member of its Presidential Council.

It is important on this occasion to state the facts concerning his academic career, but there are personal recollections we will not indulge here; Professor Alvisi would not have approved. Be it enough to say that he is still present not only in our thoughts, but in the work that we carry out every day.

Domenico Regazzi

PREFACE

This book contains a selection of the papers presented at the Joint Conference on Food, Agriculture, and the Environment, which was held in Bologna, Italy, on June 12-14, 2001. This was the seventh gathering of a biennal meeting born from a cooperation agreement between US and Italian academic and research institutions.

This round of the Conference was organized in the Faculty of Agriculture in Bologna by the Dept. of Agricultural Economics and Engineering (DEIAgra) and the CNR Land and Agri-System Management Research Centre (GeSTA-CNR) of Bologna. There were two main reasons for the choice of this location: first, the Conference was dedicated to Maurizio Grillenzoni and Franco Alvisi, two colleagues and friends who passed away in recent years, and who committed themselves and played an important role in developing the collaboration agreement and promoting the past Conferences; second, in the year 2000 the Faculty of Agriculture in Bologna celebrated its first centennial, and this Conference was part of a wide set of events organized to highlight the relevant role of the Faculty in the research activity, both at an Italian and international level.

The Conference papers were articulated both in plenary and concurrent sessions, dealing with key topics for agricultural economists. A structure similar to the Conference was adopted for grouping the papers into the four sections contained in this book:

- food, nutrition, and quality, focusing i.e. on aspects related to concentration and vertical coordination in the food industry, or food safety and quality issues;
- land and resource assessment, taking into consideration the water resource management and the recreational use of land and many other issues;
- agriculture and rural development, discussing several issues related to the chances in development of rural areas;
- environment and markets, coping with problems involving the effects of agro-environmental policies and market mechanisms to improve or protect the environment.

Thanks to the efforts of all the contributors and participants, every session was intense and very productive, so that the end result was a successful Conference. This book gives the readers an opportunity to benefit from the thorough analysis provided in the papers and to gain new insights concerning alternative approaches to dealing with important policy issues.

These and several other questions may represent the challenge that we should take up for the next meetings, trying to seize the opportunity offered by the chance to meet and collaborate with scholars and researchers coming from different countries.

We hope that publishing this book shall stimulate a wider participation and contribute to the improvement of cooperation in our disciplines.

The editors
Maurizio Canavari
K. William Easter
Paolo Caggiati

ACKNOWLEDGEMENTS

Many people and institutions were involved in organizing and funding the Conference.

A grateful thanks goes to the Committee for Celebrations of the Centennial of the Bologna Agricultural Faculty. The President of the Committee, Giorgio Stupazzoni, together with the Cassa di Risparmio in Bologna Foundation, assured the moral and financial support that made this event possible.

Significant financial and organizational support also came from the Banca Popolare dell'Emilia-Romagna, the Commune of Mirandola, the Raganella Association, and the Focherini Farm, which helped in financing and offering interesting sites to visit during the Conference's closing day.

We also wish to express our appreciation and heartfelt friendship to Danilo Agostini, whose commitment in promoting the cooperation, and whose efforts to stimulate and encourage us during the organization of the Conference was of great help.

Invaluable assistance before, during and after the conference was offered by many collaborators and staff members of the Dept. of Agricultural Economics and Engineering and of the CNR Land and Agri-System Management Research Centre (GeSTA-CNR). We would particularly like to mention Alessandra Castellini, Sabrina Di Pasquale, Maria Rosa Lelli, Lamberto Marchesi, Andrea Pancaldi, Daniele Levorato, Eva Righi, and Angelo Tummarello.

Davide Viaggi, Mayrah I. Rodriguez, Alessandro Ragazzoni and Mara Thiene were also involved in the organization of the Conference, and their contribute was effective and highly appreciated.

Regarding the editing of this book, a special thanks goes to Mary Beth Lake, who carried out a crucial, precious and fatiguing task, helping us to thoroughly review and revise many papers; this book could not have been printed without her help. Many thanks also to Giuseppe Nocella for helping in the first review of the papers. The collaboration of Alessandra Castellini was also essential for transferring ideas to paper, which we all know is not a trivial issue.

Obviously, our thanks are also directed to all the contributors, and especially to the publisher's editor Susan Safren, whose patience and collaboration helped us to carry out our editing job.

CONTENTS

FOOD, NUTRITION & QUALITY

                    LAND & RESOURCE ASSESSMENT

6. WATER SCARCITY: INSTITUTIONAL CHANGE, WATER
       MARKETS, AND PRIVATIZATION... 91
Cesare Dosi and K. William Easter

    1. Introduction   91
    2. Water Scarcity and Social Capital   92
    3. Improving Water Management through Institutional Changes   93
        3.1. Institutional changes: objectives and constraints   94
        3.2. Water markets   95
        3.3. Privatization of water services   96
        3.4. Regulatory issues   99
        3.5. The intersection of markets and privatization   101
    4. US and EU: Experience and On-going Developments   102
        4.1. The U.S. experience with privatization and markets in the public water
              sector   102
        4.2. The European Union   107
    5. Final Remarks   112
    6. References   114

7. EVALUATION OF THE RECREATIONAL USES OF RURAL LAND:
       A CASE STUDY... 117
Guido Maria Bazzani, Davide Viaggi, and Giacomo Zanni

    1. Introduction   117
    2. Demand, Supply, and Policies for Recreational Services Produced by
        Agriculture   118
    3. Methodology   122
    4. Results of the Survey   123
    5. Relationships with Agro-environmental Policies   126
    6. Discussion   129
    7. References   130

8. THE VALUE OF LICENSES FOR RECREATIONAL USE OF
       NATURAL RESOURCES.. 133
Michele Moretto and Paolo Rosato

    1. Introduction   134
    2. The Model   135
        2.1. The value of the license   137
        2.2. The value of the option to purchase the license   138
        2.3. Optimal purchase time   139
    3. The Optimal Duration of the License   141
        3.1. The trade-off between cost and duration of the license   144

FOOD, NUTRITION & QUALITY

DIFFERING U.S. AND EUROPEAN PERSPECTIVES ON GMOs
Political, economic and cultural issues

C. Ford Runge, Gian Luca Bagnara, and Lee Ann Jackson [*]

SUMMARY

Genetically Modified Organisms (GMOs) burst onto the scene in 1996 with the rapid commercial introduction in the United States of genetically engineered corn (maize), cotton, and soybeans. By 1998, more than 500 genetically modified plant varieties were available in the United States, accounting for 28 percent of the areas (2.57 million hectares) planted to maize, soybeans and cotton. Argentina and Canada had each planted an additional 100,000 hectares to GMOs and other countries (South Africa, Spain, France, Mexico, China, Australia, Brazil) had planted less than 100,000 hectares each (James, 1999). Perhaps more significant to consumers, these crops rapidly entered the supply chain for processed foods using corn, soybean, or cotton seed oils, with some estimating that between 70-100 percent of processed foods now contain GMOs (*The Economist*, 1999).

In response to this rapid expansion, countries have developed diverse regulatory approaches to the production, marketing and development of these products. The US has been a particularly strong advocate for the biotechnology industry in terms of intellectual property protection and limited governmental regulatory oversight in the production and marketing of these products. In contrast, Europe has approached biotechnology with skepticism and has been slow to grant regulatory approval for new products. These differences, which grow out of variations in political, economic and cultural characteristics, have raised challenges for trade, agricultural and consumer policy.

The purpose of this paper is to examine the underlying factors that have contributed to divergent US and European views on GMOs. After a brief overview of the historical

[*] C. Ford Runge, distinguished McKnight University Professor of Applied Economics and Law, University of Minnesota. Gian Luca Bagnara, Certified Management Consultant. Lee Ann Jackson, Graduate Research Assistant, Departments of Applied Economics and Law, University of Minnesota. Our thanks to John Kasson and Richard Leppert for comments and suggestions.

Economic Studies on Food, Agriculture, and the Environment
Edited by Canavari *et al.*, Kluwer Academic/Plenum Publishers, 2002

3

and cultural differences dividing US and EU attitudes, we focus primarily on consumer labeling strategies and international trade.

1. HISTORICAL AND CULTURAL DIFFERENCES

To our knowledge, no comprehensive assessment has been undertaken of the differences in US and European technology policy in food and agriculture (for an impressionistic assessment, see Zechendorf, 1998). However, it is clear that since the Treaty of Rome, the EU has taken a different approach to food policy and regulation. Here we suggest several differences in the US and European approaches, some historical and some contemporary.

The deep roots of European and U.S. attitudes toward agricultural science and technology relate in part to C.P. Snow's distinction between humanistic and scientific culture (Snow, 1964). In the EU, food, seen through a humanistic and aesthetic lens, is part of what differentiates what is Germanic from Latin, and French from Italian or Iberian. Some of these distinctions fall down on close inspection, with the origins of French *haute cuisine* traceable to the influence of Italian, especially Florentine, court-cooking, or the large number of "European" foods which actually have origins in colonial expansion in the 16th, 17th and 18th centuries (Field, 1982). The potato and tomato, for example, were New World foods, brought to Europe without fanfare (and in the face of considerable suspicion) early in the colonial period. Many classic European seasonings such as nutmeg and pepper were Oriental in origin. Even so, these and other foods have created a rich tapestry of national and regional European dishes that clearly demarcate European food from foods in other parts of the world, notably the US.

In the United States, while some regional variations in cookery exist, most food bears the stamp of European (and later Hispanic and Far Eastern) ethnicity. Examples include the French Creole cooking of Louisiana, the seafood stews of New England, the piquant foods of Texas and New Mexico, and the Germanic and Scandinavian meat and potatoes of the Upper Midwest. In general, Americans' perception of food is more utilitarian and less aesthetic than Europeans'. In contrast to a humanistic or aesthetic bias, Americans' think of food in nutritional terms linked to science and sustenance, with quantity often preferred to quality.

A second difference in European and U.S. attitudes toward food relates to science itself. From the earliest enthusiasm of US Founding Fathers such as Benjamin Franklin (influenced profoundly by his exposure to British and French enlightenment scientists), Americans have thought of themselves as both a democratic and scientific culture, with a special mission for human development and improvement. This was especially evident in the mechanical revolutions of the 19th century, when doubters and agnostics such as Nathaniel Hawthorne (see his "The Celestial Railroad") were dominated by celebrants and technological optimists. Noble (1999) has noted that American technological optimism has deep roots in a Christian perception of a new and better world to follow the millennial reckoning predicted in the Book of Revelations. As Milton stated, nature "would surrender to man as its appointed governor, and his rule would extend from command of the earth and seas to dominion over the stars" (quoted in Noble, p. 48). Such technological optimism was supported in America by the Freemasons, who became technological evangelists. Benjamin Franklin, a lifelong Freemason active in the movement in America,

England and France, and grand master of *La Loge des Neuf Soeurs*, was among the main evangelicals (Hans, 1953). His "Proposals Relating to the Education of Youth in Pennsylvania" led to the establishment of the Academy (later University) of Pennsylvania, and his precepts were followed by other Masons including De Witt Clinton, father of the Erie Canal, Stephen Van Rensselaer, founder of the first engineering school in America, and industrial inventor Robert Fulton (Noble, p. 79). All but a few signers of the Declaration of Independence were Freemasons. At the 1876 Exposition in Philadelphia, huge crowds gathered to stare in amazement and rapture at the giant Corliss engine, which had a presence and power that dominated descriptions of the event (Kasson, 1976).

In Europe, by contrast, the technological revolution of the 19th century was met by an aesthetic and cultural reaction to science and technology. The Arts and Crafts movement led by Morris and Ruskin in England, the Jugendstil in Germany, and the Romantic poetry of France all signaled a revolt against mechanical and technological advance. In the hiatus before the First World War, the application of science and technology to armaments set the stage for the massive destruction of European cities and landscapes and the annihilation of millions of soldiers and civilians, confirming the nightmarish visions of technological pessimists.

America, by contrast, remained largely untouched by the First World War, and retained a sense of the possibilities of technology and science, particularly related to improvements in agriculture and food production. The number of tractors on American farms jumped from 1000 in 1910 to 246,000 in 1920 (Cochrane, 1979). Herbert Hoover was elected President of the United States in 1928 largely on the strength of his record in charge of the U.S. Food Administration during the First World War. Luther Burbank and Henry Wallace (later Vice President) became folk heroes for their early work on hybrid horticultural and corn crops, respectively.

The application by Nazi Germany of scientific principles drawn from American Frederick Winslow Taylor's 1911 book, *Principles of Scientific Management* (the basis of Henry Ford's assembly line, thus "Fordism") led to the even more wholesale destruction of modern Europe in the Second World War. Taylor and his followers, using "time and motion" studies, inspired an efficiency craze in American culture. As Surowiecki (2000) notes:

> Agriculture, schools, churches, homes, government - almost every area of life became a candidate for Taylorization. Progressives like Walter Hippmann argued that scientific management would help replace "drift" with "mastery". In 1914, sixty-nine thousand people attended an efficiency exposition in New York City. *Ladies Home Journal* ran a series on Taylorism in the household…Soon, visionaries were telling us that Taylorist engineers held 'the material welfare of all the advanced industrial peoples' in their hand.

Adolf Hitler, an admirer of Henry Ford's, and of Taylorism, took these ideas seriously. In the Nazi case, the assembly line was applied to the extermination and cremation of Jews, Gypsies and the handicapped and disabled.

In the wake of World War II, Europe also confronted a food crisis. Unwilling to face the vagaries of the world market, the European Coal and Steel Community began to construct what would ultimately become, under the Treaty of Rome, a frankly protectionist policy which strove (all too successfully, it turned out) to guarantee self-sufficiency in basic foodstuffs. Scientific and technological progress in agriculture, much of it borrowed or imported from the U.S., combined with guaranteed high prices to push European production possibilities upward dramatically.

Still, Europeans remained deeply ambivalent toward science and technology, and fiercely devoted to the aesthetic component of food and rural culture. Given the terrible history of Europe in the 20th century, it is unsurprising that when genetic engineering burst onto the agricultural scene in the 1990s, it should trigger European misgivings. The experience with Mad Cow Disease in England during the early 1990s also deepened European suspicion of government regulation. Europe's intellectual and media elites distained the corporate capitalistic culture of America, and feared for European cultural sovereignty in the face of the juggernaut of America technology, now in the form of GMOs.

For all of the reasons noted above, the notion that regulatory differences over GMOs should be resolved by "sound science," the position aggressively advocated by U.S. trade negotiators, sounded far less convincing to European ears. In part as a function of their own sense of cultural superiority, Europeans also remained surprisingly vague over the facts of genetic engineering, which sounded uncomfortably like eugenics. As Regazzi, Senauer and Kinsey (2000) note in their contribution to this conference, nearly half (44 percent) of those surveyed in Germany and Austria thought that "ordinary tomatoes do not contain genes while genetically modified ones do," compared with 10 percent of those surveyed in the US. Yet the unattractive characteristics that European consumers attributed to GMOs clearly derived from the broad cultural and historical factors noted above.

Ultimately, the stance of the EU respecting GMOs reflects a non-scientific, even anti-scientific, sensibility. As Konrad von Moltke (2000) writes in a recent paper detailing the history and basis of the "precautionary principle":

> Scientific research is not designed to provide answers to questions of policy. It is a method to validate or invalidate hypothesis about natural phenomena, and only hypothesis susceptible to validation can form useful questions for scientific research. The questions policy makers typically ask are rarely appropriate for science. (p. 3).

Placed in this context, it is easier to appreciate the differences over regulatory policies toward GMOs. The European traditions of an aesthetic appreciation for food, a skepticism toward science wrought of destruction by military and Fascist technology, and a protection of domestic markets, stand in marked contrast to scientific and utilitarian attitudes toward food, a scientific optimism unscarred by war, and general support for free markets and trade in the US.

In biotechnology, the historical American evangelical belief in science as savior is also evident, and the connection to eugenics is rather disturbingly explicit. Molecular geneticist and biotechnology pioneer Robert Sinsheimer in 1969 declared that biotechnology would create "a new eugenics," which would rise above the "old eugenics," which "was limited to a numerical enhancement of the best of our existing gene pool...The new eugenics would permit in principle the conversion of all the unfit to the highest genetic level" (Sinsheimer, 1969). More than 25 years later, as an enthusiast for the Human Genome Project, Sinsheimer described its significance:

> From the time of the invention of writing, men have sought for a hidden tablet or papyrus on which would be inscribed the reason for our existence in this world...How poetic that we now find the key inscribed in the nucleus of every cell of our body. Here in our ge-

nome is written in DNA letters the history, the evolution of our species...when Galileo discovered that he could describe the motions of objects with simple mathematical formulas, he felt that he had discovered the language in which God created the universe. Today we might say that we have discovered the language in which God created life...After three billion years, in our time we have come to this understanding, and all the future will be different." (Sinsheimer, 1994, quoted in Noble pp. 189-190).

With this backdrop, we turn now to consider some specific issues of labeling policy, harmonization and trade.

2. GMO LABELING CONTROVERSY IN U.S. – EUROPEAN TRADE

Because the US and the EU have developed regulatory schemes for GMO's based on their unique national perspectives and divergent traditions, it is unsurprising that these have led in turn to international conflicts over GMO policy, particularly in international trade. Recently the US and the EU have clashed over the validity of national labeling requirements on GMO agricultural products. The most controversial of proposed regulations would mandate segregation of GMOs from traditional crops, requiring labels on all seed products containing greater than a specified amount of GMO product. Although some consumer groups have criticized the lack of national GMO regulatory policies in the US, the US government continues to maintain that GMO corn is "substantially equivalent" to non-GMO corn, and should be traded freely along with it. US producers have contended that labeling requirements would force industry to segregate crops at the field level and require monitoring of crop content along the chain of production from farmers to processors, costing billions of dollars. Hence many experts in the industry interpret the EU regulatory stance as a form of non-tariff trade barrier (NTB) (see discussion below). The EU maintains that these cost estimates are exaggerated, and that reasonable concerns about the safety of these organisms justify mandatory labeling of all agricultural products containing GMOs.

Although in general US citizens are more favorably inclined towards GMOs than are EU citizens, a broad array of advocacy groups in both countries support labeling policies. Due to the rapid introduction of GMOs in the food supply, consumer advocates and a wide range of environmental and food safety groups have mounted an active labeling campaign. In response to these concerns food distribution companies around the world have begun calling for segregation of GMOs from traditional crops and indicated their willingness to avoid GMO products in their food marketing efforts. In May 1999, major food chains in England (Sainsbury, Tesco, Marks and Spencer, Burger King, McDonalds) announced their intention to avoid GMO ingredients. In Spain, Pryca, which had been the largest importer of soy-based GMOs and a producer of GMO corn, announced it would no longer use GMO ingredients in its branded products. The French food company Carrefour instituted a similar policy. In Switzerland, Nestlé announced a temporary halt on GMO product use, and Russia announced that after July 1, 1999, any imported GMO product would require testing and licensing (Kinsey, 1999). Most recently, US food companies including Archer Daniels Midland and ConAgra have signaled their farmer suppliers of the possible future need to segregate GMO crops from conventional varieties, indicating the possibility of segregated product streams as well.

Numerous parties to the debate over GMOs have proposed labeling food products containing genetically modified material and/or segregating seeds or GMO products in supply streams allowing "identity preserved" products. Labeling has particular appeal as a market based alternative to those who believe that consumers, once informed of the presence of GMOs in food or seed, will choose to purchase (or not to purchase) them based on this information. Labels might, in fact, result wholly from voluntary decisions by firms to offer such information to consumers. However, for a variety of reasons, notably those of uniformity and coordination across both private firms and national regulatory regimes, it is probable that some international standards or norms will be necessary (Runge and Jackson, 2000).

A labeling strategy could either highlight GMO content or spotlight those products that do not contain GMOs. The first type of labeling, which we will call positive labeling, might involve the statement: "This product may contain GMOs." Given the extent to which GMOs have already entered the food and fiber chain, such a label would convey relatively little information. In contrast to a cigarette health warning labels, a simple indication of possible GMO content may imply risk but does not accurately reflect health consequences of consuming GMO products. Moreover, it is unclear from such a label how much GMO content is implied or whether the GMOs in question are specifically identified (as, for example, Bt corn or Round-Up Ready© soybeans).

Nonetheless, some advocates of positive labels assert that they would help steer consumers (or, in the case of seed, farmers) away from GMO products. Since this would now include the vast majority of processed food products (and a large share of the seed market) their motivation for positive labels may lie in a desire to reduce consumption of and trade in these products, and thus the revenues of their manufacturers, large or small. In addition, it is possible, that the threat of such a label, or the label itself, would lead investments in research and marketing into GMOs and their products to wither and eventually to die. In short, positive labels could impose costs on the agricultural industry without providing equivalent compensating benefits to consumers.

Before rejecting positive labels as purely destructive, however, we should note that positive labels may have certain advantages in the future that current efforts to limit trade in GMOs obscure. We may distinguish current, market-limiting labels "as positive labels with negative intent," from positive labels that may indicate an attractive nutritional or pharmaceutical property of a particular GMO food or seed, such as the vitamin enhanced rice varieties now being promoted by the International Rice Research Institute (IRRI) (Ye et al., 2000). In these cases a "positive label with positive intent" would convey these attractive properties, potentially enlarging the market for the seed or food product.

In contrast, a negative label would read: "This product (or seed) contains no GMOs." Such a label has certain requirements. One requirement is to define "no." "No" would necessarily imply a minimum threshold approaching zero. Once agreement on such a threshold is reached, it would need to apply across firms and national boundaries. Another requirement would be to carefully define "GMO," so that only "transgenics" in which some form of gene "splicing" had occurred would be included. Assuming such agreement could be reached, the effect of such a label would be to create niche markets for those choosing to purchase, process, segregate and sell no-GMO food or seed products. Each action will likely entail additional costs or effort at some point in the food or seed supply chain. In the case of food products, purchasing no-GMO ingredients will entail monitoring inputs closely, and requiring farmers and suppliers to conform to no-GMO practices.

Processing would need to be in separate lots, or even separate facilities, to guarantee against co-mingling. In the case of seed, similar restrictions would apply to growing, processing, segregating and selling no-GMO varieties. Assuming these extra costs are not prohibitive, firms will bear them if the market for no-GMO products or seed is perceived to be large enough, and the price elasticity of demand adequate to support product or seed prices sufficient to cover the variable costs of production.

This type of labeling strategy would have advantages. There is already evidence that such markets are perceived to be worth the effort to segregate product and processing methods. Apart from the examples of large food companies noted above, several less visible instances suggest the emergence of market opportunity for products carrying a no-GMO identity. Natural Products Inc., a Grinnell, Iowa company that processes unmodified soybeans (sold to Ben and Jerry's Homemade Inc. for use in its ice cream products), expects sales to triple in 2000 to about $10 million. The Hain Food Group of Uniondale, New York, is labeling its organic snacks as nonbiotech. Hain's method, for the moment, is to switch from frying the snacks in corn oil to safflower oil, which has not yet been genetically engineered (Kilman, 1999). Among the most active producers of no-GMO products are American Growers Foods, of Embarrass, Minnesota. The company promotes its foods as chemical and GMO free. It independently tests and certifies a variety of breakfast cereals, snack foods, and baked goods as organically grown and containing no GMOs (American Growers Foods, 1998).

In summary, Governments and the private sector will likely need to respond to calls for the labeling of foods and seeds. A system is therefore needed that could effectively use and develop GM technologies while allowing consumers to reject them if they wish. A Canadian survey of 8 countries found significant variation in consumer attitudes. For example, although 68 percent of all respondents said they would be less likely to buy groceries labeled as GM products, national responses ranged from a low of 57 percent in the United States to 82 percent in Germany. But the combination of consumer choice with freer trade would remove the chance that the GM issue could be exploited for protectionist purposes; consumers could choose between GM food and organic products without resorting to trade discrimination. As Alexander Haslberger, a leading European expert on biotechnology, noted in a recent contribution to *Science*, the significant public opposition to GM food will require that the industry adopt honest and appropriate labeling if it wants to avoid consumer resistance. One possible multilateral response could be under the auspices of the new biosafety protocol — or the U.N. Food and Agricultural Organization's Codex Alimentarius — to harmonize differing national standards (Runge and Senauer, 2000).

The Montreal talks last January, when more than 130 countries agreed on the Biosafety Protocol to the Convention on Biological Diversity, were a good start. The protocol discusses the environmental risks and benefits in biotechnology and creates a framework to protect biodiversity in developing countries. But many unanswered questions remain. Most prominent is whether the new protocol will allow a protectionist loophole for a "precautionary principle" that bars GM-food trade even if scientific evidence of harm is insufficient. Another central issue is the balance between trade restrictions justified on environmental or health grounds and the larger obligations of nations to trade without discrimination under the WTO.

3.   TRADE AND HARMONIZATION

As described in the previous section, labeling strategies include various policy options that would provide consumers with information about GMO content in their foods. Each of these options would create costs for agricultural producers and processors, while some of them might allow producers and processors to capture a price premium for segregated products. If these were the only costs associated with labeling policies, national policy makers could simply weigh the costs and benefits to their national constituents and choose an appropriate strategy. However, in this case, the US and the EU are choosing their labeling policies within a system of international trading relationships. Hence the potential costs associated with national policy choices spill over national borders.

Labeling policies have become a contentious policy issue between Europe and the US because labeling will have potentially large trade distorting effects. If labeling strategies implicitly define content standards for products containing GMOs then labels may act as non-tariff barriers, inhibiting US products from entering the European market. However, if national labeling strategies converged towards similar standards, then the non-tariff barrier effects would be minimal. Policy harmonization, in short, would help to ensure the least distortion to trade, but will be difficult to achieve (Caswell, 2000).

The US-EU trading relationship is defined within the context of rules and commitments established by the World Trade Organization (WTO). The WTO establishes reciprocal relationships among countries so that they are obligated to follow jointly accepted institutional norms for justifying their non-tariff restrictions. While in general the WTO advocates the removal of trade barriers to achieve broad societal benefits of free trade, the institution also acknowledges through agreements such as the agreement on Sanitary and Phytosanitary Standards (SPS) and the agreement on Technical Barriers to Trade (TBT) that nations have the right to establish national regulatory systems that insure a safe food supply and protect the domestic environment from pests and diseases. Nevertheless, in order for mandated crop segregation and labeling policies to be justified as a health and safety measures, the WTO requires that these policies be based on generally accepted scientific criteria.

The Agreement on the Application of Sanitary and Phytosanitary Measures and the Agreement on Technical Barriers to Trade (TBT) state that national regulations that act as non-tariff barriers are acceptable if they conform with formally recognized international standards, guidelines and recommendations, including the Codex Alimentarius.[1] These recommendations are used as benchmarks against which national food measures and regulations are evaluated within the legal parameters of the SPS and TBT Agreements. Thus, they contribute to the facilitation of international trade and resolution of trade disputes in international law.

The Codex Alimentarius provides one mechanism by which to judge the justifiability of national food safety standards. However, the concern over GMO crops and their introduction into the environment is also related to the potential environmental impacts of these crops. Because the long-term environmental consequences of these genetically

---

[1]   The Codex Alimentarius was established in May 1963 when the Sixteenth World Health Assembly approved the establishment of the Joint FAO/WHO Food Standards Programme and adopted the statutes of the Codex Alimentarius Commission. The Codex Alimentarius system is based upon the ideal that countries should come together to set mutually agreed upon food standards and to ensure the global implementation of and compliance with these standards.

modified products are uncertain, scientific consensus over the risks associated with their production has not emerged. Scientists continue to debate these issues and the controversy surrounding these regulations remains highly politicized. As Roberts (1998) notes in an evaluation of the Uruguay Round Sanitary and Phytosanitary Standards (SPS):

> Regulations rationalized on technical grounds seem to lack firm scientific foundations and, at least from the perspective of exporting countries, seem to be imposed primarily to thwart the commercial opportunities created by other trade liberalization policies.

Unfortunately, no simple mechanism exists for uncovering the underlying motives for policy choices.

The United States and Europe differ on their regulatory approach to managing environmental risk. European regulation, particularly in the environmental areas, is influenced by the precautionary principle. The precautionary principle states that in areas where science is limited and outcomes are unpredictable, regulatory authorities are justified in taking action to avoid possible negative outcomes. In relation to the case of GMO labeling, the EU policy stance is that GMOs are new goods, not extensions of their natural counterparts, thus policy makers have the obligation of not approving products for release until they are shown not to pose a danger to human or environmental health. The US, on the other hand, treats GMO products as extensions of existing products and thus these products simply must satisfy the same safety requirements as their natural counterparts. Thus, the US does not agree with the application of the precautionary principle in this area, and perceives the EU policy makers as succumbing to interest group pressure (Perdikis, 2000).

Adding to the confusion about potential harmonization of labeling strategies, some experts maintain that the labeling controversy does not fall under the auspices of the SPS or TBT agreements. Rather, they argue, it is based on a fundamental right of consumers to know the content of their food. Current WTO agreements have not addressed the possibility that consumer interest may represent an alternative, and equally valid, justification for regulations that have trade inhibiting impacts.

Harmonization of labeling strategies would facilitate trade and contractual relationships by ensuring that importers and exporters face similar economic conditions. However harmonization could also impede trade by requiring compliance at levels that impose differential costs and burdens on importers or exporters. These costs and benefits are difficult to determine without access to specific producer cost information. Therefore, even relatively unbiased national policy makers may ultimately be swayed by the political pressure within their countries for or against labeling.

Disaggregating the effects of labeling requirements on national level costs and benefits suggests a credible political explanation for the current controversy over legitimacy of labeling. Consider a compromise to harmonize around a labeling strategy, requiring that Europe drop its standards for segregating GMOs, but also requiring the U.S. to raise its labeling requirements. In Europe, producer surplus initially falls as competitive U.S. exports of GMO oilseeds and feedgrains enter the European market, less impeded by EU standards. However, European producers gain access and the ability to produce GMOs themselves, quickly regaining a competitive edge. Consumer surplus increases with less expensive food, but may fall if consumers fear GMOs as a form of health or environmental risk. Hence the EU stance on the proposed labeling harmonization depends on the perceived competitiveness effects on producers and the perceived risks to consumers. In the U.S., raised labeling requirements may raise costs and reduce producer surplus, but these costs are offset by expanded market access to the EU. Consumers gain if the labeling requirements do not appreciably affect food costs and increase the perception of food safety.

The future of the GMO labeling controversy is unclear. Within the context of the institutional obligations defined by the WTO, the US and EU may be willing to deviate from their optimal divergent strategies. However, until the WTO formally addresses this issue the definition of acceptable environmental and health risks and the consumer right-to-know issue will challenge existing WTO structures and will continue to act as stumbling blocks to US-EU labeling policy harmonization.

4. CONCLUSION

The EU and the US would benefit from a constructive dialogue over labeling policies and harmonization. If they continue to disagree and choose widely divergent strategies, the impacts on international agricultural trade, as well as on the biotechnology industry, will be immediate and large. In addition, a continued impasse in this area threatens to stall the development of new biotechnology products that offer significant global health benefits in the future. Although the WTO has not explicitly stated how it would address a trade dispute over labeling as a non-tariff barrier, such a dispute would certainly be costly for both countries in terms of resources and time. A more productive approach would be for the two to begin a dialogue about labeling options, including negative labeling, in order to develop a mutual understanding of the potential international benefits and costs associated with divergent and harmonized policies. Although such a constructive dialogue may be difficult to achieve in the context of widely divergent views concerning GMOs, without this type of dialogue potential global social benefits of harmonization will be forfeited. However, such a discussion must proceed with due recognition of the cultural differences in attitudes toward food and its role in national life and identity, and an appreciation for the evangelical zeal with which American science and industry approaches technological change.

5. REFERENCES

Bagwell K. – Staiger R.W. (1999), An Economic Theory of the GATT, *American Economic Review*, **89**(1): 215-248.
Caswell J.A. (2000), An Evaluation of Risk Analysis as Applied to Agricultural Biotechnology (with a case study of GMO Labeling), *Agribusiness*, **16**(1): 115-123.
Cochrane W.W. (1979), *The Development of American Agriculture*, University of Minnesota Press, Minneapolis, MN.
Economist (May 1999), *Sticky Labels*, 1: 75.
Field M. (1982), *All Manner of Food*, Ecco Press, New York.
Hans N. (1953), *UNESCO in the 18th Century: La Loge des Neuf Soeurs and Its Venerable Master Benjamin Franklin*, Proceedings of the American Philosophical Society, 9(5).
Jackson J.H. (1997), *The World Trading System*, The MIT Press, Cambridge, MA.
James C. (1999), *Global Review of Commercialized Transgenic Crops*.

Kasson J. (1976), *Civilizing the Machine: Technology and Republican Values in America 1779-1900*, Viking Press, New York.

Kilman S. (October 7, 1999), Food Fright: Biotech Scare Sweeps Europe, and Companies Wonder if U.S. is Next, *Wall Street Journal*, A1.

Kinsey J.D. (1999), Genetically Modified Food and Fiber: A Speedy Penetration or a False Start?, *Cereal Foods World*, **44**(7): 487-489.

Maggi G. (1999), The Role of Multilateral Institutions in International Trade Cooperation, *American Economic Review*, **89**(1): 190-214.

Noble D.F. (1999), *The Religion of Technology: The Divinity of Man and the Spirit of Invention*, Penguin Books, Harmondsworth, Middlesex, England.

Perdikis N. (2000), A Conflict of Legitimate Concerns or Pandering to Vested Interests?, *The Estey Centre Journal of International Law and Trade Policy*, **1**(1): 51-65.

Regazzi D. – Senauer B. – Kinsey J.D. (2000), *Food Quality and Safety: A Comparison of Europe and the U.S.*, Paper prepared for the 7th Joint Minnesota -Italy Research Conference, Bologna.

Roberts D. (1998), *Implementation of the WTO Agreement on the Application of Sanitary and Phytosanitary Measures: The First Two Years*, International Agricultural Trade Research Consortium Working Paper, #98-4.

Runge C.F. – Jackson L. (2000), Labelling, Trade and Genetically Modified Organisms: A Proposed Solution, *Journal of World Trade*, **34**(1): 111-122.

Runge C.F. – Senauer B. (2000), *A Removable Feast*, Foreign Affairs, pp. 39-51.

Sinsheimer R. (1969), The Prospect of Designed Genetic Change, *Engineering and Science*, **32**: 8-13.

Sinsheimer R. (1994), *The Strands of Life*, University of California Press, Berkeley.

Snow C.P. (1964), *Two Cultures*, Cambridge University Press, Cambridge.

Sugden R. (1984), Reciprocity: The Supply of Public Goods through Voluntary Contributions, *Economic Journal*, **94**: 772-787.

Surowiecki J. (2000), *The Financial Page: The Visionaries of the New Economy Dream On*, New Yorker, May 29: 50.

Taylor F.W. (1911), *The Principles of Scientific Management*, Harper, New York.

von Moltke K. (2000), *The Precautionary Principle, Risk Assessment, and the World Trade Organization*, International Institute for Sustainable Development, Manuscript.

Ye X. et al. (2000), Engineering the Provitamin A (ß-Carotene) Biosynthetic Pathway into (Carotenoid-Free) Rice Endosperm, *Science*, **287**: 303-305.

Zechendorf B. (1998), *Agricultural Biotechnology: Why do Europeans Have Difficulty Accepting It?*, Ag Bio Forum, 1(1), http://www.agbioforum.org/vol1no1/zechen.html.

THE ITALIAN LEGISLATION ON INTELLECTUAL PROPERTY RIGHTS RELATED WITH GMOs

Eleonora Sirsi [*]

SUMMARY

The European Union Directive 44/98 on the legal protection of biotechnological inventions establishes the patentability of products consisting of or containing biological material, and the process by means of which biological materials are produced, processed or used. The member states of the European Union (EU) are required to implement this directive in national law by 30 July 2000.

1. PREMISE

In both Italy and Europe, developments in the technology of living organisms have posed a problem with respect to the adaptation of the existing regulations. These problems are starting to be solved following the issuing of the European Union Directive 44/98.

Before describing the main points of this regulatory intervention, at least two aspects are worth mentioning. First, the long and difficult ratification stage of this Directive, lasting ten years and costing numerous re-drafts and repeated interventions by the various EU bodies, is noteworthy. This reflects the doubts of public opinion and resistance to the development and use of biotechnology, especially in primary sector. There are signs of this debate in the text of the Directive in the section on the premises of the considerations, where some ethical doubts are reported. These doubts arise as a result of the potential impact of the inventions in this sector because of their intrinsic relation with living matter. There is specific reference to some impacts on man and animals and to the problem of respecting the rights of weak subjects related to farmers and countries holding germ plasm, usually developing countries. The references appear in the text of the articles where the exclusions of patentability in relation to the contravention of public interest and decency are stressed, with a listing of typical exclusion cases and a limit to the extension

---

[*] Faculty of Law, University of Pisa.

Economic Studies on Food, Agriculture, and the Environment
Edited by Canavari *et al.*, Kluwer Academic/Plenum Publishers, 2002

of the patent with reference to the case of a farmer (the so-called "farmer's privilege"). Finally, an important role has been assigned to the European Group on Ethics in Science and New Technology (EGE) for evaluating the ethical aspects with specific reference to patent materials (cons. 44, art.7).

The other technical aspect that deserves mention refers to the nature of the Directive, which is aimed at harmonising national regulations. This requires acts of implementation in the regulations of each of the nations that form the European Union. By means of these acts, the solutions defined by the Directive will become part of the regulations governing patents in every European country. These regulations, that currently control the release of patents for industrial inventions, have been progressively harmonized within Europe over the last thirty years, following a series of international conventions. However, the Directive must be evaluated considering both the internal system of each country and the international system, as indicated by the presence of many acts of differing importance and the numerous references contained in the Directive to international conventions/agreements: the TRIPs agreement [1], the Convention on Biological Diversity, the European Convention for Defending Human Rights and Basic Freedoms, and the Munich Convention on the subject of plant varieties. The 1973 convention on European patents (signed in Munich by 18 states) was very important as, among other things, it instituted the so-called "European patent", released by the European Patent Office (EPO) on the basis of the European Patents Convention (EPC) regulations.

2. EU DIRECTIVE 44/98 AND PATENTING SYSTEM

Formally the patenting system appears to be a system of general application, that offers undifferentiated protection to the inventions of any technical sector; in reality it was developed to protect mechanical inventions. The surfacing of the chemical, electronic, computing, bioengineering sectors has prompted a new question of protection. The response to inventions in these new sectors has differed: by means of case law adaptation (e.g. chemical); through the creation of special legislation (topography for semiconductors, new plant varieties); and with the technique of adapting the base-regulation for the needs of the new sector. This last technique is the method chosen by directive 44/98 on the legal protection of biotechnological inventions, which has been inserted in the patent legislation without creating a specific statute. This presents a need for clarification of some points that were of uncertain interpretation.

As this is the method of intervention chosen by the European legislator it would be useful to briefly mention the elements of the patenting system that cause concern over the application of patenting rights to biotechnological inventions.

In the European and national regulations governing patents, there is no explicit exception to the patenting of living things; on the contrary, statements of maximum opening of the patenting instrument "in all fields of technology" are found in important international conventions ratified by Italy and Europe (from the Paris Convention in 1883 to the TRIPs agreement in 1994). Since the 1970's, new plant varieties have been protected in

---

[1] TRIPs Agreement: Agreement on Trade-Related Aspects of Intellectual Property Rights. It is one of the WTO's agreements, which is often called the Final Act of the 1986–1994 Uruguay Round of trade negotiations, although strictly speaking the Final Act is just the first of the agreements.

the national and EU members' regulations, and microbiological inventions are patented under the existing regulation. On the other hand, for years the European Patent Office judiciary – that is currently of great importance for the interpretation of the rules on patents – has made pronouncements that allow the patenting of organic material and has formed the basis of many of the regulations in the European directive.

The system of patent release is based on the differentiation of invention from the concept of discovery, as the requisites for patenting (newness, inventive activity, possibility of industrial use) and on the exclusion of some materials, possible objects of invention, from patentability. The regulation of the European patent excludes patenting of plant varieties and animal breeds – in Italian law the ban is limited to animal breeds – and essentially biological procedures for plant and animal production, while it is possible to patent microbiological procedures and the products obtained. The patenting of inventions whose exploitation would contravene *"ordre public"* and morality is also excluded. The latter type of limitation has been shown to be of great importance for biotechnological patents, and many argue that it should form the basis for regulating biotechnological inventions.

3. THE EU DIRECTIVE 44/98

We now come to the main contents of the Directive.

3.1. Requirements of patentability. Gene sequences

The Directive establishes the patentability of products consisting or containing biological material (i.e. product inventions) and the process by means of which biological material is produced, processed and used (i.e. process invention). It stresses that the requirements of novelty, inventive step and possibility of industrial application must also be valid for these inventions. Thus, it must be stressed that the concept of invention, instead of those of discovery, is applied. This is very important, considering that the concept of discovery is excluded from patentability. In particular, with reference to a request to patent biological material isolated from its natural environment and therefore previously occurring in nature, the European directive states explicitly that in this case, the term invention can be used, provided the compulsory requirements are met. The same is true for elements, particularly sequences or partial sequences of a gene, that have been isolated from the human body. In the same way, there is no hindrance to invention consideration where the biological matter or the element of the human body, obtained with technical procedures, are the product, when they are identical to the natural ones. These evaluations are based on the necessity of human intervention and the impossibility of obtaining the same result "naturally".

Considerable argument was triggered by the question of the patentability of gene sequences or partial sequences. The directive has resolved this in a positive way, with the stipulation that industrial application must be empirically stated. In the case of a partial gene sequence used to produce a protein, it is necessary to indicate what the protein is and what function it will perform (cons. 24). Great importance is attributed to the regulatory guideline of art.5.3 of the Directive. If it is interpreted in the sense of giving a generic explanation, the tactic of premature patenting will prevail (i.e. data tagging) and the patent will have a "blocking" function. If, on the other hand, a concrete and detailed explanation is required, this will leave space for the so-called inventions of use and/or improvement. The latter are held to be particularly important in this sector where inventions, such as chemical ones, are characterised by an ambiguous relationship between structure and function (e.g. more than one single function corresponding to a product that is structurally similar or identical).

3.2.   The patentability of transgenic plants

In recent years, important case law has brought to light the problem of the patentability of bio-engineered plants. The question posed was: do bio-engineered plants come under the concept of plant varieties, which are excluded from patenting, or rather are they closer to the concept of production of microbiological process, a situation expressly allowed by the patenting regimen? The response from the EPO has upheld the "non-patentability" of the "product-plant variety" independently of the method with which it was created. This response was also guided by the rules that regulate plant variety patents (i.e. reg. 2100/94 and d.lg. 455/98 that incorporates the last version of the UPOV Convention), which expressly cite the hypothesis of the use of genetic engineering techniques for the development of varieties. The same regulatory definition of variety, which is also referred to in the text of the directive, appears entirely compatible with any production technique. By "variety" it is intended "a plant grouping within a single botanical taxon of the lowest known rank, which [...] can be defined by the expression of the characteristics resulting from a given genotype or combination of genotypes, distinguished from any other plant grouping by the expression of at least one of the said characteristics, and considered as a unit with regard to its suitability for being propagated unchanged".

In the case of inventions that involve more than one variety without expressly claiming a variety (because, for example, three genetic groups are referred to that alone do not characterise the entire genome of a given variety, but lend themselves to identifying more than one type of variety), the Directive has ruled that biotechnological inventions shall be patentable if the technical feasibility of the invention is not *confined* to a particular plant or animal variety. Variety has therefore been treated as a fairly restricted and literal definition, given that variety is defined by its entire genome. A plant group characterised by a determined gene, and not by the entire genome, does not come under the notion of variety and is therefore not excluded from patenting, even if it may comprise a variety.

3.3.   "Essentially biological process" and "microbiological process"

In light of the above discussion, the question of the patentability of a biotechnological *process* for obtaining plants and animals and varieties of plants and animal breeds should get a positive response, but for this it is important to know what position the Directive has taken with regard to the "essentially biological process" of plant and animal production, which is excluded from patenting. The Directive states that a process is essentially biological if it "consists entirely of a natural phenomena such as crossing or selection" thus accepting the idea of the biological-technical contradistinction. The intervention seems appropriate as the EPO judiciary had failed to give an entirely clear response to the question.

There is nothing new on microbiological process, for which the Directive confirms the rule of patentability already made by national and international regulations, considering a "microbiological process" to be any process involving, or performed upon, or resulting in microbiological material. The EPO, on the other hand, had defined the term microorganism not only as bacteria and yeasts, but also fungi, algae, protozoa, animal and plant cells. In other words this definition includes all unicellular organisms of non visible dimensions that can be reproduced and manipulated in the laboratory, including plasmids and viruses. It should not be forgotten that regulations of apropos to microbiological process (which are an "exception to the exception" in the regulation) extended the sphere of the patentability of biological material.. It is on the basis of this regulation that the patentability of the onco-mouse was recognised, and that, to date, only transgenic mammals have been patented in Europe.

3.4. The contravention of "*ordre public*"

The exclusions that come under the contravention of "*ordre public*" have prompted, as has been said, great interest, even though the Directive repeats an analogous limit provided by national and international legislation (in particular, by the Italian act on inventions l.1127/39, and the EPC art.53, and art.27 of the TRIPs Agreement). The provision lends itself to a series of observations, the first of which derives from the way art.6 identifies the ban. After having provided, in a general way, for the exclusion from patentability of inventions whose commercial exploitation contravenes "*ordre public*" and morality, four cases are listed of non-patentability "in the senses of paragraph 1", in other words because they are presumed to contravene "*ordre public*": "a) processes for cloning human beings; b) processes for modifying the germ line genetic identity of human beings; c) uses of human embryos for industrial or commercial purposes; d) processes for modifying the genetic identity of animals which are likely to cause them suffering without any substantial medical benefit to man or animal, and also animals resulting from such processes" This four cases do not represent, by express provision of the Directive, an exhaustive list (consid.38). This provision avoids problems of effectiveness with a restrictive interpretation of cases of non-patentability that must arise from a regulation of the exception type.

This could affect the interpretative procedure of the EPO attributed to the concept of "*ordre public*". In the few cases that have been, to date, evaluated according to this criterion, the EPO has responded by refusing the exclusion of patentability for contravention of "*ordre public*". In the interpretation of the corresponding EPC regulation, the significance of protection of the physical integrity of individuals as members of society and of environmental conservation, requires that there must be proof of the harmfulness of the invention. The EPC further contends that the evaluation of dangerousness is the responsibility of national authorities in accordance with the regulations specifically provided for the introduction of the product on the market.

This last notation places the focus on an aspect that the Directive also emphasises: that it is necessary to distinguish the aspect of patentability, which refers exclusively to the determination of a monopoly in utilising a given good, from the production and marketing of that good. Art.6 of the directive (complying with the EPC art.53 and Italian Inv. Act art. 13) refers to this concept when it states that "the utilisation of an invention cannot by itself be considered as contravening "*ordre public*" or morality just by the fact that it is banned by a legislative or regulatory directive", thus distinguishing the concept of "*ordre public*" from a mere legal impediment.

The concept of "*ordre public*" remains a legal criterion, to be kept separate from the ethical opinion body, for which the Directive includes the possible intervention of a specific agent (e.g. the European Group on Ethics in Science and New Technology), independently of the fact that some conclusions can be drawn. At present, ethical thought on patents tends to dwell on aspects like the inequality in terms of trade between developing countries supplying germ plasm and nations supplying the technology, the loss of biodiversity following the success of improved cultivars as substitutes for traditional varieties, and the situation of dependence (a more "fierce" form of vertical integration) of the farmer on an increasingly powerful chemical industry. These topics are to date extraneous to the accepted idea of "*ordre public*". However, the notion that these topics may soon also become subject to evaluation by the patent offices and legal bodies cannot be overlooked.

3.5. The farmer's privilege

Within the ambits of the topic of the extension of the patent, that opens up a host of evaluations and questions arising from the typical reproducible nature of living organisms, the institution of the so-called farmer's privilege has assumed prominence (evidenced by its importance in the debate that accompanied the forming of the Directive). The farmer's privilege is already present in the European regulation on property rights for plant discoveries and expressly disciplined by a 1995 European regulation (n.1768/95). In dispensation to the regulations (arts.8 and 9), according to which the protection given by the patent extends to all biological material deriving from the reproduction or propagation in an identical or differentiated way and endowed with the same properties, art.11 of the Directive provides for the sale of propagated material to the farmer for agricultural utilisation and implies the authorisation to use the harvested product for reproduction or propagation on that farm. The same regulation is valid for livestock or other breeding material of animal origin.

3.6. Description and licences

The aspect of description, that has in the past been one of the most widely-discussed topics related to microbiological inventions, is dealt with in the Directive by transposition of the rule provided for in the Budapest Treaty for micro-organisms. This rule states that the description can be integrated (adequate for an expert on the subject to put that invention into effect) with the deposit of the biological material in one of the recognised international deposits, and that access to the biological material is guaranteed through the issuing of a sample.

Another aspect of importance is the regulation of the compulsory reciprocal secondary licenses between plant breeders and patent holders for a biotechnological invention. This requires the payment of a suitable fee exclusively in the case where it is impossible to exploit the property rights or patent without contemporary violation of a patent or property rights.

4. CONCLUSION

As it is well-known, Italy, adhering to the recourse presented by the Netherlands, has contested the European Directive pleading the formal reason of the erroneous adoption of the legal basis of art.100 A (now 95) – that requires the approval of a qualified majority – instead of that of art 235 (now 308), that provides for unanimous consensus.

This opposition does not block the adoption procedure of the Directive, and the member States are still obliged to implement it by 30 July 2000. In Italy, the act that would allow the government to implement the European Act is still under discussion, therefore it is reasonably certain that Italy will not have implemented the directive by the planned date. European Union law has provided for the directive to be applicable in these cases after the expiration date for implementation, if given in detail - certainly a requisite in this case. The Italian patent office will therefore be able to release patents on biotechnological material or processes until next August. The fact that the EPC, that regulates the European patent (the patenting system most often used nowadays in Europe), began an initial harmonization with the contents of Directive 44/98 should also forward the implementation. In effect from 1st September 1999 the modifications introduced in the implementation statute of the EPC with art. 23b, 23c, 23d, 23e are in the force, which reiterate the main points of the Directive and provide for the Convention to be interpreted according to the prescriptions in Dir.44/98.

5. REFERENCES

AA.VV. (1991), La brevettabilità delle biotecnologie: una riflessione preliminare, *Giurisprudenza italiana*, 285-288.

Caforio G. (1995), *Le invenzioni biotecnologiche nell'unità del sistema brevettuale*, Giappichelli, Torino.

Campiglio C. (1999), I brevetti biotecnologici nel diritto comunitario, *Diritto del commercio internazionale*, **4**: 849-876.

Gimelli F. (1999), Biotecnologie si; brevetti no, *Ambiente Risorse Salute*, **70**: 44-47.

Guidetti B. (1999), La direttiva 98/44 sulle invenzioni biotecnologiche, *Contratto e impresa*, **1**: 482- 491.

OEB (1998), *La Jurisprudence des Chambre de recours de l'Office européen des brevets*, 3ème édition, München.

Rambelli P. (1999), La direttiva europea sulla protezione delle invenzioni biotecnologiche, *Contratto e impresa*, **1**: 492-502.

Ricolfi M. (1998), La direttiva sul brevetto biotecnologico: efficienza allocativa, equità, potere, *Quaderni di sociologia*, **XLII**(18): 167-177.

Salza G. (1998), Etica, politica e rappresentazioni del mondo nella direttiva europea sul brevetto biotecnologico, *Quaderni di sociologia*, **XLII**(18): 178-186.

Sena G. (2000), L'importanza della protezione giuridica delle invenzioni biotecnologiche, *Rivista di diritto industriale*, **I**: 65-76.

CHILD NUTRITION AND ECONOMIC GROWTH IN VIETNAM IN THE 1990s

Paul Glewwe, Stefanie Koch, and Bui Linh Nguyen *

SUMMARY

Rapid economic growth in the 1990s has led to a huge decline in poverty and an increase in household incomes in Vietnam. According to many economists economic growth can further lead to a better nutritional status of a population. Malnutrition rates in Vietnam, in terms of stunting (low height-for-age) in children under 5, have been reduced from 50% in 1992-93 to 34% in 1997-98. Disparities, however, exist between different regions, urban and rural areas, ethnicities and income quintiles. Given this dramatic decline, it is tempting to conclude that the nutritional improvements are due to higher household incomes. This paper therefore attempts to investigate the impact of household income growth on children's nutritional status in Vietnam. Different estimation methods applied to the 1992-93 and 1997-98 Vietnam Living Standards Survey data find that growth in household expenditures accounts for only a small proportion of the improvements in children's nutritional status.

1. INTRODUCTION

Child malnutrition is pervasive in almost every low income country. Among all developing countries, about 30% of all children under age 5 have abnormally low weight given their age (UNDP, 1998). For the least developed countries, this figure rises to 39%. Most economists would agree that economic growth can lead to better nutrition among children in these countries. However, the size of this impact is uncertain and will probably vary across countries. If the impact is small, policymakers in developing countries may

* Paul Glewwe, University of Minnesota and the World Bank. Stefanie Koch, Gerhard-Mercator-Universität. Bui Linh Nguyen, General Statistical Office of Vietnam. The findings, interpretations, and conclusion expressed in this paper are entirely those of the authors. They do not necessarily represent the views of the World Bank or the Government of Vietnam.

Economic Studies on Food, Agriculture, and the Environment
Edited by Canavari *et al.*, Kluwer Academic/Plenum Publishers, 2002

23

These issues are of crucial importance for Vietnam. It is one of the poorest countries in the world, with an estimated annual per capita income of about $330 in 1998. It has a very high incidence of child malnutrition; in 1993 50% of Vietnamese children under age 5 were stunted (abnormally low height given their age), although the situation has improved since that time. The role of economic growth in improving children's nutritional status is of particular interest in Vietnam because it enjoyed very rapid economic growth in the 1990s. Its annual rate of real economic growth since 1988 has been about 8%, or about 6% in per capita terms, yet at the same time it remains a very poor country with high rates of child malnutrition.

The objective of this paper is to estimate the impact of economic growth on child nutrition in Vietnam. It does so using data from two household surveys recently completed in Vietnam, the 1992-93 and the 1997-98 Vietnam Living Standards Surveys (VNLSS). A recent study of child nutrition in Vietnam, based on the 1992-93 data, found only a weak relationship between household income and child nutrition (Ponce, Gertler and Glewwe, 1998). This suggests that Vietnam's rapid economic growth in the 1990s resulted in little improvement in children's nutritional status, yet the 1997-98 data show that the incidence of stunting (low height for age) for children under 5 declined from 50% in 1992-93 to 34% in 1997-98. Given this apparent contradictory evidence, this paper seeks to clarify the role of economic growth.

This paper is organized as follows. Section II presents some basic information about child nutrition and economic growth in Vietnam in the 1990s. The data used and the analytical framework employed are discussed in Section III. Estimates of the impact of household income on child nutrition are given in Section IV. Section V provides a brief summary of the results and several concluding comments.

2. CHILD NUTRITION AND ECONOMIC GROWTH IN VIETNAM IN THE 1990s

This section provides a broad overview of the nutritional status of Vietnamese children in the 1990s, along with some information on Vietnam's economic performance during that time period. Before examining the data, it is useful to discuss briefly how children's nutritional status can be measured.

A. Measurement of Children's Nutritional Status. The nutritional status of children can be assessed using data on their age, sex, height and weight. In particular, such data can be used to calculate three indicators of children's nutritional status: 1) stunting (low height-for-age), 2) wasting (low weight-for-height) and 3) underweight (low weight-for-age). Each indicator describes different aspects of malnutrition.

Stunting is defined as growth in a child's height that is low compared to the growth in height of a reference healthy population. Slow growth in height over long periods of time causes children to fall further and further behind the height of the reference population. Thus stunting is a *cumulative* indicator of deficient physical growth. In developing countries stunting is caused primarily by repeated episodes of diarrhea, other childhood diseases, and insufficient dietary intake.

In contrast to stunting, wasting is an indicator for current severe malnutrition, which leads to rapid loss of weight. Thus it indicates *current* nutritional problems, such as diarrhea, other childhood diseases, and insufficient dietary intake. While stunting is usually not reversed – children who become stunted remain so throughout their lives, as opposed to "catching up" – the low weight associated with wasting can be restored quickly under favorable conditions.

The third indicator, underweight, can reflect stunting, wasting, or both. Thus it does not distinguish between long-term and short-term malnutrition.

All three measures are commonly expressed in the form of Z-scores, which compare a child's weight and height with the weight and height of a similar child from a reference healthy population. More precisely, the stunting Z-score of a child i is the difference between the height of that child, H_i, and the median height of healthy children of the same age and sex from the reference population, H_r, divided by the standard deviation of the height of those same children (same age and sex) from the reference population, SD_r:

$$Z - score = \frac{H_i - H_r}{SD_r}$$

Relatively short children have negative height for age Z-scores, and thus stunted children are commonly defined as those that have Z-scores of –2 or lower.

Z-scores for low weight for age (underweight) are calculated in the same way, using the weight of the child (instead of height) and the median weight (and standard deviation) of children of the same age and sex from a healthy reference population. Finally, Z-scores for wasting (low weight for height) are obtained by comparing the weight of the child with the median weight (and standard deviation) of children from the reference population who have the same height as that child. The reference population was selected by the National Center for Health Statistics (NCHS), in accordance with the WHO recommendations (WHO, 1983).

The two preferred anthropometric indices for the measurement of nutritional status of children are stunting and wasting since they distinguish between long-run and short-run physiological processes (WHO, 1986). The wasting (low weight for height) index has the advantage that it can be calculated without knowing the child's age. It is particularly useful in describing the current health status of a population and in evaluating the benefits of intervention programs since it responds more quickly to changes in nutritional status than does stunting. A disadvantage of this index, however, is that it classifies children with poor growth in height as normal (Gibson, 1990). Stunting measures long-run social conditions because it reflects past nutritional status. This is why the WHO recommends it as a reliable measure of overall social deprivation (WHO, 1986).

B. Children's Nutritional Status in Vietnam. Children that receive sufficient breastmilk, infant foods and "adult" foods grow quickly and attain their potential weight and height, unless disease or other illnesses intervene. In developing countries, children that fail to attain their potential growth typically suffer from inadequate dietary intake, illness, or both. During the first years of life, the single most important factor is the incidence of diarrhea. Children who are exclusively breastfed are much less likely to be exposed to pathogens that lead to diarrhea and other gastrointestinal diseases, and they also receive immunitive agents from breastmilk. Yet when weaning foods are introduced, typically in the first 3-6 months of life, infants are exposed to many pathogens that often lead to diarrhea and other diseases. This typical pattern is found in the 1992-93 survey data from Vietnam. The first column of Table 1 shows that less than 1% of children age 3 and older are reported to have had diarrhea in the past 4 weeks, while the incidences in the first, second, third and fourth six-month periods of life are 2.6%, 4.5%, 5.8% and 4.3%, respectively.

Repeated bouts of diarrhea interfere with human growth, leading to low weight gain. This is seen in the second column of Table 1, which shows that wasting (defined as a weight for height Z-score below -2) is relatively rare among very young children, who are still primarily breastfed. Specifically, only 1.7% of children age 0-5 months are wasted, but this figure increases to 6.1% for children age 6-11 months, then 7.9% for children 12-17 months, and finally peaks at 10.9% for children age 18-23 months. For children 2 years and older wasting fluctuates between 4% and 7%, but this is more likely caused by inadequate food intake since diarrhea is relatively rare by that age.

The long-run consequence of diarrhea, other illnesses and inadequate food intake is stunting (low height for age). As seen in the last column of Table 1, stunting (defined as a height for age Z-score below –2) rises dramatically during the first two years of life, from 13.6% for children age 0-5 months to 65.3% for children age 18-23 months, and then "settles down" to about 55% for children aged two to nine years. In other words, Vietnamese children follow the typical pattern in that their worse bouts of malnutrition occur during the first two years of life, and as a consequence slightly more than half of them become stunted during these years and remain so for the rest of their childhood.

The previous paragraphs described the situation in the early 1990s. The situation in the late 1990s shows substantial improvements, at least in terms of stunting. As seen in Table 2, stunting in 1997-98 was less common for all age categories, relative to 1992-93. At the same time, one sees the same general pattern that wasting (low weight for height) peaks in the second year of life, as does the incidence of diarrhea, so that stunting again typically develops during the first two years of life, after which it remains relatively high.

While Tables 1 and 2 demonstrate typical patterns of malnutrition and show that stunting declined during the 1990s, two anomalies stand out. The first is that the incidence of diarrhea seems to have increased dramatically for almost all age groups. This apparent increase is spurious because the way the question was asked changed between the two surveys. In the 1992-93 survey each person was asked (or, for small children, parents were asked) whether they had been sick in the last 4 weeks and, if so, what illness they suffered from. This underestimates the incidence of diarrhea for two reasons. First, some people may think of diarrhea as "normal" and so would answer that they had not been sick during the past 4 weeks. Second, persons suffering from more than one illness in the past 4 weeks were allowed to report only one illness. Thus if a child had diarrhea and another illness, the other illness may have been reported instead of diarrhea. In the 1997-98 survey all individuals were asked directly whether or not they had had diarrhea in the past 4 weeks, which resulted in a much higher reported incidence of diarrhea.

The second anomaly in Tables 1 and 2 is that wasting (low weight for height) appears to have increased, which is inconsistent with the dramatic decline in stunting. More specifically, the data show that, for each age category, average weight and height increased substantially between 1992-93 and 1997-98, both of which indicate that the nutritional status of Vietnamese children greatly improved in the 1990s. However, height increases were larger than weight increases, so that weight for height indicates a worsening condition (wasting). This suggests that examining changes in weight for height over time may provide a misleading picture of changes in children's nutritional status when household

Table 1. Stunting and Wasting in Vietnam, 1992-93

Age	Percent Diarrhea During Last 4 Weeks	Percent Wasted	Percent Stunted
0-5 months	2.6	1.7	13.6
6-11 months	4.5	6.1	25.3
12-17 months	5.8	7.9	54.6
18-23 months	4.3	10.9	65.3
2 years (24-35 months)	1.7	4.4	52.8
3 years	1.0	4.7	55.4
4 years	0.8	6.7	59.2
5 years	0.5	5.7	55.3
6 years	0.7	4.5	57.5
7 years	0.5	6.1	53.4
8 years	0.4	7.0	53.4
9 years	0.3	6.5	55.7

Source: Authors' calculations based on 1992-93 and 1997-98 Vietnam Living Standards Survey

Table 2. Stunting and Wasting in Vietnam, 1997-98

Age	Percent Diarrhea During Last 4 Weeks	Percent Wasted	Percent Stunted
0-5 months	2.3	7.1	4.4
6-11 months	14.2	11.4	7.9
12-17 months	16.3	16.1	38.1
18-23 months	10.2	13.2	49.9
2 years (24-35 months)	4.6	13.7	31.1
3 years	4.9	9.2	36.4
4 years	3.3	8.8	46.9
5 years	2.9	7.2	41.6
6 years	0.9	8.6	43.1
7 years	2.2	7.7	38.2
8 years	0.7	9.6	36.8
9 years	1.9	7.7	38.1

Source: Authors' calculations based on 1992-93 and 1997-98 Vietnam Living Standards Survey

incomes are rising. Similar contradictory results over time have been found in sub-Saharan Africa (Sahn, et al. 1999).

More information about the nature of malnutrition (as measured by stunting and wasting) is provided in Table 3. The first three columns present data from the 1992-93 survey. At that time the overall incidence of stunting for children age 0-9 years was about 53%, while the incidence of wasting was about 6%. Stunting is most prevalent in rural areas, affecting about 56% of the population, while only about one third of urban children (35%) are stunted. This is not surprising since real per capita expenditures in urban areas are almost double those of rural areas (1,897,000 vs. 1,012,000 Dong), as seen in the third column. The figures on wasting are somewhat surprising in that it is slightly more common among urban children than among rural children (6.8% vs. 5.7%).

Regional rates of stunting and wasting are also instructive. Vietnam can be divided into seven regions. The two regions with the highest incidence of stunting in 1992-93 are

Table 3.   Stunting and Wasting by Region, 1992-93 and 1997-98 (All children 9 years or younger)

	1992-93			1997-98		
Region	**Stunting** %	**Wasting** %	**Per Capita Expend.**	**Stunting** %	**Wasting** %	**Per Capita Expend.**
Northern Uplands	62.2	6.4	827	43.5	8.2	1629
Red River Delta	55.5	6.2	1099	32.2	8.6	2473
North Central	63.8	4.5	884	44.2	10.9	1879
Central Coast	48.7	5.0	1229	43.8	8.0	2193
Central Highlands	56.7	3.2	988	47.6	7.0	1781
Southeast	33.0	6.3	1802	19.7	9.3	4575
Mekong Delta	44.9	7.0	1277	33.5	10.9	2191
All Urban	34.9	6.8	1897	20.3	9.7	4406
All Rural	55.7	5.7	1012	40.9	9.2	1884
All Vietnam	52.5	5.9	1147	37.3	9.3	2305

Note: Per capita expenditures are given in thousands of current Dong
Source: Authors' calculations based on 1992-93 and 1997-98 Vietnam Living Standards Survey

the Northern Uplands and the North Central region. As one would expect, these two regions also have the lowest average per capita expenditures. Stunting is least common in the Southeast region, which has the largest city in Vietnam (Ho Chi Minh City, formerly known as Saigon) and by far the highest per capita expenditures. This strong correlation of stunting and per capita expenditures is not found in the data on wasting. The Mekong Delta has the highest incidence of wasting (7.0%) even though it has the second highest per capita expenditures. The Central Highlands have the lowest incidence of wasting (3.2%) even though it has the third lowest per capita expenditures. Overall, there is no clear correlation between wasting and per capita expenditures, which casts doubt on its use as an indicator of nutritional status.

Columns 4-6 of Table 3 present information from the 1997-98 survey. Incomes have increased in all regions (although deflated numbers are not presented) and the incidence of stunting has declined by almost one third, from 53% to 37%. This decline is found in both urban and rural areas and in all seven regions. Indeed, the region with the largest percentage increase in per capita expenditures – the Red River Delta, which moved from 4th highest to 2nd highest per capita expenditures – has the largest decline in stunting. from 56% to 32%.

In contrast, the wasting data are rather puzzling. The incidence of wasting increased in urban and rural areas of Vietnam and in all seven regions. It shows no clear relationship with income or with changes in income. Given that other indicators of child health also show improvement over this time period, for example the infant mortality rate dropped from 44 to 39 (World Bank, 1999a), the rest of this paper will focus on the stunting data.

C. Vietnam's Economic Performance in the 1990s. In the 1980's Vietnam was one of the poorest countries in the world. A rough estimate of its GNP per capita in 1984, in 1984 U.S. dollars, is $117. This would place it as the second poorest country in the world, barely ahead of Ethiopia and just behind Bangladesh (the poorest and second poorest countries in the world as reported in World Bank, 1986). By 1998, Vietnam's GNP per

Table 4. Malnutrition By Expenditure Quintiles, 1992-93 and 1997-98

	Stunting		
Quintile	1992-93	1997-1998	1997-98 with 1992-93 Quintile
1	63.4	47.8	50.9
2	59.0	41.9	45.9
3	50.2	36.8	41.7
4	45.7	29.5	36.7
5	31.5	15.4	20.4
	Wasting		
Quintile	1992-93	1997-1998	1997-98 with 1992-93 Quintile
1	4.9	9.5	11.8
2	6.5	10.0	8.2
3	6.1	9.5	10.0
4	6.3	8.6	9.3
5	6.0	7.6	8.0

Source: Authors' calculations based on 1992-93 and 1997-98 Vietnam Living Standards Survey

capita had increased to $350 (1998 U.S. dollars), so that instead of being second to last it ranked 173 out of 206 countries (World Bank, 2000).

This rapid improvement in Vietnam's economic performance dates back to 1986, when a series of decrees transformed Vietnam from a planned to a market-oriented economy. In particular, the government dissolved state farms and divided agricultural land equally among rural households, removed prices controls, legalized buying and selling of almost all products by private individuals, stabilized the rate of inflation and opened up the economy to foreign trade. In the 1990s Vietnam was one of the ten fastest growing economies in the world, with an average real GDP growth of 8.4% per annum from 1992 to 1998.

This rapid economic growth has led to a dramatic decline in the rate of poverty, from 58% in 1992-93 to 37% in 1997-98 (World Bank, 1999b). As seen in Table 3, it also appears to have led to large decreases in the rate of stunting among Vietnamese children. Are the dramatic increases in the incomes of Vietnamese households the main cause of the large decreases in stunting among young children? Table 4 provides a first glance of the evidence. For each survey, households were divided into five groups of equal size on the basis of their per capita expenditures. The first group, quintile 1, is the poorest. In 1992-93 about 63% of the children in that group were stunted. The second poorest group, quintile 2, had a somewhat lower rate of about 59%. Quintiles 3, 4 and 5 had steadily lower rates of 50%, 46% and 32%, respectively. The same pattern is seen in the 1997-98 survey; the incidence of stunting among the poorest quintile is 48% and steadily drops to 15% for the wealthiest quintile. This pattern, based on cross-sectional data, strongly suggests that higher incomes reduce child malnutrition. In contrast, the data on wasting (low weight for height) show no such pattern, raising further doubts about the informational content of this nutritional indicator, at least in the Vietnam context.

Returning to the stunting data in Table 4, note that stunting rates decline over time within each quintile. This suggests that something else in addition to income growth was leading to reduced malnutrition in Vietnam in the 1990s. Yet these quintiles are not strictly comparable because the poorest 20% of the population in 1997-98 had a higher income than the poorest 20% in 1992-93. The last column in Table 4 adjusts for this difference, classifying households in the 1997-98 survey according to the quintile categories used in the 1992-93 survey. Even after this adjustment is made there are still dramatic declines in stunting for households in the same income group.

3.   DATA AND ANALYTICAL FRAMEWORK

A. Data. All of the analysis done in this paper uses the 1992-93 and 1997-98 Vietnam Living Standards Surveys (VLSS). The 1992-93 survey covered 4800 households, while the 1997-98 survey covered 6000 households. Both surveys are nationally representative. About 4300 households were interviewed in both surveys and thus constitute a large, nationally representative panel data set. In both surveys, the household questionnaire covered a wide variety of topics, including education, health (including anthropometric measurements of all household members), employment, migration, housing, fertility, agricultural activities, small household businesses, income and expenditures, and credit and savings.

These two household surveys are particularly useful for examining the impact of economic growth on children's nutritional status. All household members, both children and adults, were measured for height, weight and arm circumference. The vast amount of information on the households surveyed, including detailed income and expenditure data, reduces problems of omitted variable bias. Finally, the panel data allow for estimation that controls for unobserved household fixed effects.

B. Analytical Framework. The data presented in Section II show changes over time but they do not attempt to explain what caused those changes, or more generally what determines children's nutritional status. Such causal analysis is much more difficult and requires a clear analytical framework to avoid drawing false inferences from the data.

The starting point for thinking about the determinants of a child's nutritional status is a health production function, since nutritional status is a major component of child health. In general, a child's health status (H) is determined by three kinds of variables, health inputs (HI), the local health environment (E) and the child's genetic health endowment (ε):

$$H = f(HI, E, \varepsilon) \tag{1}$$

The child's health endowment (ε) is defined as all genetically inherited traits that affect his or her health. It is exogenous (cannot be altered by the child or anyone else), but is rarely observed in any data. The local health environment (E) consists of the characteristics of the community in which the child lives that have a direct effect on his or her health, such as the prevalence of certain diseases and the extent of environmental pollution. It is also exogenous, although one could argue that it is endogenous to the extent that households migrate to areas with healthier environments or take measures to improve the local health environment. Finally, there are a wide variety of health inputs (HI) that are provided by the household to the child, including prenatal care, breastmilk, infant formula, all other foods, medicines and medical care. In addition, the quality of the household's drinking water, toilet facilities and other hygienic conditions around the home can be treated as health inputs.

While researchers would often like to estimate a health production function, it is almost impossible to do so because one rarely has complete data on health inputs and the local health environment, and data on the child's genetic endowment is rarer still. This incompleteness may well lead to serious problems of omitted variable bias. This problem is further complicated by the need to have this information not only for the current time period but for all past time periods of the child's life. A more practical alternative is to consider what determines health inputs and "substitute out" that variable from equation (1). In general, the health inputs that households choose for their children are determined by the household income level (Y), the education levels of both parents (MS and FS, for mother's schooling and father's schooling), their "tastes" for child health (η), the local health environment and the child's genetic health endowment:

$$HI = g(Y, MS, FS, \eta, E, \varepsilon) \qquad (2)$$

Note that family size and the presence of other siblings are not included as determinants of health inputs. This is done because those variables are clearly endogenous, and including endogenous variables will often lead to biased estimates unless suitable estimation methods, such as instrumental variables, are used. Thus it is best to include in equation (2) only those variables that are clearly exogenous. Of course, one could rightly claim that household income is endogenous; for example, parents may change their hours of work in response to the health status of their children. However, removing this variable from (2) would preclude estimation of the key relationship of interest in this paper, so it is retained. The approach used to deal with possible estimation biases from retaining this variable is discussed below.

Substituting (2) into (1) gives the basic equation that this paper will attempt to estimate:

$$H = f(g(Y, MS, FS, \eta, E, \varepsilon), E, \varepsilon) = h(Y, MS, FS, \eta, E, \varepsilon) \qquad (3)$$

This paper will use each child's height for age Z-score as the indicator of child health, H. As mentioned above, both surveys have data on household income and expenditures. Household per capita expenditures will be used instead of household per capita income to measure Y, for two reasons. First, expenditure data are likely to be more accurate than income data (Deaton, 1997). Second, expenditure data are more likely to reflect a households "permanent income", which is more appropriate in this case because Y represents the household's income stream since the child was born, not just current income.

The remaining variables in equation (3) merit further comment. The schooling of each parent is provided in both surveys, even for children who are no longer living with one or both parents (8% of the children in the sample are not living with their father, and 3% are not living with their mother). However, parental tastes for child health, η, are difficult to ascertain in any survey and no attempt was made to do so in the Vietnamese surveys used here. Dropping this variable from the estimation altogether is risky; doing so would relegate it to the error term and there are many scenarios that suggest that it could be correlated with household income (in which case it would lead to biased estimates of the impact of household income on child health). For example, some parents may be "irresponsible", which implies low tastes for child health and low income. This would lead to overestimation of the impact of income on child health. This paper uses three approaches to deal with this problem. First, dummy variables representing different ethnic and religious groups are included to approximate, albeit only partially, tastes for child health. Second, in some estimates instrumental variables are used for household per capita expenditures, which should eliminate some or perhaps even all of the bias due to correlation between per capita expenditures and unobserved tastes for child health. Third, some estimates presented below are based on panel data, and for those parental tastes can be thought of as a fixed effect that differences out of the estimation.

The last two variables in (3) are the local health environment, E, and the child's innate healthiness, ε. The estimates presented in Section IV use community fixed effects that are used to control for all differences across communities, one of which is differences in the local health environment. Finally, consider the child's genetic health endowment, ε. In the cross-sectional estimates, this is (partially) represented by the height of each parent (which reflects both "normal" variation in height that is not associated with health status and the innate healthiness of each parent) and by the sex of the child (since girls are typically healthier than boys, but note that this masks any sex discrimination taking place). In estimates using panel data, the average healthiness of each household's children is treated as a fixed effect and thus is differenced out.

The last issue to address is the problem that household income is endogenous, which raises the possibility of simultaneity bias. In general, households make decisions about their children's health at the same time that they make decisions about income earning activities, and these two decisions could be related. For example, parents whose children are chronically ill may decide to purchase costly medicines or medical services, and to do this they may increase some household members' work hours in order to pay for those medicines. In this scenario simple ordinary least squares (OLS) estimates would tend to underestimate the impact of household income (expenditures) on child health because unobserved negative shocks to child health would be positively correlated with household income. Alternatively, households may decide to reduce hours worked in response to a child's illness, for example the mother may work fewer hours in order to spend more time caring for the child. In this case OLS estimates would overestimate the impact of household income (expenditures) on child health.

Another problem with both household income and expenditure data is that they are often measured with random error, simply because it is difficult for households to provide accurate answers to detailed questions about their incomes and expenditures. As explained above, this paper will use household expenditures instead of household income because it is likely to be more accurate. However, even household expenditures are likely to suffer from a significant amount of measurement error, much of which will be rather random. Such random measurement error will lead to underestimation (attenuation bias) of the true impact of household expenditures on child health.

Of course, instrumental variable methods can, in principle, remove the bias caused by either endogeneity or measurement error in the household expenditures variable. The task is to find plausible instrumental variables, that is variables that are correlated with household income but are not correlated with unobserved determinants of child health and are not correlated with the measurement error in the household expenditure variable. Two plausible categories of instrumental variables are types of agricultural land allocated to the household and certain sources of non-labor income. In Vietnam, agricultural land is tightly controlled by the government, and markets for land simply do not exist in most rural communities (less than 3% of households in the 1992-93 survey reported that they had bought or sold land in the previous year). Thus households' land assets are unlikely to be influenced by children's health status. Similarly, some types of non-labor income are received regardless of children's health status. Thus the following instrumental variables are used for households' per capita expenditures: irrigated annual cropland, unirrigated annual cropland, perennial cropland, water surface, income from social funds, social subsidies, dowries, inheritances and lottery winnings. Finally, the existence of relatives (more specifically, children of household members) living overseas may also indicate an additional source of income; although the amount of remittances sent by such relatives may respond to child illnesses, the existence of such relatives is unlikely to be affected by those illnesses. Two such variables are used, overseas relatives in other Asian countries and overseas relatives in Western countries.

As will be seen in the next section, while these instrumental variables have statistically significant predictive power they are rather weak in terms of the R^2 coefficient in the first-stage regressions. If the main problem is measurement error, as opposed to household expenditures being correlated with unobserved determinants of child health, then one could use household income as an instrumental variable for household per capita expenditures. In the regression results presented below, two sets of instruments are used, one without household income, which should be robust to both endogeneity and measurement error in the income variable, and another set that adds income, which is robust to measurement error but is invalid if household income is endogenous with respect to child health.

4.   INCOME GROWTH AND CHILD NUTRITION

This section presents regressions that estimate equation (3). In all regressions the dependent variable is the child's height for age Z-score. Separate estimates are presented for urban and rural areas. For cross-sectional regressions separate estimates are given for 1992-93 and 1997-98. The cross-sectional estimates are presented first, followed by estimates based on panel data.

A. Cross-Sectional Estimates. Table 5 presents estimates of equation (3), the determinants of child malnutrition (as measured by height for age Z-scores), for urban areas of Vietnam in 1992-93. The first column presents OLS estimates, which are likely to suffer from omitted variable bias due to unobserved characteristics of local communities (such as the local health environment). OLS estimates may also be biased because they do not account for endogeneity or measurement error in the household expenditure variable. The second column of estimates includes community fixed effects, which avoids bias due to unobserved community characteristics as long as those variables enter equation (3) in a simple additive form without interaction terms with household or child level variables. Yet these fixed effects estimates are not robust to endogeneity or measurement error in the expenditure variable. The third and fourth columns employ both fixed effects and instrumental variables for household expenditures. The third column does not use household income as an instrument, so it should be robust to both measurement error and endogeneity, while the fourth adds household income as an instrument and thus controls only for measurement error.

Although the OLS estimates in the first column are likely to be biased, it is useful to begin with them because the results for many variables change only slightly when other estimation methods are used. As one would expect given the results of Tables 1 and 2, the age of the child has a strong relationship to malnutrition as measured by stunting. In addition to a linear term (age in months), quadratic and cubic terms were added to allow for flexibility in this relationship. Mother's and father's height are both strongly and positively related to child health, which partially controls for unobserved children's health endowment but also reflects natural variation in height across a healthy population. For some mothers and fathers (3% of mothers and 16% of fathers), the height variable was missing. In this case the parent is assigned the average height and a dummy variable is added to indicate this type of observation. These dummy variables are never significant and thus indicate little difference between parents whose heights were and were not measured.

In all the regressions in Table 5, girls in urban areas appear to be slightly healthier than boys, but this apparent advantage is never statistically significant. The impact of mothers' and fathers' schooling is also never statistically significant, which is somewhat surprising, especially for mothers. One would think that better educated mothers are more able to care for their children's illnesses, ceteris paribus. Perhaps better educated women are also more likely to work outside the home, which could have negative consequences for their children's health, so that the net effect of mother's education is zero. Finally, there are few differences across ethnic and religious groups in urban areas (the omitted ethnic groups are Vietnamese and "no religion"), the two exceptions are that Protestants and households practicing religions other than Buddhism and Christianity had children who were significantly less healthy. Both groups are relatively rare in urban areas, and it is not clear what to make of this result; indeed the result for Protestants is based on a single child and so should be treated with caution. Since the focus of this paper is on the impact of household income and health care services, these apparent impacts of religion on child health will not be discussed further.

Turn finally to the impact of per capita household expenditures (expressed in natural logarithm) on child health. The OLS estimate is 0.302, which is fairly precisely estimated (the standard error of 0.068 yields a t-statistic is 4.47). This is a somewhat higher than the estimate of 0.22 found by Ponce, Gertler and Glewwe (1998), which may reflect small differences in sample size and specification.

Even if household expenditures were exogenous and measured without error, the OLS estimate of the corresponding coefficient is likely to be biased due to correlation between household income and unobserved community differences. The basic problem is that wealthier communities may have a better health environment, for example better sanitation and health care facilities. If these community characteristics have effects that are primarily additive, community fixed effects estimates will remove this bias. Such estimates are shown in the second column of Table 5. As expected, the impact of household per capita expenditures is much smaller, falling from 0.302 to 0.193. Yet the impact of household expenditures on child health is still statistically significant (the standard error is 0.084, yielding a t-statistic of 2.31). This result is very similar to the coefficient of 0.20 found by Ponce, et al.

Table 5. Determinants of Child Malnutrition in Urban Areas, 1992-93

	OLS	Fixed Effects	2SLSFE (1)	2SLSFE (2)	Means
Constant	-18.112	-	-	-	1.00
	(-9.67)				
Age (months)	-0.080	-0.080	-0.080	-0.080	62.3
	(-4.44)	(-4.29)	(-4.19)	(-4.29)	
Age^2	0.0013	0.0013	0.0013	0.0013	-
	(3.98)	(3.90)	(3.75)	(3.89)	
Age^3	-0.000006	-0.000006	-0.000006	-0.000006	-
	(-3.69)	(-3.67)	(-3.51)	(-3.65)	
Mother's height (cm)	0.051	0.047	0.045	0.047	152.6
	(8.05)	(6.91)	(6.52)	(6.81)	
Mother's height missing	0.333	0.178	0.483	0.196	0.03
	(1.60)	(0.85)	(1.24)	(0.77)	
Father's height (cm)	0.046	0.046	0.045	0.046	1.624
	(5.53)	(5.23)	(4.39)	(5.12)	
Father's height missing	-0.129	-0.131	-0.085	-0.128	0.16
	(-1.54)	(-1.62)	(-1.02)	(-1.63)	
Female child	0.115	0.082	0.103	0.083	0.50
	(1.55)	(1.09)	(1.23)	(1.08)	
Log mother's years schooling	0.048	0.061	-0.081	0.053	1.90
	(0.58)	(0.67)	(-0.60)	(0.51)	
Log father's years schooling	0.086	0.061	-0.088	0.053	2.07
	(1.04)	(0.69)	(-0.48)	(0.45)	
Log per capita expenditures	0.302	0.193	0.878	0.233	7.35
	(4.47)	(2.31)	(1.51)	(0.87)	
Buddhist	0.017	-0.118	-0.129	-0.119	0.30
	(0.18)	(-1.04)	(-0.90)	(-1.04)	
Catholic	0.092	-0.015	-0.010	-0.015	0.14
	(0.78)	(-0.09)	(-0.05)	(-0.08)	
Protestant	-2.312	-2.524	-2.222	-2.506	0.001
	(-13.11)	(-15.17)	(-7.16)	(-11.40)	
Other religion	-0.771	-1.026	-0.896	-1.019	0.01
	(-3.48)	(-2.96)	(-2.96)	(-3.03)	
Chinese	-0.008	-0.078	-0.159	-0.083	0.10
	(-0.06)	(-0.40)	(-0.80)	(-0.42)	
Ethnic Minority	0.096	0.027	0.424	0.050	0.01
	(0.36)	(0.09)	(0.78)	(0.16)	
R^2	0.216	0.257	0.201	0.256	
Overidentification test (p-value)			0.188	0.077	
Observations	870	870	870	870	

Note: 1. Asymptotic t-statistics in parentheses.
2. Standard errors and t-statistics adjusted for sample design.
3. Instrumental variables for (1) of 2SLSFE are copy from P.16. The estimates in (2) add per capita household issue as an instrument.

Source: Authors' calculations based on 1992-93 and 1997-98 Vietnam Living Standards Survey

The last two estimations in Table 5 attempt to correct for endogeneity and measurement error in the household expenditures variable. The third column of Table 5 presents estimates that predict household expenditures using the land assets and non-labor income variables. Although these instrumental variables have strong explanatory power in the sense that they have high F-statistics, they do not by themselves explain a large percentage of the variation of per capita household expenditures (the R^2 coefficient of a regression of the expenditure variable on the excluded instruments is only 0.08). Thus, although the coefficient on per capita expenditures appears to have increased tremendously, from 0.193 to 0.878, it is not statistically different from zero because the standard error has increased from 0.084 to 0.582. This imprecision means that one can say almost nothing about the impact of household expenditures on child health in urban areas of Vietnam in 1992-93.

Somewhat higher precision can be obtained if one assumes that the only problem with household expenditures is measurement error, i.e. if one assumes that it is exogenous. This allows one to use per capita income as an instrumental variable, which greatly increases the precision of the estimates. When this is done the coefficient falls to 0.233. Although the standard error falls from 0.582 to 0.267 the coefficient is still quite imprecisely estimated and thus not significantly different from zero. To make matters worse, the standard overidentification test (see Davidson and MacKinnon, 1993) indicates that one or more of the instrumental variables is correlated with the residual (this could well be per capita income, since the estimates in the third column easily pass this test), although this is only significant at the 10% level. Overall, it is difficult to estimate with any precision the impact of household expenditures on children's nutritional status in urban areas of Vietnam in 1992-93 once one accounts for the possibility that the expenditure variable may be endogenous and may be measured with a significant amount of error.

Turn now to the cross-sectional results for rural areas of Vietnam in 1992-93, which are reported in Table 6. The age and parental height variables show the same patterns as in urban areas. As in urban areas, girls are somewhat healthier than boys, and this time this difference is significant at the 5% level. Mother's schooling has a significantly negative effect in the OLS results, but this counterintuitive finding is not found in the fixed effects or 2SLS estimates. Father's schooling occasionally has a significantly positive effect, but this result is not very robust. No estimates regarding religious and ethnic groups are consistently statistically significant except that again Protestant children appear to be more malnourished, though the magnitude of the impact is much smaller than in urban areas (perhaps because it is based on a larger number of children).

Focusing on the (log) household expenditure variable, the OLS results show a precisely estimated impact of 0.365 (the standard error is 0.053). As in urban areas, this figure declines when community fixed effects are introduced, to 0.227 (with a standard error of 0.62). This is similar to the fixed effects result of 0.25 in rural areas found in the Ponce, et al. study.

The third column in Table 6 specifies the expenditure variable as endogenous, using the land and non-labor income variables as instruments. The point estimate is quite large, at 0.467, but the precision of the estimate is quite low because the standard error increases to 0.318. This imprecision is not surprising because a regression of household expenditures on the excluded instruments alone yields an R^2 coefficient of only 0.060. When (log) per capita income is added as an instrument the coefficient drops to 0.281; although the standard error is smaller (0.206) this estimate is still not significantly different from zero. A final point to notice is that both 2SLS specifications easily pass the overidentification test, which suggests that it may well be legitimate to use household income as an instrumental variable (that is, the expenditure variable can be treated as exogenous).

Table 6. Determinants of Child Malnutrition in Rural Areas, 1992-93

	OLS	Fixed Effects	2SLSFE (1)	2SLSFE (2)	Means
Constant	-15.004	-	-	-	1.00
	(-14.85)				
Age (months)	-0.091	-0.090	-0.091	-0.090	60.9
	(-14.49)	(-14.21)	(-14.01)	(-14.08)	
Age^2	0.0014	0.0014	0.0014	0.0014	-
	(12.75)	(12.59)	(12.50)	(12.47)	
Age^3	-0.000006	-0.000006	-0.000006	-0.000006	-
	(-11.24)	(-11.16)	(-11.12)	(11.05)	
Mother's height (cm)	0.042	0.040	0.039	0.039	151.7
	(11.49)	(9.99)	(9.74)	(9.96)	
Mother's height missing	-0.008	-0.018	-0.025	-0.019	0.03
	(-0.07)	(-0.14)	(-0.20)	(-0.15)	
Father's height (cm)	0.035	0.031	0.029	0.030	161.9
	(7.27)	(6.39)	(5.53)	(6.02)	
Father's height missing	-0.056	-0.39	-0.034	-0.038	0.11
	(-0.69)	(-0.49)	(-0.42)	(-0.47)	
Female child	0.078	0.076	0.079	0.076	0.48
	(2.31)	(2.25)	(2.29)	(2.26)	
Log mother's years schooling	-0.072	0.005	-0.031	-0.003	1.60
	(-2.07)	(0.13)	(-0.53)	(-0.06)	
Log father's years schooling	0.038	0.085	0.049	0.077	1.81
	(1.14)	(2.32)	(0.90)	(1.77)	
Log per capita expenditures	0.365	0.227	0.467	0.281	6.80
	(6.94)	(3.67)	(1.47)	(1.37)	
Buddhist	0.049	0.053	0.074	0.057	0.25
	(0.84)	(0.73)	(0.87)	(0.75)	
Catholic	-0.168	-0.132	-0.127	-0.131	0.08
	(-1.62)	(-1.31)	(-1.25)	(-1.29)	
Protestant	-0.304	-0.512	-0.483	-0.505	0.01
	(-2.37)	(-12.61)	(-8.23)	(-10.09)	
Other religion	0.341	0.268	0.253	0.265	0.02
	(1.36)	(0.94)	(0.87)	(0.92)	
Chinese	-0.023	0.138	0.098	0.129	0.01
	(-0.18)	(1.00)	(0.66)	(0.92)	
Ethnic Minority	-0.247	-0.144	-0.112	-0.137	0.18
	(-3.37)	(-1.61)	(-1.10)	(-1.49)	
R^2	0.175	0.220	0.216	0.220	
Overidentification test (p-value)			0.507	0.485	
Observations	4787	4787	4787	4787	

Note: 1. Asymptotic t-statistics in parentheses.

2. Standard errors and t-statistics adjusted for sample design.

3. Instrumental variables for (1) of 2SLSFE are copy from P.16. The estimates in (2) add per capita household issue as an instrument.

Source: Authors' calculations based on 1992-93 and 1997-98 Vietnam Living Standards Survey

The 1997-98 survey had a larger sample size, which may add more precision to the estimates. The results for urban areas are presented in Table 7 while those for urban areas are presented in Table 8. Many of the results for the urban areas in 1997-98 are very similar to those for 1992-93. The age effects and parental height impacts are similar, the sex of the child and parental schooling again have no significant effects, and the impacts of the religion variables are similar. The one change is that the Chinese and ethnic minority variables now have positive effects that are statistically significant at the 5% level, but this is of little interest for the purposes of this paper.

The OLS estimate of the impact of household expenditures is somewhat higher in 1997-98 than in 1992-93 (0.431 and 0.302, respectively), but this difference is not so large after fixed effects are introduced (0.260 in 1997-98 vs. 0.193 in 1992-93). The first set of 2SLS estimates shows an effect of almost zero (0.024), which is quite different from the comparable 2SLS estimate of 0.878 in 1992-93. However, in both cases the standard errors of the estimates are huge (0.582 for 1992-93 and 0.590 for 1997-98), so these differences are not statistically significant. If one assumes that household expenditures are exogenous, so that one can add household income as an instrumental variable, one obtains a much more precise and indeed a statistically significant estimate of 0.413 (standard error of 0.163). Unlike the estimate from the 1992-93 data, this easily passes the overidentification test, although one should bear in mind that the power of this test to detect bad instrumental variables may not be very high.

In rural areas in 1997-98, again there are few differences in most variables. The only differences are that the Catholic dummy variable is significantly negative, and the Protestant variable is less negative, though still statistically significant. Turning to the variable of primary interest, household expenditures, the OLS and fixed effect estimates are very similar across the two years. However, the first set of 2SLS estimates is very different: in 1992-93 the estimated impact was 0.467 while in 1997-98 it was –0.430. Yet when one recalls that both of these estimates have very large standard errors (0.318 in 1992-93 and 0.303 in 1997-98) the difference between them is barely statistically significant at the 5% level (t-statistic of 2.04). The second set of 2SLS estimates, which adds household income as an instrumental variable, are much closer to each other, with point estimates of 0.281 in 1992-93 and 0.155 in 1997-98. Again, neither of these are very precisely estimated (with standard errors of 0.206 and 0.161, respectively), so they are not statistically different from each other.

Overall, the cross-sectional estimates settle on impacts of about 0.2 in rural areas, while the estimates for urban areas may be as high as 0.4. Before assessing the policy significance of these estimates, it is worthwhile to examine some estimates based on panel data.

B. Panel Data Estimates. In principle, there are two possible ways to use panel data to estimate the impact of household expenditures on children's nutritional status in Vietnam. First, one could examine data on the same children over time, and estimate the impact of changes in income on changes in their height for age Z-scores. However, this is rather problematic because, as seen in Tables 1 and 2, stunting develops in the first two years of life, after which there is little change. Thus any child who was covered in the 1992-93 survey was already at least 5 years old in the 1997-98 survey, and the impact of the household's expenditure levels in the latter survey should have almost no effect on the stunting of those children because their stunting developed 3 or more years prior to the time of that survey.

Table 7.   Determinants of Child Malnutrition in Urban Areas, 1997-98

	OLS	Fixed Effects	2SLSFE (1)	2SLSFE (2)	Means
Constant	-16.947	-	-	-	1.00
	(-12.95)				
Age (months)	-0.062	-0.064	-0.063	-0.065	64.9
	(-5.26)	(-5.31)	(-5.24)	(-5.21)	
Age²	0.0009	0.0009	0.0009	0.0009	-
	(4.09)	(4.30)	(4.42)	(4.22)	
Age³	-0.000004	-0.000004	-0.000004	-0.000004	-
	(-3.45)	(-3.69)	(-3.81)	(-3.61)	
Mother's height (cm)	0.044	0.045	0.046	0.044	152.7
	(6.24)	(6.07)	(6.53)	(5.84)	
Mother's height missing	-0.137	-0.096	-0.110	-0.086	0.04
	(-0.82)	(-0.62)	(-0.660)	(-0.56)	
Father's height (cm)	0.041	0.041	0.043	0.040	162.7
	(8.12)	(7.05)	(6.51)	(7.00)	
Father's height missing	0.036	-0.062	-0.064	-0.060	0.13
	(0.31)	(-0.53)	(-0.57)	(-0.51)	
Female child	0.052	0.073	0.076	0.071	0.47
	(0.69)	(0.99)	(1.04)	(0.95)	
Log mother's years schooling	0.053	0.037	0.093	0.001	1.99
	(0.96)	(0.059)	(0.58)	(0.02)	
Log father's years schooling	-0.005	-0.034	0.029	-0.074	2.10
	(-0.08)	(-0.43)	(0.18)	(-0.84)	
Log per capita expenditures	0.431	0.260	0.024	0.413	8.19
	(5.15)	(3.03)	(0.04)	(2.53)	
Buddhist	-0.148	-0.151	-0.161	-0.144	0.24
	(-1.77)	(-1.61)	(-1.67)	(-1.55)	
Catholic	-0.006	0.005	-0.015	0.019	0.08
	(-0.06)	(0.04)	(-0.10)	(0.12)	
Protestant	-0.433	-0.428	-0.396	-0.448	0.004
	(-3.42)	(-4.52)	(-2.44)	(-4.58)	
Other religion	-0.541	-0.528	-0.610	-0.475	0.01
	(-3.39)	(-2.52)	(-2.08)	(-2.04)	
Chinese	0.441	0.413	0.415	0.411	0.07
	(2.32)	(2.23)	(2.32)	(2.17)	
Ethnic Minority	0.618	0.807	0.780	0.825	0.01
	(1.75)	(2.12)	(2.05)	(2.13)	
R^2	0.244	0.322	0.315	0.319	
Overidentification test (p-value)			0.493	0.557	
Observations	1066	1066	1066	1066	

Note: 1. Asymptotic t-statistics in parentheses.

2. Standard errors and t-statistics adjusted for sample design.

3. Instrumental variables for (1) of 2SLSFE are copy from P.16. The estimates in (2) add per capita household issue as an instrument.

Source: Authors' calculations based on 1992-93 and 1997-98 Vietnam Living Standards Survey

Table 8.   Determinants of Child Malnutrition in Rural Areas, 1997-98

	OLS	Fixed Ef-fects	2SLSFE (1)	2SLSFE (2)	Means
Constant	-15.096	-	-	-	1.00
	(-12.87)				
Age (months)	-0.095	-0.094	-0.094	-0.094	66.9
	(-14.05)	(-14.47)	(-14.31)	(-14.47)	
Age^2	0.0014	0.0014	0.0014	0.0014	-
	(11.93)	(12.17)	(12.19)	(12.19)	
Age^3	-0.000006	-0.000006	-0.000006	-0.000006	-
	(-10.47)	(-10.63)	(-10.69)	(-10.65)	
Mother's height (cm)	0.041	0.040	0.044	0.041	152.1
	(10.22)	(11.32)	(10.41)	(11.27)	
Mother's height missing	0.143	0.098	0.097	0.098	0.03
	(1.40)	(1.04)	(0.95)	(1.04)	
Father's height (cm)	0.041	0.036	0.040	0.036	161.9
	(8.02)	(7.79)	(8.24)	(7.62)	
Father's height missing	0.022	0.052	0.045	0.052	0.10
	(0.34)	(0.94)	(0.81)	(0.92)	
Female child	0.146	0.145	0.133	0.144	0.50
	(4.08)	(4.10)	(3.65)	(4.05)	
Log mother's years schooling	-0.093	-0.048	0.036	-0.041	1.62
	(-2.43)	(-1.36)	(0.68)	(-1.15)	
Log father's years schooling	0.051	0.035	0.123	0.043	1.81
	(1.00)	(0.76)	(1.92)	(0.77)	
Log per capita expenditures	0.312	0.207	-0.430	0.155	7.47
	(6.67)	(3.86)	(-1.42)	(0.96)	
Buddhist	-0.015	-0.018	-0.008	-0.017	0.15
	(-0.25)	(-0.27)	(-0.11)	(-0.26)	
Catholic	-0.240	-0.260	-0.231	-0.258	0.10
	(-3.49)	(-3.78)	(-3.30)	(-3.75)	
Protestant	-0.414	-0.385	-0.358	-0.382	0.02
	(-2.15)	(-2.38)	(-1.80)	(-2.36)	
Other religion	0.011	0.049	0.063	0.051	0.03
	(0.06)	(0.28)	(0.34)	(0.29)	
Chinese	0.299	0.394	0.517	0.404	0.004
	(0.92)	(2.38)	(1.41)	(1.27)	
Ethnic Minority	-0.091	0.019	-0.083	0.010	0.20
	(-1.19)	(0.19)	(-0.72)	(0.10)	
R^2	0.215	0.280	0.250	0.280	
Overidentification test (p-value)			0.732	0.223	
Observations	4086	4086	4086	4086	

Note: 1. Asymptotic t-statistics in parentheses.
        2. Standard errors and t-statistics adjusted for sample design.
        3. Instrumental variables for (1) of 2SLSFE are copy from P.16. The estimates in (2) add per cap-ita household issue as an instrument.
Source: Authors' calculations based on 1992-93 and 1997-98 Vietnam Living Standards Survey

The alternative possibility, which is pursued in this paper, is to compare children five years or younger in the first survey to children who were five years or younger in the second survey. This can be done using panel data for those households in the panel data that had children of that age in both 1992-93 and 1997-98, which occurs for 1663 of the 4300 households in the panel data set. For households that had two or more children in this age range in either year (or both years), all variables used are averages over those children.

Before examining the estimates, a discussion of their usefulness is in order. Recall that parental tastes for child health (η) and child innate ability (ε) are unobserved variables that could be correlated with household income. One way to try to get around this problem is to use instrumental variables for income that are not correlated with these variables. The approach with panel data is somewhat different. Instead, one assumes that the impact of these variables on child health takes an additive form and that these additive components do not change over time. In this case, *changes* in household income will be uncorrelated with these household fixed effects, so regressing changes in height for age Z-scores on changes in household expenditures (and other variables that may change over time) should eliminate bias due to these two types of household unobserved characteristics.

While this sounds like a promising approach, there are at least two problems with it. First, child innate ability varies at the child level, not the household level, so that although a household's *average* child health endowment differences out, the variation across different children within the household does not, and this could (at least in principle) lead to biased estimates for the same reason that it would do so in OLS estimation of cross-sectional data. Second, regressing differences in variables on each other greatly exacerbates measurement error, as stressed in Deaton (1997). Thus one would like to find instrumental variables that can predict changes in household income over time. This excludes many of the instrumental variables used above. In this subsection we simply try one instrumental variable, changes in household income over time.

Table 9 presents panel data estimates for urban and rural areas. The only variables that change over time are the age and sex of the children and (log) per capita expenditures. The sex dummy variable has no significant impact in any of the regressions. The age variable is again specified in a flexible way, with linear, quadratic and cubic terms. The coefficients on these age terms are quite similar to those seen in Tables 5-8.

The three urban regressions (OLS, fixed effects, and 2SLSFE) reveal a rather odd finding: negative point estimates for the impact of household expenditures on child health. However, note that the standard errors on these coefficients are very large (0.168, 0,185 and 0.920, respectively), which reflects the small sample size. Thus the positive estimates in Tables 5 and 7 are not necessarily inconsistent with these results. On the other hand, these standard errors are so large that it is probably not possible to infer much of anything from them.

In rural areas, the sample sizes are much larger. In the OLS and fixed effects estimates, the estimated impacts of household expenditures are very close to zero, and the standard errors are small enough (0.077 and 0.084) to exclude the point estimates in Tables 6 and 8 from the associated confidence intervals. However, recall that such differenced estimates may suffer from considerable attenuation bias due to increased bias due to measurement error, The final column of estimates in Table 9 uses household income to correct for measurement error in the household expenditures variable (but recall that this

Table 9.   Determinants of Child Malnutrition: Panel Data Estimates

	Urban			Rural		
	OLS	Fixed Effects	2SLS	OLS	Fixed Effects	2SLS
Sex	0.033	0.032	-0.037	0.017	-0.003	0.004
	(0.10)	(0.09)	(-0.09)	(0.11)	(-0.02)	(0.03)
Age (months)	-0.059	-0.067	-0.054	-0.083	-0.086	-0.087
	(-2.68)	(-2.99)	(-1.90)	(-9.06)	(-9.07)	(-9.29)
Age^2	0.0009	0.0010	0.0008	0.0011	0.0011	0.0011
	(2.10)	(2.35)	(1.61)	(6.51)	(6.32)	(6.13)
Age^3	-0.000004	-0.000004	-0.000004	-0.000005	-0.00005	-0.000005
	(-1.71)	(-1.92)	(-1.29)	(-5.47)	(-5.29)	(-5.20)
Log per capita			-0.939			
expenditures	-0.016	-0.106		0.004	-0.046	0.376
	(-0.10)	(-0.57)	(-1.02)	(0.05)	(-0.55)	(0.73)
R^2	0.068	0.239	0.178	0.138	0.240	0.226
Sample Size	237	237	237	1426	1426	1426

Note: 1. Asymptotic t-statistics in parentheses.

2. All variables were differenced for estimaton.

3. Sample includes all panel households who had at least one child age 0-60 months in both surveys.

Source: Authors' calculations based on 1992-93 and 1997-98 Vietnam Living Standards Survey

assumes that expenditures can be considered exogenous). The point estimate of 0.376 is much larger and comparable with estimates in Tables 6 and 8. Unfortunately, this point estimate also has a very large standard error (0.512), so even in rural areas the panel data estimates are probably too imprecise to add anything to what has been learned from the cross-sectional estimates.

C. Impact of Income Growth on Child Nutrition. Given the estimates in Tables 5-9, what can be said about the impact of Vietnam's economic growth on child nutrition? More precisely, is the rapid increase in household incomes and expenditures the main cause of Vietnam's substantial decrease in child stunting?

This question is examined in Table 10, which shows changes in mean height for age Z-scores and in the percent of children who are stunted from 1992-93 to 1997-98. The first three lines of the table show the actual changes, for rural and urban areas separately, while the rest of the table uses the estimated impacts of household expenditures from Tables 5-9 to examine how much of the change was brought about by directly raising households expenditure levels.

Table 10 shows that the mean height for age Z-score in urban areas of Vietnam increased by 0.49 standard deviations, while the mean in rural areas increased by 0.38 standard deviations. These increases are quite dramatic over a period of only five years; they correspond to a drop of almost 15 percentage points in the incidence of stunting in both urban and rural areas. Given the high rate of income growth over this time period, it is tempting to conclude that this large improvement in children's nutritional status is due to higher household income.

Table 10.   Role of Economic Growth in Reducing Child Malnutrition

	Mean HAZ		Percent Stunted	
Actual Figures:	**Urban**	**Rural**	**Urban**	**Rural**
1992-93	-1.566	-2.108	34.9	55.7
1997-98	-1.073	-1.727	20.3	40.9
Change (over 5 years)	+0.493	+0.381	-14.6	-14.8
Estimates of change due to economic growth:				
OLS	0.175	0.088	-6.6	-3.4
	[0.248]	[0.114]		
FE	0.108	0.057	-3.5	-2.2
	[0.191]	[0.086]		
2SLS (1)	0.216	0.005	-8.1	-0.0
	[0.783]	[0.165]		
2SLS (2)	0.154	0.057	-5.1	-2.2
	[0.363]	[0.152]		
OLS (panel)	-0.008	0.001	+0.4	-0.0
	[0.156]	[0.041]		
FE (panel)	-0.051	-0.013	+2.4	+0.7
	[0.130]	[0.031]		
2SLS (panel)	-0.449	0.098	+16.8	-3.5
	[0.451]	[0.363]		

Note: 1. Cross-sectional estimates are based on the mean of the 1992-93 and 1997-98 estimates.
2. Increase in real expenditures per capita was 29.8% in rural areas and 61.3% in urban areas (GSO, 1999). This implies that the change in real exp. per capita was +0.261 in rural areas and +0.478 in urban areas.
3. Numbers in brackets are upper bounds of 95% confidence intervals.

Source: Authors' calculations based on 1992-93 and 1997-98 Vietnam Living Standards Survey

The remaining lines of Table 10 assess whether this conclusion is valid. For each estimator the predicted change in the mean height for age Z-score is given, which is simply the estimated coefficient of the impact of household expenditures multiplied by the change in (the log of) average household expenditures. For estimates not based on panel data, the estimated impact used is a simple average of the 1992-93 and 1997-98 estimates. In addition, these estimated impacts are added to each child's Z-score in 1992-93 to see how they change the incidence of stunting (low height for age). Those calculations are reported in the third and fourth columns of Table 10.

The clear conclusion to draw from the results in Table 10 is that growth in household expenditures accounts for only a small proportion of the improvement of children's nutritional status in Vietnam from 1992-93 to 1997-98. In urban areas, the mean height for age Z-score increased by 0.49, but the highest predicted change among seven different specifications is only 0.22 (from the 2SLS specification without income as an instrumental variable), less than half the total amount. Similarly, the incidence of stunting dropped by 14.6 percentage points, but the predictions from the econometric estimates are much smaller, the highest one showing a drop of only 8.1 percentage points. The same conclusion holds even more forcefully for rural areas; the mean height for age Z-score dropped by 0.38 standard deviations but the largest predicted drop is only 0.10 standard deviations, and the incidence of stunting dropped by 14.8 percentage points while the largest predicted drop is only 3.5 percentage points.

Given the imprecision of the estimated impacts, it is useful to check the upper bound of the 95% confidence interval of the estimated impacts, since it is possible that even though the point estimates are low the actual change may still lie within that confidence interval. The mean changes in height for age Z-scores using the upper bounds of the 95% confidence intervals are shown in brackets in the first two columns of Table 10. In only one out of fourteen cases does the actual change lie within that confidence interval. Thus one must conclude that growth in household incomes accounts for only a proportion, and probably a rather small proportion, of the improvement in children's nutritional status in Vietnam during the 1990s.

5.   SUMMARY AND CONCLUSION

This paper has investigated the impact of household income growth, as measured by household expenditures, on child nutritional status in Vietnam in the 1990s. Vietnam was fortunate in this decade in that both household incomes rose dramatically while children's nutritional status improved rapidly. While one may conclude that the former caused the later, the estimates presented here do not support such a conclusion. Using many different estimation methods, this paper has shown that the impact of household expenditures on children's nutritional status (as measured by height for age Z-scores) is not necessarily significantly different from zero. More specifically, the impact may well be positive, but it is not very large. In particular, none of the estimates is large enough to account for even half of the measured improvement in children's nutritional status from 1992-93 to 1997-98.

While some observers may argue that this finding casts doubt on the benefits of economic growth for children's health status, such a conclusion would be premature. This is because economic growth may lead to other changes in society, such as improved health care services. That is, economic growth should increase government budgets through increased tax revenues, some of which can then be used to provide better health care services. The question then becomes: What kinds of health projects are most effective at raising child (and adult) health? This is an issue that requires further investigation and thus is beyond the scope of this paper. Fortunately, the data available from Vietnam, and from other countries, may well have the answer to this question.

6.   REFERENCES

Davidson R. – MacKinnon J. (1993), *Estimation and Inference in Econometrics*, Oxford University Press, Oxford.
Deaton A. (1997), *The Analysis of Household Surveys*, Johns Hopkins University Press, Baltimore.
Gibson R. (1990), *Principles of Nutritional Assessment*, Oxford University Press, Oxford.
GSO-General Statistical Office (1999), *Viet Nam Living Standards Survey 1997-98*, Ha Noi, Viet Nam.
Ponce N. – Gertler P. – Glewwe P. (1998), Will Vietnam Grow Out of Malnutrition?, in: D. Dollar, P. Glewwe and J. Litvack, eds., *Household Welfare and Vietnam's Transition*, The World Bank.

Sahn D. – Stifel D. – Younger S. (1999), *Intertemporal Changes in Welfare: Preliminary Results from Nine African Countries*, Cornell Food and Nutrition Policy Program, Working Paper #94, Cornell University. Ithaca, NY.

UNDP-United Nations Development Programme (1998), *Human Development Report*, Oxford University Press, Oxford.

World Bank (1986), *World Development Report*, Oxford University Press, New York.

World Bank (1999), *Vietnam Health Sector Review*, The World Bank, Hanoi.

World Bank (1999b), *Vietnam: Attacking Poverty*, Joint Report of the Government of Vietnam and the Donor-NGO Poverty Working Group, Hanoi.

World Bank (2000), *World Development Indicators*, The World Bank, Washington, D.C.

WHO-World Health Organization (1983), *Measuring Change in Nutritional Status*, Geneva.

WHO-World Health Organization (1986), *Use and interpretation of anthropometric indicators of nutritional status*, Bulletin of the World Health Organization, Geneva.

LARGE SCALE RETAILERS AND DIFFUSION IN THE AGRI-FOOD SYSTEM OF THE ISO 9000 CERTIFIED QUALITY MANAGEMENT SYSTEM

Maurizio Canavari, Roberta Spadoni, Domenico Regazzi, and Federica Giacomazzi [*]

SUMMARY

This paper reports the initial results of a research program examining the relationship between large retail chains and the certification of quality management systems. The aim is to analyze the role of large retail firms in the development of the ISO 9000 standard model of Quality Management System (QMS) in the agri-food environment, through the pressure exerted by the retailer on its suppliers.

The analysis is based on surveys of the buyers for the main large retail firms in Emilia-Romagna.

The results seem to confirm the positive role of large retail business, but fails to show a clear strategy for these operators in implementing their own QMS. The influence of large scale retailers may be the only factor that will definitely promote a dramatic increase in the adoption of such management systems in the agri-food market.

1. INTRODUCTION AND OBJECTIVES

Quality in food and agricultural production has been a major topic in recent times, particularly in relationship to aspects such as genuineness and safety.

The term "quality" has gained more importance as an assurance for a predetermined set of characteristics rather than as the definition of excellence. The first concept does not refer to better or worse characteristics of the produce, but to a particular quality level obtained with a high degree of reliance. Several factors favour this interpretation, including the qualitative change and growth of supply in comparison to demand, an increased

[*] Maurizio Canavari (sections 3 and 4), Roberta Spadoni (sections 1 and 2), and Domenico Regazzi (coordinator), Department of Agricultural Economics and Engineering, University of Bologna. Federica Giacomazzi (responsible for interviews, data collection and data processing), Agri2000, Bologna.

Economic Studies on Food, Agriculture, and the Environment
Edited by Canavari *et al.*, Kluwer Academic/Plenum Publishers, 2002

competition and, consequently a higher attention to customers' needs, and an internation-alization and widening of markets, which then become more competitive.

In light of these factors, it has been shown that a higher quality assurance level can-not be achieved solely through controls on final products. The more promising method to achieve these assurance levels is through process control, and more recently, through management control. There are several models for quality system management that could be used, but a standard reference model within an international context, is described in the rules of the ISO 9000 series, which applies to all activities, firms and countries.

This study is part of a research project whose aim is to clarify the role of large retail on spreading Quality Management Systems and on related auditing and certification ac-tivities in the agri-food market environment.

Together with this goal, advantages flowing to a production industry willing to have direct commercial relations with large retail firms will be also investigated.

The study arised from the concern that, from a superficial view of the market, certi-fied QMSs seem to have become crucial to competitiveness. The issue is whether adop-tion of a QMS is an unavoidable requirement to compete.

Another question arises as a part of this evaluation: Is large retail appropriately ori-ented to prefer certified suppliers? Large retail is fully included in the agri-food sup-ply-chain, and like the other operators is forced to provide suitable safety conditions and adequate and efficient control systems. Moreover, since large retail acts as a contact point between consumers and the supply-chain, it takes on the final responsibility for quality assurance for goods sold in its stores, and on any food problems that occur, including quality or safety concerns. This is now particularly true, because after a period in which the major commercial brands have gained a prominent role in assurance and information to consumers, large retail tries to make up for the information delay by using the private labels. This strategy gives several benefits to large retail, but obviously it also includes a greater responsibility, considering that large retail can not (and will not) produce its own products. In this case, reliable strategies for selection, control and testing of goods, to-gether with selection of suppliers able to satisfy requirements for quality assurance, be-come fundamental.

Large retail must introduce refined control systems, but this is not sufficient to assure quality and safety to an acceptable level against rare but crucial flaws. That's why the process control on suppliers (giving for example the production rules determined by the retail chain) goes beyond the simple itemized list of characteristics requested in the con-tract. In this way, the retail chain provides rules for the production method, or for the raw materials, or for the type of workers that can be used. Thus, the retail chain becomes one with the producer, tending to invade its field. In this context, it seems that a supplier hav-ing a certified quality management system could better respond to the above mentioned requirements, control the "invasion", and provide a simpler way to give the second party the requested assurance on reliability of food quality and safety.

2. METHOD OF ANALYSIS

Certified quality systems are still a debated issue. In this study, the analysis is carried out by using a questionnaire submitted during interviews to the buyers of the food division and to the Quality Assurance Managers of the large retail firms. They have been chosen because of their role in defining and maintaining relations with suppliers.

The target group has been selected from among the major large retail firms operating in the Emilia-Romagna region, and is listed as follows:

- Coop Italia
- Conad
- Crai
- Rinascente
- Gruppo GS S.p.a.
- Finiper

- MDO S.p.a.
- C3
- Sisa
- Sigma
- Colmark
- Fininvest - Standa

- Despar Italia
- Interdis
- A & O - Selex
- Pam
- Esselunga
- Lidl

The questionnaire has been submitted directly or sent by fax, and includes two sections, divided by specific topics.

First section:

1. Identification of survey respondent.
2. General information on the large retail firm, including turnover, number and types of stores, etc.
3. Purchasing policy, types of goods sold, supply source, etc.

Second section:

1. Purchase of "controlled" food, information on various types of certification, including product, process and system certification.
2. ISO 9000 series certification: behavior of the retail chain towards the certified suppliers.

The second section is aimed at verifying whether the choice of a supplier is somewhat influenced by the presence of a certified quality management system.

Finally, the role of quality systems inside of the large retail firm is investigated.

3. RESULTS

Although efforts were made to make questionnaire concise, returned surveys were not always properly filled out, particularly the section related to data in which the interviewed persons were not directly involved.

Moreover, two food safety emergencies occurred during the period of the survey (Coca-Cola™ and poultry meat dioxin crisis), and many of the persons addressed were very busy solving the related problems. However, the second section of the questionnaire was generally more thoroughly completed, and it is the most interesting part of the survey. The main difficulty in data collection was encountered in scheduling meetings with the addressed persons.

Only 6 out of 18 retail chains contacted completed the questionnaire, mainly during an interview with a buyer. The six respondents were:

1. COOP Italia   (hypermarket Grand'Emilia Modena - Coop Estense)
2. CONAD   (headquarters + hypermarket Pianeta of Bologna)

3. SIGMA   (headquarters)
4. CRAI   (headquarters)
5. INTERDIS   (headquarters)
6. A & O - Selex (headquarters)

The return rate is unsatisfactory for a general analysis, even though two of the above mentioned retail chains represent a very high market quota in the Emilia-Romagna region. The survey is expected to be continued, and product certification will also be considered because of the great interest arising in the topic during this period. The outcome of the current survey is outlined in the following section, though it is necessarily just a partial and rough picture.

3.1   Retail chains types

Four out of the 6 above mentioned retail chains are co-operatives, while the other two are joint-stock companies. With respect to product turnover classes (Table 1) the sample is quite differentiated, involving national leading companies and smaller companies. The types of stores are mainly represented by supermarkets, but larger and smaller stores and discount stores are also included.

In the sample, the retail chains normally manage more than 2,000 stores each (Table 2). Only one of them encompasses less than 1,000 stores. The overall surface managed by each retail chain is always larger than 700,000 square meters. Among the sample, the share for food sales of total sales is always higher than 60% and a private label product assortment is set up for each retail chain. Purchases in the fresh produce division are always less than in the canned and frozen food division. Purchases are made in a differentiated way: in one case, the share managed at the headquarters level reaches 75%, while in two other cases, this share is much smaller and limited to the fresh produce division.

The form of supply is rather homogeneous. Supply for canned and frozen food is mainly carried out by the producer industry or farm (Table 3). A similar situation occurs in the other divisions. The minor role played by wholesalers should be pointed out: the retail chains use them primarily for smaller purchases. In the remaining cases, the high volumes managed in large retail allow retail chains to negotiate a better contract and to set up a partnership with the producer.

3.2   Request for "controlled" products

No retail chain was able to state the share of "controlled" and certified products they request. There is a rather homogeneous situation in relation to the type of requested certification, considering that, except for integrated farming for fruit juices, very few large retailers explicitly request certification for specific foodstuff items, as can be seen in the following:

- Integrated farming:
  3 retail chain for fruit juice
  1 retail chain for fruit and vegetables
- Organic farming:
  1 retail chain offers the entire range of such items

Table 1.   Product turnover classes

Turnover in Italian currency (billions, 1998 or average 1996/98)	Number	%
< 1,000	1	17
1,000 – 5,000	1	17
5,000 – 10,000	2	33
> 10,000	2	33
Total	6	100

Source: survey data.

Table 2.   Distribution of retail chains, by number of stores and managed surface

no. of stores	retail firms	%	Overall sale surface (sq.m.)	%
< 1,000	1	17	< 500,000	-
1,000 – 2,000	-	-	500,000 – 700,000	-
2,000 – 3,000	2	35	700,000 – 1,000,000	67
> 3,000	3	50	> 1,000,000	33
Total	6	100	Total	100

Source: survey data.

Table 3.   Canned and frozen food division by source

% of total supply	Industry	Co-operatives	Wholesalers
1 – 20	-	1	1
21 – 40	-	-	-
41 – 60	-	-	-
61 – 80	-	-	-
81 - 99	1	1	-
100	4	-	-

Source: survey data.

- Denominations of Origin:
  2 retail chains request it for wines, sausages and cheeses (normally for local or regional traditional products)
- ISO 9000 series:
  1 retail chain requests it for egg-noodles
- Other: for many products, internal rules of production are set in the contracts, particularly for private label products

3.3   ISO 9000 certification

In the following section, issues related to ISO 9000 certification are examined. First, the number and share of certified suppliers was requested in the second section of the questionnaire. The number of certified suppliers of the total varies from about 10 to 30% (Table 4). In almost all cases, certified suppliers are used in order to supply the private label assortment products. The prevailing certification scheme is the ISO 9002 (Table 5), while no supplier declared an ISO 9003 certificate.

Table 4.   Number and share of certified suppliers of total suppliers

nr. of suppliers	no. retail chains	% on total suppliers
1 – 10	1	N.A.
11 – 20	1	22
21 – 30	-	-
31 – 40	1	11
> 40	1	30

Source: survey data

Table 5.   Suppliers following the certificate types

certified suppliers adopting ISO 9001 %	No. retail chains	certified suppliers adopting ISO 9002 %	Number retail chains
0	4	< 50	-
1 – 10	2	51 – 99	2
51 – 100	-	100	4

Source: survey data

Table 6.   Motivations for choice of certified suppliers

Reasons	no. of choices	%
Better control/assurance	6	36
Increased consumer confidence	3	17
Increased competition between suppliers	3	17
Better image	2	12
Decreased control costs	1	6
Consumers' request	1	6
Other	1	6
Total	17	100

Source: survey data

Finally, motivations driving the retail chains to promote the quality management systems have been investigated using an open-ended scheme. In Table 6, we can clearly see that a better control/assurance is the main motivation behind the choice of certified suppliers. Beyond the given answers, the major part of the interviewed persons agree that supplier certification does not cause a reduction in control costs. The role of price in choice of certified suppliers remains in many cases a crucial factor. In fact, most of retail chains are not willing to pay more for products coming from certified suppliers, even if they consider certification a decisive factor in the case of slight price differences. Only one of the retail chains is willing to recognize a premium price for certification, while two other retail chains consider price as a secondary factor.

In addition, we asked how the retail chains in relation to their customers view certification. Most of the retailers have never publicized the system certification of a supplier, while a smaller number did it for specific goods, particularly for some private label products.

Retailers were also asked if they thought certification had an impact on consumers (Table 7), using an open-ended multiple choice scheme. One retailer expressed a certain

Table 7.   Impact of the ISO 9000 certification on retail chain customers

Answer	no. of choices
Increase of trust/assurance	3
Not perceived by customers	3
Not addressed to customers	2
Increases fidelity	-
Is considered a price booster	-
Other	-
Total	8

Source: survey data

Table 8.   Trends for quality management system certification

Answer	no. of choices
Positive - Encounters market needs	5
Positive - Firm imagine improvement	2
Positive - Decreases costs/complaints	3
Positive - Diminishes risks	3
Negative - The market is not interested	-
Negative - Is expensive	-
Negative - Is not enough perceived by customers	-
Negative - It is just bureaucracy	-
Other: Gives a better service	1
It will never be crucial	1
Total	15

Source: survey data

confidence in the role of system certification, but the others clearly assert that certification does not have a significant impact on consumers.

Yet the question regarding trends (Table 8) suggests that all the interviewed buyers forecast an increasing role for certification in the future. Several reasons for this trend were offered, but market needs dominated.

These answers tend to present a picture of reluctant retailers accepting the inevitability of certification, possibly driven by a general market belief, more than by a real need.

In the last question, our aim was to briefly verify this issue by asking about the intention to implement a quality management system in the retail firm (Figure 1). Presently no retail chain has started up the certification procedure, and only one of them declared that it planned to implement one. This situation seems to show little confidence in the usefulness of a quality management system certification. In some cases, something like a quality system management is already active, but the retail chains do not want to submit it to an external auditing process.

4.   FINAL REMARKS

The survey carried out among the retail chains of the Emilia-Romagna region allows a partial assessment of the attitude of such firms towards certified quality management systems. Further analyses are necessary in order to complete the picture. Enormous inter-

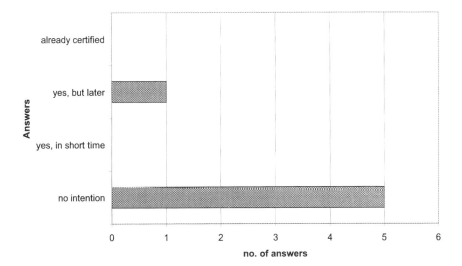

Figure 1. Retail chains planning to have their QMS certified according with the ISO 9000 (Source: survey data)

est in this issue has arisen among operators, not only with regard to the future of QMS certification, but also because of the product certification issue and the increasing importance recently acquired by this topic in Italy and in the modern distribution system. The study generally shows a vivid interest in the quality issue in the large retail sector, and a more positive approach towards system certification, especially in those chains that are managing a private label.

A large retailer, that wants to maintain its information and assurance role, tries to do that by offering private label products, as well as minor brand products, beside the main commercial brands. With the exception of major brands, consumers strictly associate the producer with the retail chain and tend to ascribe satisfaction or discontent after consumption to the latter. This phenomenon is clearer in the case of private label products, where the link between product and retailer is immediate. That explains why quality system certification may be more interesting for suppliers providing private label products.

The trend towards concentration of retail chains and the resulting reduction of purchasing centers also lead to higher competition, a process of elimination of intermediary passages, and to direct relationships between producers and retailers. Besides this, in many retail chains a trend towards the reduction of suppliers to a restricted number is clearly expressed. The accredited supplier must pass through a severe auditing process in order to demonstrate his reliability, and that process is much simpler when a certified quality management system is already set up.

However, many retail chains are not willing to pay a premium price on products for system certification. This is important information for those firms aiming to implement a QMS and gain the certificate in order to get higher market prices. This is likely the wrong motivation in approaching quality management systems.

The ISO 9000 rules are often criticized for the confusion concerning the role of system certification, which is the main obstacle to its diffusion. An ISO 9000 certificate also does not have a significant impact on consumers, and is difficult to promote using non equivocal and intelligible terms. It is mainly intended to be a business-to-business tool.

Still, all the contacted buyers agree that system certification will gain in importance. The reasons vary, but many think that this is inevitable, and that in the long run the market will prefer certified firms. The third party validation of the quality management system implemented by the company will be essential for a foodstuff producer in order to have access to the large retail market.

On the other hand, the retail chains themselves have no intention of certifying their quality management system. This sounds a little inconsistent with their quest for quality assurance, but some retail chains still do not have a working HACCP (Hazard Analysis Critical Control Point) plan [1], that recently became compulsory in Italy.

From this survey we do not find large retailers playing a decisive role in expanding system certification. This role appears to be real and positive, however it seems to be determined not so much by a clear strategy, but in a passive way - analogous to swimming along with the tide.

In the future, it will be important to analyze the impact derived from developments in the ISO model for quality management systems after the periodic revision scheduled at the end of year 2000. This will require several changes in the approach to quality and a deeper awareness by the actors of the real objectives of this tool.

Finally, an interesting issue involving large retailers could be the integration of several certification types, for instance, organic products or GMO-free foodstuff, that could (or should) be inserted in a quality management system built in accordance with the ISO 9000 model.

5. REFERENCES

Bonato P. (1996), Situazione normativa ed esperienze, in: *Qualità e certificazione dei sistemi e dei prodotti agro-alimentari*, I Georgofili - Quaderni 1996-II, Firenze, pp. 15-23.

Bassi M. (1998), Warranties on production have become a real must, *GDO Week*, **25**: 34-38.

Calderoni O. (1996), La diversificazione delle aziende italiane di distribuzione, *Commercio*, **56**: 51-84.

Canavari M. – Spadoni R. (1997), I "Sistemi Qualità" nell'agro-alimentare dell'Emilia Romagna, *Economia Agro-Alimentare*, **3**: 73-100.

Cantarelli F. (ed., 1998), *Rapporto sull'agroalimentare in Italia nel 1998*, Franco Angeli, Milano.

Cremonesi G. (1996), Se la qualità esige un attestato, *Largo Consumo*, **4**: 118-121.

De Jacobis G.M. (1998), Quality, safety and controls meeting European standards, *GDO Week*, **25**: 16-19.

Ferraris de Celle B. (1999), Dalla qualità dell'impresa alla Qualità del Sistema, *De Qualitate*, **1**: 21-23.

Fraïoli P. (1996), Normativa e prospettive comunitarie in materia di valorizzazione e protezione dei prodotti agricoli e delle derrate alimentari di qualità specifica, in: *Qualità e certificazione dei sistemi e dei prodotti agro-alimentari*, I Georgofili - Quaderni 1996-II, Firenze, pp. 63-75.

Martani G. – Salomone L. (1998), Private label verso l'eccellenza, *Largo Consumo*, **7/8**: 105-113.

---

[1] The Hazard Analysis Critical Control Point (HACCP) system identifies specific hazards and preventative measures for their control to ensure the safety of food. HACCP is a tool to assess hazards and establish control systems that focus on preventative measures rather than relying mainly on end-product testing, thus being very similar (in principles) to a QMS. Its adoption has been fostered in the European Union by the Directive 43/1993, and the need of a HACCP plan has been introduced in Italy as a compulsory requirement for all firms handling and processing foodstuff by a recent decree (D.lgs. 155/1997).

Mirandola R. – Righi R. (1996), Glossario, in: *Qualità e certificazione dei sistemi e dei prodotti agro-alimentari*, I Georgofili - Quaderni 1996-II, Firenze, pp. 117-178.

Olivieri O. (1996), *Stato della legislazione sulla certificazione di qualità in Italia - interventi regionali*, in: *Qualità e certificazione dei sistemi e dei prodotti agro-alimentari*, I Georgofili - Quaderni 1996-II, Firenze, pp. 25-40.

Peri C. (1994), *Qualità, concetti e metodi*, Franco Angeli, Milano.

Pieri R. – Venturini L. (1996), *Cambiamenti strutturali e strategie nella distribuzione alimentare in Italia: le conseguenze per il sistema agro-alimentare*, Franco Angeli, Milano.

Pontiggia C. (1997), Quando la certificazione diventa un must, *Largo Consumo*, **9**: 202-203.

Spadoni R. (1997), Qualità nell'agro–alimentare: problematiche e opportunità, *Economia Agro-Alimentare*, **1**: 211-224.

VERTICAL COORDINATION AND THE DESIGN PROCESS FOR SUPPLY CHAINS TO ENSURE FOOD QUALITY

Luciano Venturini and Robert P. King [*]

SUMMARY

There is growing interest in vertical integration and supply chain management in the food system. Several dimensions of supply chain performance have been analyzed such as logistic efficiency, quantitative input control, risk reduction. This paper focuses on supply chain design strategies for ensuring food quality. We present a series of brief descriptions of food supply chains in the U.S. and Europe. These help illustrate both the common features and the diversity of emerging supply chains in the food system. We review key theories that can serve as the conceptual foundation for supply chain analysis and design, including transaction cost economics, agency theory, property rights theory, and the resource-based theory of the firm. We assess the usefulness of these theories in explaining similarities and differences in the illustrative supply chains with respect to chain leadership, choice of mechanisms for quality assurance, the distribution of margins among chain participants, and adaptability in the face of change. In our concluding remarks, we look ahead to challenges and opportunities for future work on supply chain design in the food system.

1. INTRODUCTION

The food system, like many other key sectors in the world economy, is undergoing fundamental, far-reaching change. New mechanical, biological, and information technologies; new institutions and business arrangements; and new national and international policies are driving this change. These innovations are the basis for new products, new ways of doing business, new competitors in once stable industries, and a redefinition of horizontal and vertical firm boundaries.

---

[*] Luciano Venturini, "Sacro Cuore" Catholic University of Piacenza. Robert P. King, Department of Applied Economics, University of Minnesota.

Economic Studies on Food, Agriculture, and the Environment
Edited by Canavari *et al.*, Kluwer Academic/Plenum Publishers, 2002

One important aspect of the transformation of the food system is the shift away from broadly defined commodities to products with differentiated characteristics that add value when product identity can be preserved. The shift to mass production and mass distribution at the end of the nineteenth century made identity preservation a defining feature of the post-processing end of the food supply chain. Processing, packaging, and brand marketing turned commodities into differentiated food products but did not require identity preservation at the primary production end of the chain. Now, as primary product attributes substitute for or facilitate processing, identity preservation is moving to the upstream end of the system, creating new opportunities for enhancing food quality and safety. At the same time, this creates new challenges, since identity preservation and maintenance of food quality requires a higher degree of vertical coordination.

Supply chain concepts are also playing an important role in the analysis, management, and redesign of key sectors of the economy, including the food system. A supply chain is a linked set of value creating activities encompassing product design, input procurement, primary production and processing, marketing, distribution, and service. As such, this is hardly a new concept. What is new, however, is that supply chain thinking encourages participants to view the chain as a unified system, focusing first on improving system-wide performance and then on the distribution of gains from improvements.

There is a growing literature on vertical integration and supply chain management in the food system (e.g. Galizzi and Venturini; Hughes; Ziggers, Trienekens, and Zuurbier). In this paper, we focus on supply chain design strategies for ensuring food quality. After defining the concept of food quality and describing general categories for food quality assurance mechanisms, we present a series of brief descriptions of food supply chains in the U.S. and Europe. These help illustrate both the common features and the diversity of emerging supply chains. Next we review key theories that can serve as the conceptual foundation for supply chain design, and we explore the usefulness of those theories in explaining similarities and differences in the illustrative supply chains. In our concluding remarks, we look ahead to challenges and opportunities for future work on supply chain design in the food system.

2. THE CONCEPT OF FOOD QUALITY AND MECHANISMS FOR FOOD QUALITY ASSURANCE

The term "food quality" has many meanings. In this paper we take a very broad view of food quality. Within this concept we include sensate attributes, such as taste, smell, appearance, and texture. We also include health and safety attributes, such as nutritional content, therapeutic benefits, and freedom from pathogens and contamination. Product origin, either with respect to geographic location or production practices is another dimension of our concept of food quality. Finally, we include a service dimension in our concept of food quality. These attributes include not only factors such as convenience for the final consumer but also levels of service provision between trading partners as reflected in practices such as electronic data interchange. An effective supply chain design enhances, protects, and preserves the quality attributes of the final product.

Supply chain participants' efforts to improve food quality are motivated, in part, by public policies and regulations regarding health claims, safety, and labeling. As Segerson notes, public policies and regulations often focus on ensuring quality attributes that are difficult or costly for consumers to assess, such as food purity, absence of food-borne pathogens, or product origin. Conforming with these policies and regulations is mandatory. Often this increases the need for coordination and information exchange across the supply chain, since one participant's failure to comply can jeopardize the viability of the entire chain. For example, coordination among farm producers, processors, wholesalers, and retailers may be necessary to ensure that requirements are met for a product that has a protected denomination of origin designation.

Market forces may also motivate voluntary efforts to improve food quality, especially in cases where quality attributes are easily observed (Segerson). For example, participants in fresh fruit and vegetable supply chains are highly motivated to take steps to improve the taste, smell, appearance, and convenience of their products. Again, these efforts may require increased coordination, since a failure at one point may lower revenues for and damage the reputation of all firms in the chain.

Mechanisms for ensuring food quality are numerous, but they can be grouped into four general categories.

1. Procedures involving direct inspection or measurement of intermediate or final products make up the first category. These include laboratory tests for the presence of pathogens or contaminates, for varietal integrity or the presence of genetically modified organisms, or for other measurable characteristics associated product grades and standards. They also include direct visual inspection, such as carcass inspection in a slaughter plant or visual sorting of fruit in a packing plant. Finally, devices such as time-temperature integrators, which measure exposure to unsafe temperatures, also belong in this category.

2. Trace-back systems make up the second category of quality control mechanisms. Maintaining the ability to trace a product back through the supply chain to its origin provides strong incentives for firms in each segment of the chain to take steps to ensure quality. Trace-back systems also make it possible to selectively remove products from the supply chain when a quality problem is discovered in an upstream segment.

3. Quality control technologies represent a third category of quality control mechanisms. Examples include packaging for intermediate products as well as for final goods, physical processes such as pasteurization or irradiation, and temperature-controlled containers or vehicles. All these are tools for maintaining or restoring sensate and/or health and safety attributes.

4. The final category includes quality assurance mechanisms such as ISO certification, organic certification, and proprietary quality systems, which involve "... (1) a reliance on documentation of production processes and practices and (2) third-party auditing and certification." (Holleran, Bredahl, and Zaibet, p. 671) We also include HACCP in this category. In all these systems, the focus is on designing and maintaining processes that will ensure quality rather than on direct measurement of quality attributes. Of course there is some overlap with other categories, since direct measurements, trace-back systems, and quality control technologies are often elements of quality assurance processes.

One of the most important challenges in supply chain design is that of selecting a set of quality control mechanisms that ensures quality will be maintained at desired or mandated levels while keeping costs for quality assurance at a minimum.

3. EXAMPLES OF EMERGING SUPPLY CHAINS FOR AGRICULTURAL PRODUCTS

While the concept of supply chain management is being adopted rapidly in the food system, the implementation of supply chain management practices differs considerably across products and geographical locations. One of the difficult challenges in developing conceptual foundations for supply chain design is to identify theories that can explain similarities and differences in supply chain structures. In this section we illustrate the magnitude of that challenge by presenting brief descriptions of supply chains for specific food products within each of six stylized product types: (1) branded products, (2) genetics-based products, (3) production practice-based products, (4) retailer private-label products, (5) protected denomination of origin (PDO) products, and (6) products produced under a national quality assurance system. For each supply chain, we describe relevant quality attributes, key supply chain segments, and the flows of products and information across those segments. We also characterize the distribution of returns and costs across the supply chain. We then identify the chain leader – the supply chain participant primarily responsible for coordinating product flows and ensuring quality – and the quality assurance mechanisms chain participants use. Finally, we briefly discuss how the basic supply chain structure differs for other products in the category.

3.1. A branded product supply chain – Sourcing wheat for Wheaties breakfast cereal [1]

Wheaties is a popular breakfast cereal made from whole wheat. It has been manufactured and marketed by General Mills since 1921. Until recently, General Mills procured wheat for Wheaties in traditional commodity markets. However, results from in-house research revealed that wheat variety has subtle but important effects on product quality. More specifically, researchers discovered that cereal flakes made from particular wheat varieties are more curly, crispy, and resistant to breakage than flakes produced with other varieties. In addition, flakes produced with a special variety are consistently preferred by consumers in taste tests. Therefore, General Mills decided to use variety specific wheat exclusively in the production of Wheaties – a decision that led to development of a supply chain to ensure an adequate supply of identity preserved grain to Wheaties manufacturing plants.

The Wheaties wheat supply chain has five potentially separable segments prior to the cereal manufacturing plant: seed production, seed distribution, farm production, storage, assembly, and transportation. While the functions associated with each segment in this chain could be performed by a separate firm, the fact that General Mills owns a line of grain elevators makes it possible for them to integrate all the segments except farm production into their own operations. The General Mills elevators that handle the specific wheat are located in Idaho and contract with farmers for the production. Provisions in the one-year contracts require the farmer to use seed distributed through the elevator to produce identity preserved wheat on a specified number of acres, following production practices that ensure product quality and safety. General Mills agrees to purchase all produc-

---

[1] Much of this supply chain description is based on a presentation made by Ronald D. Olson, Vice President Grain Operations, General Mills.

tion from the contract acres at a price equal to the local price for commodity wheat plus a premium that varies from $0.05 to $0.25 per bushel depending on agronomics and functionality. The identity preserved wheat is delivered to the elevator at harvest, stored in the elevator, and transported to Wheaties manufacturing plants to meet ingredient needs. New seed is grown every year to preserve the stability of performance. In total, General Mills contracts through two elevators with up to twenty-five producers for production on approximately 8,000 acres for this specific product. They also contract through their line of elevators for other products. The fact that contracts are offered only through a controlled number of elevators makes it easier for General Mills to efficiently manage the logistics of transport to manufacturing plants. The supply chain structure also facilitates information flows for General Mills, since it gives them access to up-to-date information on acreage under contract prior to the growing season, crop condition during the growing season, and the location and quality of supplies after harvest.

General Mills is clearly the chain leader in this supply chain. Because they own the Wheaties brand, an asset of considerable value, they are able to derive added returns through increased sales of a higher quality product, though to date they have not tried to convey information about these product quality improvements directly to consumers. The supply chain structure also lowers costs for General Mills through improved ingredient quality, better information flows, and more efficient logistics. While General Mills retains some of the net benefits from the value added by the supply chain, the premium they pay farmers helps assure product segregation and quality. Quality assurance mechanisms in the system include monitoring crop production practices, varietal integrity checks at key points in the chain, and standard grain storage and handling practices designed to prevent contamination. It is noteworthy that General Mills' ownership of elevators significantly lowers monitoring costs, since there is no incentive to misrepresent product quality when the product moves across separable chain segments that are under common ownership. The effect of vertical integration on adaptability is less clear. On the one hand, General Mills can quickly change to a new wheat variety by working through its own elevators. On the other hand, their significant investment in elevators in a single region may make it more difficult to shift production to another region. It is highly unlikely that such a shift in the location of production would be necessary, however, because growing conditions in Idaho are ideally suited for high yields and excellent grain quality.

Of course, many other branded products have ingredient supply chains designed to ensure both quality and quantity of key ingredients. The fact that most food manufacturing firms do not own the infrastructure required for product assembly, storage, and transportation is an important difference, which makes it necessary to work through independent intermediaries that can provide these services. Nevertheless, the Wheaties wheat supply chain is representative of the general structure of branded product supply chains.

3.2. A genetics-based supply chain – LoSatSoy™ oil supply chain [2]

Over the past decade seed companies have placed increased emphasis on developing varieties with traits well suited for special uses. For example, the DuPont Company has developed corn, soybean, and sunflower varieties that have unique nutritional attributes as ingredients in animal feeds and food products for human consumption. In some cases, these varieties are developed through conventional breeding methods, but increasingly they are being developed through new methods based on advances in biotechnology. Regardless of how they are developed, seed companies face two difficult challenges in commercializing these new products. First, these varieties have value only if they can be segregated from other varieties as they move from the farm to the manufacturer, and they often change hands several times as they move through the supply chain. Second, while seed companies typically operate at the upstream end of the supply chain, the added value from these products may not be realized until they reach downstream users. This can make it difficult for the seed companies to capture a share of the added value.

LoSatSoy™ cooking oil is manufactured from a low-palmitic-acid soybean developed at Iowa Sate University using traditional breeding methods. Pioneer Hi-Bred International has a non-exclusive license for development and marketing of high yielding varieties of these soybeans.[3] Oil produced with low saturate soybeans has a low level of saturated fat – half that in conventional soybean oil and comparable to that in canola oil (Iowa State University Office of Biotechnology). The LoSatSoy™ oil supply chain preserves the identity of the low saturate soybeans through seven potentially separable supply chain segments: seed production and distribution, crop production, storage, assembly and transportation, crushing, refining, and distribution to the supermarket shelf. Iowa State University has retained ownership of the LoSatSoy™ trademark. Optimum Quality Grains, L.L.C. (OQG) and Protein Technologies International (PTI) work closely together and play key integrative roles in the supply chain. OQG coordinates activities in the upstream segments of the supply chain (seed production and distribution through assembly and transportation), while PTI manages the downstream segments (crushing through retail distribution). Both companies are owned by the DuPont Company, which also owns Pioneer Hi-Bred International.

PTI serves as the primary contact for LoSatSoy™ oil customers, including several food retailers and the USDA School Lunch Program. Based on knowledge of contractual commitments and demand forecasts, PTI works with OQG to project the quantity of low saturate soybeans that will be needed for a crop year and the number of acres required to produce that quantity. OQG then establishes a network of elevators that have grain handling facilities suitable for identity preserved soybeans and will serve as delivery points for contracts with farmers.[4] For the 2000 crop year, all contract delivery points for low saturate soybeans are in Michigan. This decision reflects both agronomic and logistical

---

[2] This supply chain description is based on information from the Web sites of Optimum Quality Grains, L.L.C. (http://www.oqg.com) and Protein Technologies International (http://www.protein.com) and from personal communication with Robert E. Kennedy at Optimum Quality Grain. This case is based on a more complete description in King (pp. 2-6).

[3] Kalaitzandonakes and Maltsbarger provide an excellent description of a similar supply chain for high-oil corn as a feed ingredient.

[4] OQG organizes and supports an identity preserved marketing system for corn, soybean, and sunflower varieties offered by Pioneer Hi-Bred. OQG works with elevators throughout the Midwest.

considerations. The low saturate soybean variety yields well in Michigan, and the crusher currently used by PTI is located there.

Acting as an agent for PTI and working through its network of local elevators, OQG offers contracts to growers. Contracting opportunities are posted on OQG's Internet-based contracting system called OSCAR™, which allows farmers to identify elevators near their farms that are offering contracts for identity preserved products. Participating farmers finalize contract terms with an employee of a participating elevator or a Pioneer sales representative, who enters the contract information into the OSCAR™ system. Low saturate soybean seeds are offered only to farmers who have contracts to produce for the supply chain.

Contracts for the 2000 crop year are on a per-acre basis. Subject to quality standards, the participating elevator agrees to purchase all production from the contract acres and to pay a premium of $0.25 per bushel over the market price for commodity soybeans. The grower is free to use the full range of standard pricing tools to manage price risk on the base price to which premiums are added. The contract also stipulates that the grower must purchase seed for the contract acreage from an approved seed dealer, must allow PTI and OQG to inspect the condition of the crop, and must provide a representative sample prior to delivery so that soybeans can be tested for oil content and varietal integrity. To qualify for premiums, soybeans must be less than 8% saturate and less than or equal to 0.1% GMO. An attachment to the standard contract provides a checklist of practices that are required for grower certification.

At harvest, the grower delivers the low saturate soybeans to the elevator or places them in on-farm storage, depending on the delivery terms for the contract. On delivery, the elevator pays the farmer and takes title to the soybeans. Working through OQG, PTI reimburses the elevator for the premium. The elevator also receives a small payment from PTI for maintaining identity preservation.

Based on oil sales, PTI and OQG jointly manage the flow of beans from participating elevators to the crusher. The buyer call contracts give them flexibility in adjusting the movement into the downstream segments of the chain and in scheduling efficient transportation to the crushing plant. When directed by PTI, the elevator sells the beans at the regular commodity price to a designated crushing firm, which processes the beans and receives a small quantity-based premium from PTI for maintaining identity preservation. The crusher sells the meal in the regular commodity market, since it does not have attributes that add value. The oil is sold at the regular commodity price to a refinery designated by PTI. The refiner receives a small quantity-based premium from PTI for identity preservation but also pays a royalty to PTI for each unit of LoSatSoy™ oil it sells. The refiner, in turn, is able to recover the cost of the royalty because its customers pay a premium price for LoSatSoy™ oil. The oil then moves through normal supermarket and food service channels to end-users.

In their coordinating roles, OQG and PTI bring many independent actors together into a well integrated identity preserved marketing channel. In the process, they gather valuable information on end-user demand, projected seed requirements for the next growing season, and the spatial pattern of production and stocks. Finally, they help DuPont capture a larger share of the margins generated by the supply chain by effectively negotiating the premiums received by farmers, elevator handling fees, and the royalties paid by the refiner. It is remarkable that they accomplish all this without ever taking title to the low saturate soybeans or the products derived from them. This keeps capital requirements for OQG and PTI to a minimum and allows these companies to focus their energy on maximizing returns from the intellectual property embodied in the low saturate soybean seeds and their knowledge of the supply chain.

King (p. 5) estimates that the retail price premium for LoSatSoy™ oil translates to $2.21 per bushel of added value for low saturate soybeans relative to conventional soybeans. This is distributed among supply chain participants as a reward for identity preservation and as compensation for added costs they incur. Relative to the supply chain for conventional soybean oil, most of the added costs are in the segments prior to processing. Farmers pay more for seed and incur some added labor cost for production and postharvest handling. OQG bears the transaction costs associated with establishing a network of elevator delivery points, contract design, and negotiation. Participating elevators also bear some transaction costs in finalizing contracts with growers, and they have higher labor and logistics costs for assembly and transportation to the processor. Even after considering these costs, it appears that a substantial addition to net margin is generated by the LoSatSoy™ oil supply chain. Much of that margin remains with Pioneer Hi-Bred, OQG, and PTI.

Intellectual property rights (based on license agreements with Iowa State University) are a key factor in holding the LoSatSoy™ oil supply chain together. They give DuPont (through Pioneer Hi-Bred, OQG, and PTI) control at both ends of the supply chain. At the same time, these intellectual property rights are a significant source of market power. Overall, they provide strong incentives for developing an efficient supply chain that can achieve the identity preservation that is a precondition for consumers to enjoy the benefits of LoSatSoy™ oil.

In summary, DuPont Company, working through Pioneer Hi-Bred, OQG, and PTI, is the leader for the LoSatSoy™ oil supply chain. Laboratory testing each time product changes ownership and financial incentives for identity preservation are key mechanisms for quality assurance. The net benefits created by this supply chain are substantial, and most of those accrue to DuPont, the chain leader. Finally, this chain design is highly adaptable in at least two respects. First, from year to year the elevators, crusher, and/or refiner can be changed. This is still another factor that strengthens the bargaining power of DuPont. Second, this basic chain structure can and has been adapted for other genetics-based products. Both OQG and PTI have developed significant knowledge and expertise that can be readily applied in establishing other supply chains.

3.3.   A production practice-based supply chain – Organic dry beans [5]

Demand for organically produced food products in Europe and the United States has been growing rapidly at an annual rate of 20 to 30 percent over the past several years, and organic products typically command a retail price premium of from 10 to 50 percent over conventional products (Lohr). This segment of the food system poses interesting challenges for supply chain design. The quality attributes that distinguish organic products – production without the use of synthetic fertilizers and pesticides or genetically modified organisms – are highly valued by the consumers who purchase them. However, it is generally not possible to use laboratory tests to determine whether or not a retail product

---

[5] This supply chain description is based on material presented in William Chambers' Ph.D. dissertation titled *Changes in the Structure of the US Food System: Evidence from the Dry Bean Industry*.

conforms to organic standards. Therefore, quality control systems for organic food product supply chains must be built around mechanisms that rely on third-party certification of adherence to strict production and handling standards.

A recent study by Chambers traced supply chains for twenty-three processed dry bean products upstream from the canner back through elevators to farmers. This research was designed to investigate the impact increasingly idiosyncratic customer standards are having on the structure of previously undifferentiated commodity supply chains. In general, Chambers found that standard market mechanisms work well for even highly idiosyncratic dry bean ingredients when market transactions are used in combination with elevator site visits by canners and laboratory testing by both elevators and canners. He also observed that reputation plays an important role in motivating suppliers to meet or surpass their customers' quality standards.

Six of the twenty-three products in Chambers' study – all manufactured by a single firm – were organic. The supply chains for these products differ from those for conventional products in several ways. Most important, the canner who produces organic products buys directly from individual farmers or from small farmer-owned and managed elevators rather than using large elevators as intermediaries. This eliminates one layer of transactions and makes it easier for the canner to use a combination of third party certification and site visits to ensure that the ingredients it purchases are produced organically. Also, in accordance with standards for handling organic products after they leave the farm, beans purchased by the organic canner are transported in bags rather than in bulk, and the production from each farmer supplier is segregated until it is processed. This makes it possible to reject beans even after they have been purchased if a farmer is found to have violated organic production standards.

The canner shares the chain leader role with third party certification agencies in this supply chain. While the canner sets its own quality standards for each type of bean it purchases and initiates product flows, the overall structure of the supply chain is determined to a large extent by certification agency standards, which specify practices that must be followed, records that must be kept, and procedures for site visits and inspections. In the U.S., there are several certification agencies. These are non-profit or public organizations that represent the interests of all stakeholders in the supply chain. In effect, these agencies are mechanisms for broad-based collaboration in developing and managing a system designed to foster flows of products that meet clearly defined quality standards. This has the potential to ensure a more equitable distribution of costs and benefits across key chain participants, but concrete evidence on this is lacking. The collaborative structure of the supply chain may also facilitate innovation and adaptation, since the flow of information and ideas among participants makes it easier to identify and resolve negative externalities. On the other hand, the fact that no single party takes the lead in making decisions may slow down the decision process and make it more difficult to react quickly to unexpected developments.

3.4.   Retailer-led supply chains – Private Label beef in the U.K. [6]

Food retailing in the U.K. is competitive and increasingly concentrated. Supermarkets there have been leaders in the development of high quality private label products that

---

[6] This supply chain description is based primarily on material presented by Fearne.

shift brand identity from the food manufacturer to the retailer. British retailers have been innovative in designing supply chains for perishable private label products that facilitate strict quality assurance while guaranteeing a steady flow of product (Harvey).

Supply chains for beef products sold by the three largest British retailers – Marks & Spencer, J. Sainsbury, and Tesco – serve as excellent examples of these retailer-led supply chains.[7] As Fearne notes, the structures of these chains have been shaped by competitive pressures, government regulation, and public concerns about food safety. The 1990 Food Safety Act has had especially important impacts on supply chain design. It required retailers to take primary responsibility for ensuring the safety of the food products they receive from their suppliers. This forced retailers to pay greater attention to the upstream segments of the supply chain and led to the development of farm quality assurance schemes that set standards for traceability, feeding, animal health and welfare, and transportation and handling (Fearne, p. 220).

Each of the retailer-led chains described here has three major segments: farm production, slaughter/processing, and distribution/retailing. There has been a move toward disintermediation, with slaughter/processing firms bypassing traditional livestock auctions to purchase directly from farmers and retailers bypassing independent wholesalers by integrating back into warehousing and distribution. Fearne also notes that all chains involve some level of horizontal partnership among farmers (allowing them to realize the scale economies and cost sharing benefits of collaborative marketing) and that all rely on strong vertical partnerships to link segments of the chain.

Despite these similarities, there are also important differences among the beef supply chains for the three retailers. Marks & Spencer procures exclusively through Scotsbeef, a family-owned slaughter/processing firm that buys all its beef from Scottish producers who are approved Marks & Spencer suppliers. Marks & Spencer maintains a database with information on all its producers and conducts regular taste tests that are used to provide feedback to individual farmers and so promote learning throughout the system (Fearne, pp. 222-223). J. Sainsbury procures its beef through Anglo Beef Producers (ABP), the largest meat processor in the U.K. ABP purchases cattle from 500 farmers who are members of the "Partnership in Livestock" that links producers to ABP and Sainsbury, though Sainsbury does not maintain any direct contact with farmer-suppliers. The relationship between Sainsbury and ABP has a high level of interdependence. Sainsbury relies on ABP to assume much of the responsibility for quality assurance, while ABP has made relationship specific investments in plants and depends on Sainsbury to keep them in operation (Fearne, p. 228). Finally, the Tesco supply chain is built around "Producer Clubs" that are farmer organizations established and coordinated by Tesco. Club members submit to a nationally recognized quality assurance scheme and are subject to independent audits. They are not contractually obligated to sell to Tesco, but selling in the chain does guarantee market access, favorable transportation rates, and valuable feedback on product quality. For slaughter/processing activities, Tesco, works exclusively with a small number of regional firms (Fearne, p. 229).

Farm level quality assurance schemes designed to guarantee traceability and adherence to production standards are key quality assurance mechanisms in these chains. The use of "approved suppliers" is also a common feature. This helps create a sense of loyalty within the chain and makes it more difficult for trading partners to defect to a competing

---

[7] See Fearne and Hughes for a comprehensive description of the fresh produce supply chain in the U.K.

chain. Mechanisms for facilitating and coordinating information flows among trading partners differ across chains, with Marks & Spencer maintaining the most direct contact with its farmer-suppliers and making the most systematic use of feedback based on taste testing.

Fearne notes that upstream participants in each of these chains receive financial benefits in the form of price premiums and benefits associated with improved access to information and technology. It is the retailers that hold most of the power, however. As an increasing proportion of beef in the U.K. is marketed through these retailer-led supply chains, there is concern that price premiums for farmers and slaughter/processing firms will fall as simply gaining access to the retail market becomes the primary motivator for chain membership.

Finally, the ability to innovate is noteworthy in this example. The fact that the chains are retailer-led fosters more rapid learning about consumer demand and desired product characteristics by all chain participants. The strong competition among chains, both for consumers and for suppliers, also provides strong incentives for innovation. The U.K. approach to supply chain management and quality assurance is having impacts elsewhere in Europe and in the United States.

3.5. A Protected Denomination of Origin product – The Parmesan cheese supply chain

In July 1992 the European Union introduced legislation on geographical indications and designations of origin for agricultural products and foodstuffs through Regulation N. 2081/92. Protected Denomination of Origin (PDO) products are produced, processed and prepared within a particular geographical area. They have features and characteristics which must be due to the geographical area, and the methods used to produce the product must be unique to the area. Protected Geographical Indication (PGI) products must be produced or processed or prepared within a geographical area and have a reputation, features or certain qualities attributable to that area. PDO and PGI designations are a key element of European food quality policy.

Parmigiano-Reggiano cheese has been a geographically-based brand name for many centuries. It was one of the first Italian cheeses to be awarded the Italian denomination of controlled origin (DOC) title introduced by Italian law n. 125 of 10 April, 1954. According to this law, the brand name Parmigiano-Reggiano is applied exclusively to cheese produced within the confines of a zone comprising the territory of the provinces of Parma, Reggio Emilia, Modena, Mantova on the right bank of the Po river and Bologna on the left bank of the Reno river and aged for minimum 12 months. Since 1996, the name "Parmigiano-Reggiano" has also been registered as a PDO by the European Union. Parmigiano-Reggiano cheese, hereafter referred to as Parmesan cheese, is recognized around the world as one of the finest high-quality Italian food products.

The Parmesan cheese supply chain has three major segments: milk production, milk processing, and aging. The milk production and processing segments are highly fragmented, with approximately 8,000 milk producers and 596 milk processing facilities, most of which are cooperatively owned. A few wholesalers control the aging segment of the chain, and they are able to exert some market power. Vertical coordination of the chain is not achieved through long-term contractual relationships between the parties. Rather, spot markets still dominate as the mode of transaction between segments. Therefore, the development of the supply chain and the design process to ensure quality is strongly dependent on the existence of institutions which have resources and capabilities to assume the leadership to integrate the various parties and the many agents operating in the chain.

This leadership role in the Parmesan cheese chain is played by an association of producers called "Consorzio di Tutela del Parmigiano Reggiano" (CTPR). CTPR plays several functions. These include traditional functions associated with the enforcement of technical requirements, such as grading milk to evaluate whether it meets the standards for Parmesan cheese or should be destined to other uses and performing inspections to ensure that the trademarks are applied to the wheels of Parmigiano-Reggiano according to the norms set forth in the specific regulations. Other functions of CTPR are more innovative and strategic. These include R&D, quality assurance, promotion, and the provision of consumer information. In the area of R&D, CTPR pays increasing attention to understanding how the chemical, biochemical and physical characteristics of the raw material relate to processing requirements and to the final quality and safety of the product and to exploring how scientific advances in animal nutrition, breeding techniques, and genetics can be used to improve the quality of milk. In the area of quality assurance, CTPR encourages awareness of quality issues and adoption of better production practices, such as the implementation of HACCP. Finally, CTPR offers consumers clear information and guarantees on how Parmesan cheese is produced at all stages of the supply chain from the farm to the point of sale. Overall, there is a shift in emphasis from a reputation for quality simply based on the denomination of origin to a reputation firmly based on a well-defined and enforced quality assurance system

Quality assurance mechanisms for the chain are embodied in the program CTPR developed in 1993 to assist farm producers and processors in improving quality through a coordinated approach involving the entire chain. At the farm level, CTPR has increased pressure on producers to join on-farm HACCP-type systems for assuring quality and reducing food safety hazards. About 100 processors achieved HACCP credentials under a voluntary certification program. The adoption of HACCP along the supply chain and more formal procedures for quality control allowed a first certification of the chain in 1998 under the norm UNI EN ISO 9002 (Verrini). As a result, the Parmesan cheese supply chain is designed in order to provide customers and final consumers with a double quality assurance. The first assurance is given by the European label of PDO. In addition to this logo, the CTPR provides a private assurance through the certification of the chain. This demonstrates that long-established practices already meet high standards and that it is possible to incorporate HACCP principles into traditional artisanal production processes. The next challenge for CTPR is to move toward a culture of continuous improvement for the functioning of the supply chain and the quality of the product, while maintaining the traditional image of Parmesan cheese.

Parmesan cheese sells at a considerable premium over the price of competing cheeses. In 1999, the average retail price of Parmesan cheese was 25,000 lire per kilogram, while the price of the main substitute cheese, Grana Padano, was 17,000 lire per kilogram. Thus, the price premium for Parmesan cheese was nearly 50%. In that same year, farmers received an average price of 800 lire per liter (after adjustment for the lag in payment) for milk entering the Parmesan cheese chain, while the average farm-gate price for fluid milk was about 730 lire per liter. At the farm level, then the price premium was only about 10%. As already noted, control of the aging segment of the chain is rather concentrated in the hands of a few large wholesalers. Though further analysis is needed to confirm this statement, it appears that these firms may use their market power to capture a disproportionate share of the added value created by the Parmesan cheese supply chain.

With regard to adaptability, the Parmesan cheese supply chain serves as a truly re-markable illustration of how modern concepts of quality assurance and continuous im-provement can be applied to the design and management of a traditional product's supply chain. The CTPR plays a key role in fostering this adaptability. It is also important to note, however, that there are limits to the success of CTPR and some weakness in the chain. Further improvements are needed in areas such as promotional strategies, provision of information to retailers and final consumers, and export strategies. One weakness of the chain is given by the fact that breeding is not yet well coordinated and oriented to the specific needs of the Parmesan chain. In fact, breeding continues to be more oriented to the needs of fluid milk producers which are concerned essentially to improve milk outputs rather than to upgrade the content in protein of milk which is a crucial determinant of cheese quality. Thus, to make progress, CTPR must continue to focus on working closely with supply chain members, building a culture of quality and cooperation, fostering ini-tiatives for continuous improvement, and developing human resources throughout the supply chain.

Finally, chains of other PDO Italian products face similar challenges. In fact, the Italian experience with producers' associations shows that, more often than not, these associations are rather slow in evolving towards the more innovative and strategic func-tions (Canali).

3.6. Products produced under a national Quality Assurance System – Label Rouge poultry [8]

The French Label Rouge system is a national program for food quality assurance that differentiates products of artisanal farming with a well-defined geographical origin from industrial products. As such, the system is consistent with the protected geographical indication (PGI) designation, and the EU treats Label Rouge as a form of PGI. However, as Westgren (1999) notes, Label Rouge goes beyond PGI by placing greater emphasis on system-wide quality assurance and on proactive branding and product promotion. Label Rouge products are known for their high quality with respect to taste and appearance, for their safety and wholesomeness, and for the environmentally friendly practices used in producing them. In addition, they have a product image that is very attractive to consum-ers.

The poultry sector serves as an excellent example of the Label Rouge approach. La-bel Rouge poultry products have been available since 1965, and Westgren (1999, p. 1108) reports that Label Rouge sales accounted for more than half of French household poultry purchases in 1998. Label Rouge poultry products are produced by *filière*, supply chain networks that include genetics development organizations, hatcheries, feed mills, farms, and slaughter/processing facilities. The structure of the *filière* is established by a farmer-centered "quality group" (QG), which develops a supply chain design built around a HACCP plan for assuring quality all the way to the retail shelf and can be viewed as the chain leader. The *filière* design is presented in a document called a *cahier des charges*

---

[8] Much of this supply chain description is based on material presented by Westgren (1994, 1999).

that is submitted to a national commission for approval. Only after a group's *cahier des charges* has been approved can it use the Label Rouge. All Label Rouge products must originate from a clearly defined geographic area, and there are minimum standards for genetics, production facility design, production practices, sanitation period, and frequency of inspections that all *filière* must meet. Nevertheless, there is also some latitude with respect to chain structure and product attributes.

Quality assurance in the Label Rouge system rests on HACCP principles and third party certification by a private sector *organisme certificateur* (OC). The OC performs the inspections required under the *cahier des charges* developed by the QG. The OC has the power to order destruction of birds at any point in the production process or decertification of a feed mill, slaughter/processing facility, or transport firm. Westgren (1999) notes that supply chain participants responsible for destruction of a group of birds or for loss of the Label Rouge designation for a group of birds are required to compensate other parties in the chain. This provides strong incentives for compliance to quality assurance procedures in each segment of the chain.

It is important to note that the OC must be recognized by COFRAC, a national government agency, and that COFRAC not only certifies the OC but also has the power to close down any OC which fails to rigorously oversee its quality groups. A further important aspect is that the QGs become members of SYNALAF, which is, in turn, affiliated with CERQUA, a quality certification agency funded by the agriculture ministry whose main task is the maintenance of strict industry standards regarding the use of quality designations (Westgren, 1999). This provides a clear example of the complex network of agencies at work in ensuring quality control in these chains.

Westgren (1994, p. 572; 1999, p. 1109) reports that the retail price for Label Rouge products is from 50% to 150% higher than that for non-Label Rouge products and that prices received by Label Rouge producers carry a similar percentage premium. The QGs and their associated *filières* have distinct brand identities, and a single large supermarket may carry several Label Rouge brands. This provides a strong positive incentive for participants in a *filière* to work together not only on quality assurance but also on product promotion and innovation.

The Label Rouge poultry supply chains are similar in many respects to the Parmesan cheese supply chain.[9] In both, the product is partly defined by its place of origin and use of the quality designation is to some extent governmentally sanctioned and regulated. Also, primary production and at least some processing stages are highly fragmented in both, and a farmer-controlled umbrella organization establishes and maintains the overall structure for the supply chain and the system for quality assurance. The Label Rouge supply chains are also similar in some respects to the organic dry bean supply chain. In both settings, at least some key product attributes cannot be verified through testing, and chain participants rely heavily on third party certification and inspection to implement the quality assurance system. However, chains in the Label Rouge system differ from these other cases in at least one important respect. By allowing strong brand identities to develop at the *filière* level and by vesting control of those brands with farmers, the Label Rouge system seems to more strongly encourage collaborative relationships among trading partners. Westgren (1999, p. 1108) observes that there is a very high "... degree of

---

[9] We use the plural in referring to Label Rouge supply chains to emphasize that there are numerous *filières* that have unique characteristics and compete with each other.

interrelatedness and jointness of strategic and operational issues ..." within the *filières*. This high level of cooperation makes it possible to implement systemic innovations that require high levels of trust, coordination, and information sharing. As Sylvander notes, the Label Rouge system, which was originally designed to protect the traditional artisanal sector, has made important contributions to the evolution of an innovative, forward-looking new segment of the French food system.

4.   CONCEPTUAL TOOLS FOR SUPPLY CHAIN ANALYSIS AND DESIGN

The literature on vertical coordination and supply chain management is growing rapidly. Work in this area draws on a range of theoretical frameworks, and there is not yet any emerging consensus on a unified theory for study of these issues. In this section, we briefly review four important theoretical frameworks that we believe are especially relevant for supply chain analysis and design. They are: transaction cost economics, agency theory, property rights theory, and the resource-based theory of the firm. We assess the usefulness of each theory by discussing the insights it provides regarding the four key issues that were the focus for our supply chain case descriptions: determination of chain leadership, choice of mechanisms for quality assurance, the distribution of margins among chain participants, and adaptability in the face of change.

4.1.   Transaction cost economics

The concept of "transaction cost" was first suggested by Coase to explain the boundaries of a firm. His key insight is that there are costs to using the market mechanism, including costs for informational search, negotiation and monitoring/enforcement. The core concepts of what is known as transaction cost economics (TCE) were developed by Williamson (1979, 1993a, 1993b). TCE examines and predicts how, given the presence of bounded rationality and opportunism, organizations choose governance structures in order to minimize transaction costs.

The choice of an appropriate governance structure depends on the characteristics of transactions involving the organization. These differ in four aspects: 1) the presence of asset specificity, (2) uncertainty about other parties' actions, (3) the frequency with which the transactions occur, and (4) the complexity of the exchange arrangement. Of these, asset specificity is the most important characteristic. Williamson (1979) identifies four types of asset specificity: (1) site specificity, (2) physical asset specificity, (3) human asset specificity and (4) dedicated assets.

The main proposition of TCE is that differences in transaction costs explain the advantages associated with different governance structures. The firm and the market are identified as the primary governance alternatives for organizing economic activity. When asset specificity is low, markets provide enough discipline and information to eliminate the need for safeguards against opportunism. However, as asset specificity increases, the potential for opportunism may rise to the point that internalization through the use of hierarchy represents the most efficient solution. Williamson (1991) suggests hybrid forms that are intermediate between the polar extremes of markets and vertically integrated firms as a third alternative. There are several forms, including: alliances, partnerships, franchising arrangements, interfirm cooperation, and networks. The existence of different forms is a reflection of the variety of coordination mechanisms available to organizations. Hybrid mechanisms may be formal, as in the case of joint ventures, or informal. They may be based on contracts that range in formality from verbal agreements to long, complex written contracts that remain incomplete because not all contingencies can be included. Hybrid forms are the most cost-efficient when the cost of opportunism requires greater protection than is provided by market, while the threat of opportunism is not serious enough to require a hierarchical solution that may have its own significant transaction costs.

TCE has been applied to explain vertical coordination in industries where expensive and highly specific capital investments are common. Several studies have indicated that this is the case for food chains in the poultry and hog industries in several countries (Royer and Rogers; Galizzi and Venturini). TCE provides a particularly useful conceptual framework for analyzing the determinants of supply chain structure in relation to quality issues. In fact, quality may be a crucial source of uncertainty in transactions.

An important prediction of TCE is that vertical coordination is more likely to be observed when the level of uncertainty is high. There are, of course, several forms of uncertainty. TCE considers a very broad notion of uncertainty, including uncertainty about contingencies that can or cannot be anticipated *ex ante,* as well as informational uncertainty when one party has information that the other lacks (information asymmetry). TCE suggests that the need to reduce quality uncertainty for key inputs is an important factor leading to integration and closer vertical coordination (Mahoney, 1992). It also suggests that there may be an interaction between asset specificity and uncertainty. Vertical integration or coordination is most likely to be chosen when both uncertainty and asset specificity are high (Williamson, 1985).

As transaction costs arising from uncertainty about food quality attributes increase, vertical coordination between input suppliers, producers and processors tends to become closer, and an increasing role is played by non-market coordinating mechanisms and institutions. The empirical evidence is consistent with this prediction. Several quality related factors, in fact, contribute to an increase in transaction costs in these chains. Food quality and safety attributes may be particularly difficult to observe and measure, as food products are increasingly characterized by non-visual quality attributes and "hidden" characteristics. Usually sellers have more information about the true quality attributes of the product than buyers, and the information asymmetry between successive sellers and buyers along the supply chain complicates the buyer's problem of identifying quality. The problem tends to become more serious as food products are no longer "search" goods, whose quality characteristics can be determined easily prior to purchase through visual inspection. Increasingly, food products are "experience" goods, whose quality characteristics can only be determined after consumption, or "credence" goods, whose quality cannot be determined even after consumption (Darby and Karni; Nelson).

TCE predicts that, in a world of boundedly rational consumers, firms face three interrelated information problems. First, they need to make consumers aware of the distinctive quality attributes of their products. Second, they need to provide an institutional infrastructure that will prevent these attributes from being degraded. Third, they need to accomplish both goals in an economical fashion (Williamson, 1985). Thus, TCE explains why mechanisms to signal quality to buyers are necessary to counteract the effects of quality uncertainty and why higher transaction costs due to quality uncertainties imply greater incentive to implement quality assurance systems.

In addition to predicting and explaining the increased adoption of quality assurance systems to mitigate transaction costs, TCE also has something to say about chain leadership and the mechanisms used for quality assurance. Since quality and food safety assurance affect the cost of carrying out transactions, there may be private incentives for adopting proprietary quality assurance programs, and these incentives are strongest for those agents who are affected most by transaction costs. Thus, TCE predicts the chain leader will be the firm in the chain facing the strongest incentives to act in order to reduce transaction costs.

Leadership in the supply chains described in Section 3 is consistent with the predictions of TCE. In the case of a branded product such as Wheaties, for example, uncertainty about input quality is an important problem, since the ability to sell products with constant quality and very precise characteristics is absolutely necessary. The leader's reputation is a signal of quality for consumers, there is a strong incentive for the leader to adopt well-designed quality assurance systems. Close vertical coordination also reduces transaction costs and input uncertainty. This may help explain General Mills' decision to own elevators located in a region where production uncertainty and quality problems are minimal.[10]

Similar reasoning, based on TCE, helps explain chain leadership by retailers in the case of British beef. Retailers in the U.K. face strong incentives to establish quality assurance systems that ensure consistent quality of supplies and provide traceability as a guarantee that agreed-upon requirements have been met. This makes them quite vulnerable to opportunism from their suppliers. Largely in response to these problems, major British retailers purchase only from "approved" suppliers who will adhere to their specific proprietary quality assurance systems. However, identifying competent suppliers who will adopt "safe" production process adds to transaction costs. Once the potential suppliers are identified, their production systems need to be audited and their products and processes evaluated. Monitoring and enforcement costs after a retailer-supplier relationship has been established also add to transaction costs. They are further increased by the need for traceability, since firms are required to trace the product through the production process and to identify potential sources of contamination. Establishing a supply chain that allows for traceability requires a firm to collect information about its suppliers and their production process as well as monitoring its own production process.

These transaction costs can be mitigated through the adoption of a quality assurance system. Quality systems serve as a mechanism of communication between buyers and sellers, reducing the buyer's uncertainty regarding a product's attributes by providing information about the seller's production process. Some quality assurance systems may specify idiosyncratic production and processing methods. These requirements may operate as "insulating mechanisms" that place some firms and supply chains at a competitive advantage. They create the potential for the capture of additional rents if the participants of a closely coordinated supply chain can offer greater security and more credible guarantees to consumers than their competitors. However, as we shall see, TCE is not well-equipped to explore these aspects, and integration with other perspectives and approaches is useful.

---

[10] Concerns about input quality may also help explain why processors in the Canadian turkey industry are pressuring farmers to adopt HACCP-based on-farm quality assurance and production standards (Hobbs).

Infrequent supply chain relationships place high information and monitoring costs on retailers (and on processors) for identifying suitable suppliers and monitoring those suppliers to prevent potential food safety problems (Hobbs). Closer vertical coordination helps minimize these transaction costs. It facilitates the transfer of information along the supply chain and reduces information asymmetries along the supply chain, enabling retailers to offer quality guarantees to their customers with more confidence and at a lower cost. As we have seen in Section 2, the UK retailers provide a clear example of the relevance of well-designed chains to ensure quality. UK retailers pushed for closer supply chain partnerships. The desire to create a supply chain where transaction costs are minimized by having more frequent transactions with a smaller number of trading partners was a major factor in British retailers' decision to source beef products through partnership agreements with specific processors and groups of farmers rather than through livestock auction markets (Hobbs and Kerr).

Regarding the choice of mechanisms to ensure quality, TCE suggests supply chain participants will (or should) adopt those that are the most efficient in reducing transaction costs. This helps explain the widespread use of process-oriented quality systems and third party certification in emerging food supply chains. These systems can take many forms, including private voluntary international quality assurance standards (such as ISO 9000), national farm-level assurance systems, and proprietary quality assurance systems.

Private voluntary international standards such as ISO 9000 are internationally accepted procedures and guides that ensure the firm's output reflects the standards that have been set. They reduce transaction costs by offering a guarantee of a firm's production process that is understood by and transparent to all trading partners. In effect, use of such private international standards is a form of "outsourcing" for quality assurance. When these systems are adopted, the transaction costs of going to an outside organization are judged to be less than the cost of establishing and maintaining this function internally. In general, these approaches also require a third-party audit to evaluate the integrity of the documented production process and to review the quality system in practice. The TCE explanation for this is that it is the least costly way to establish the credibility of the system.

Farm level quality assurance systems developed by a national government or a farm organization – as in the Label Rouge and Parmesan cheese supply chains – can be viewed as public goods that emerge in cases where no single chain participant has the power and/or incentive to establish and enforce a quality system, yet all participants will benefit if it is established. Institutions play a critical role in reducing the transaction costs of establishing these systems. The laws that protect Label Rouge and PDO products provide incentives for collective action, but a strong sense of unity among farmers is also necessary. It is noteworthy that, with the exception of organic standards, farm level quality assurance systems have been slow to develop in the U.S., where there is little tradition of or legal protection for food products with a strong geographic identity. Finally, chains based on collective quality assurance systems may be particularly vulnerable to the opportunistic behavior of a small number of agents. This raises the problem of high monitoring costs to guarding against "free-riding." This explains why the implementation of quality assurance schemes is particularly difficult in these situations.[11]

---

[11] Evidence from elsewhere provides examples of both success and failures in this regard. The Danish pork industry is an example of a successful experience. Approximately 97% of Danish pigs are sold through a

When information asymmetry exists, buyers may be unwilling to pay high prices for food products given their uncertainty about quality and/or safety. The result is lower prices and few incentives for food firms to provide high quality/safety products. This is the classic situation described by Akerlof in his example of "lemons" in the used car market. The phenomenon of quality erosion depends crucially on the size of penalties and strength of enforcement mechanisms. When market-driven incentives are strong, chain participants are motivated to adopt private (and perhaps proprietary) quality assurance schemes, and the risk of quality erosion is low. In the absence of strong market incentives, the development of well-designed institutions and the active presence and cooperation of private institutions and public regulators are necessary to ensure the credibility of food safety and quality claims.

TCE provides few insights on the distribution of revenues and costs among chain participants. Its emphasis on efficiency does not help to focus this issue. For example, by developing efficient quality assurance systems, chain leaders may introduce changes that could benefit all chain participants. However, the distribution of benefits may also be affected by their strong bargaining power. For example, retailers in the U.K. have encouraged the development of private third party agencies to certify food quality and safety standards where these institutions did not preexist. However, this is explained by the fact that third party accreditation helps minimize the retailer's quality assurance costs because the cost of auditing is transferred to the supplier, who is required to pay a fee to the accreditation agency (Henson and Northen).

Finally, the static nature of most TCE models limits the usefulness of this approach for analysis of typically dynamic issues such as adaptability and innovation.

4.2. Agency theory

Agency theory focuses on situations where two or more individuals with conflicting objectives contribute to a production process. As such, it is natural to consider the applicability of work in this area to the problem of supply chain analysis and design. In their classic paper on information costs and economic organization, Alchian and Demsetz argued that problems with monitoring and metering effort in settings characterized by team production are key determinants of firm structure and firm boundaries. Normative principal-agent models (e.g., Ross; Stiglitz (1974, 1975); Holmström) are concerned with the design of incentive systems that help align the interests of employees (agents) with employers (principals) in settings like those characterized by Alchian and Demsetz.

In the most basic principal-agent model, the principal owns key capital resources used in a production process and is the residual claimant to net returns from that process. The principal hires an agent, who contributes labor and perhaps some human and/or physical capital resources to the production process. The agent's effort affects the level of

---

small number of farmer-owned cooperatives, which engage in primary and further processing. An umbrella organization, "Danske Slagterier" (DS), oversees this cooperative industry structure, coordinating industry-wide research, marketing and strategic planning. DS developed an industry-wide traceability system to satisfy the due diligence requirements of its UK supermarket customers. By adopting a single system satisfactory to all of the major UK retailers, the Danish industry avoided wasteful duplication and chain fragmentation that would have come with developing a different system for each retailer. In contrast, the national quality assurance scheme for U.K. beef and lamb failed because, with no on-farm inspections, the actions of a few producers went unchecked.

production, but it is costly or impossible for the principal to accurately observe and control that effort. The challenge is to design an incentive scheme that will maximize the principal's utility subject to participation and incentive compatibility constraints on the agent's behavior. Optimal incentive schemes typically involve some monitoring (of output if not effort) and a link between performance and the agent's compensation. When there is uncertainty in the production process, one added problem is that the link between pay and performance may also shift risk to the agent, leading to risk avoidance behavior that is not in the best interest of the principal.

Agency models have been extended to more complex, realistic situations where new monitoring technologies, information asymmetries, signaling, and reputation come into play. They have been applied to employer-employee relationships within a firm and to interfirm relationships when a dominant firm purchases from or markets through less powerful trading partners. Clearly, these models are relevant to the problem of supply chain design and managment when a supply chain leader works with its own employees and/or with trading partners to deliver high quality products to its customers as efficiently as possible. Agency problems stemming from conflicting objectives and high monitoring costs can have significant impacts on firm boundaries, the structure of contractual relationships, and the selection of quality assurance mechanisms.

Agency theory sheds little light on the question of who is (or should be) the chain leader. Principal-agent models treat the identity of the principal, or chain leader, as a prior assumption. As will be discussed later in the section on property rights theory, however, agency models can be useful in comparative assessments of supply chain structure and performance under alternative supply chain configurations.

On the other hand, agency models are very well suited for explaining the choice of quality assurance mechanisms in an existing supply chain or for identifying preferred mechanisms when a new supply chain is being established. In the LoSatSoy™ supply chain, for example, DuPont and it subsidiary companies can use relatively inexpensive laboratory tests to ensure varietal integrity each time soybeans or products derived from them change ownership. If one of their "agents" – a farmer, elevator, crusher, or refiner – fails to maintain identity preservation due to lack of effort or outright deception, that agent can be penalized by withholding the identity preservation premium and refusing to do business with the agent in the future. The Wheaties supply chain is similar in this respect, since General Mills can easily test for varietal integrity when wheat is delivered to an elevator or manufacturing plant. In both cases, the ability to use direct measurement for quality assurance reduces agency problems, and trading partner relationships are governed by short term contracts that are similar to standard commodity contracts except for one important difference. While standard forward contracts for commodities specify both the price and the quantity, contracts in these two supply chains specify a price premium and then guarantee that the principal will purchase all production from a specified planted acreage. This shifts some of the quantity risk to the buyer, giving farmers (the agents) an incentive to plant the identity preserved crop on high quality land that can be inspected during the growing season and to use management practices that ensure quality and high production.

Agency theory also helps explain the choice of quality assurance mechanisms in the retailer-led private label and organic dry bean supply chains. In both cases, quality verification by direct testing is not possible, and the focus for quality assurance shifts to process controls centered around supplier certification processes combined with regular and unannounced site inspections. Going through the certification process is expensive for the supplier (agent), and the threat of being removed from the approved supplier list as a result of an unfavorable inspection report provides incentives for the supplier to conform to quality standards established by the buyer (principal). In exchange for bearing the cost of initial certification and continuing quality control, suppliers in both of these chains receive price premiums and assurances that there is a guaranteed market for the special products they produce.

There are not clear principal-agent relationships in the Parmesan cheese and Label Rouge supply chains. While quality assurance mechanisms in both are similar to those in the private label and organic supply chains, these are established and enforced by overarching institutions rather than by a chain leader whose activities are based in a key segment of the supply chain. Therefore, agency theory adds little to our understanding of supply chain configuration in these cases.

Agency models have limited usefulness in explaining the distribution of benefits and costs among chain participants or the adaptability of chain structure to changes in technology or market conditions. Regarding distribution, the principal has nearly all the bargaining power in the typical principal-agent model, and the agent's utility or compensation is driven down to the reservation price that defines the participation constraint. This does point to the fact the distribution of net returns in a chain is strongly influenced by the attractiveness of agents' opportunities in competing supply chains. For example, concentration of retailing in the U.K. keeps margins low for producers in the retailer-led private label chains, while the rapid expansion of market opportunities for organic producers in the U.S. allows them to capture a larger share of the added value in the dry bean chain. Regarding adaptability and innovation, agency models typically treat technology and institutions as fixed and so are not an important tool for understanding this important dimension of supply chain design.

4.3.   Property rights theory

Property rights theory is concerned with the question of who should own assets in settings where two or more technologically separable production activities are vertically linked and it is not possible to write and enforce contracts that specify the actions of all parties involved in all possible situations. In a stylized food product supply chain, for example, investments made and actions taken in the primary production, processing, distribution, and retailing segments of the chain may all affect product quality and chain-wide efficiency. Property rights theory is a tool for assessing whether or not integration of two or more segments under common ownership will improve system-wide performance.

Oliver Hart's *Firms, Contracts, and Financial Structure* provides a concise introduction to property rights theory and its application to determining firm boundaries and selecting an appropriate financial structure for a firm. Earlier papers by Grossman and Hart and by Hart and Moore develop some of the key principles for this approach. Throughout this work, attention is focused on tradeoffs between the benefits and costs of concentrating ownership of relation-specific assets, where ownership is defined as having residual rights of control. The following passage from Hart (p. 33) describes the essence of this tradeoff:

> In summary, the benefit of integration is that the acquiring firm's incentive to make rela-
> tion-specific investments increases since, given that it has residual control rights, it will
> receive a greater fraction of *ex post* surplus created by such investments. On the other
> hand, the cost of integration is that the acquired firm's incentive to make relation-
> specific investments decreases since, given that it has fewer residual control rights, it will
> receive a smaller fraction of the incremental *ex post* surplus created by its own in-
> vestments.

The following example, based on the chef-skipper-tycoon example presented by Hart
and Moore (pp. 1122-1124), illustrates the basic features of property rights theory to
supply chain design. Coordination among three technologically separable activities –
production by farmers, who are organized in a cooperative and act together; assembly,
storage, and handling by a strategically located elevator; and processing by a food manufacturer – is necessary to produce a new food product using a special variety of maize.
Each activity is essential. In addition, the farmers must collectively make an investment of
$1,000,000 in human capital formation to be able to grow the special variety, and the
food manufacturer must make an investment of $1,000,000 in developing knowledge of
the market. These investments can be observed by others in the chain, but they cannot be
verified by an outside party and are not transferrable (i.e., human capital cannot be sold
along with physical assets). No unique human capital investments are needed at the elevator level, and the labor market for elevator managers functions well.

If the supply chain for this new product can be implemented, it will generate a net
present value of $2,400,000 in added revenue (relative to standard maize) over the rele-
vant planning horizon. Since this exceeds the combined cost of investments by farmers
and the food manufacturer, the first best solution is to make the investments and establish
the chain. Due to high transaction costs, however, it is not possible to write complete
contracts that will guarantee the chain will operate as expected after investments are made. One party may "hold up" the others by threatening to withdraw its assets from the
chain. Therefore, there will be *ex post* bargaining once the system is established, and it is
assumed that the $2,400,000 in added value will be divided equally among the indepen-
dent firms.

In this chain, if all three segments are independently owned, each firm will receive
$800,000 in added value. This will not be enough to compensate the farmers and food
manufacturer for their investments and so the chain will not be established. On the other
hand, if either the farmer cooperative or the food manufacturer acquires the elevator, the
added value will be divided between only two independent firms, and the supply chain
will be established. Finally, if the farmer cooperative acquires the food manufacturing
firm, it may not be possible to create an incentive scheme that will induce the manager of
the food manufacturing firm (who would be an employee rather than an owner under
these circumstances) to make the necessary investment in knowledge of the market, and
the chain will not be established. Similar arguments suggest that the chain will not form if
the food manufacturer acquires the physical assets of the farmers and hires them as employees.

This simple example shows how alternative ownership patterns can affect supply
chain performance. Property rights models can be extended to include issues such as
worker incentives, reputation effects, capital formation, and investments in physical capital. The following are some of the general propositions on optimal ownership structure
that are presented by Hart and Moore (pp. 1131-1139) and Hart (pp. 44-55).

- If only one agent has an investment, that agent should own all assets.
- If an asset is idiosyncratic to an agent (i.e., owning the asset does not affect investment incentives for any other agent), that agent should own the asset.
- If an agent is indispensable to an asset (i.e., the asset has value only if it is owned by the agent), that agent should own the asset.
- If a group of agents are needed to make investment in an asset productive, control of the asset should be governed by majority voting among them.
- When assets are economically independent, independent ownership is better than integration.
- When assets are highly complementary, some form of integration is better than independent ownership.

With its focus on the benefits and costs of alternative ownership patterns, it is not surprising that property rights theory helps to explain asset ownership patterns and chain leadership roles in the six food product supply chains described in this paper. In the Lo-SatSoy™ supply chain, for example, the overall chain coordination and leadership function is highly complementary with investments in varietal development, and these segments are jointly owned by DuPont. In contrast, the other segments of the supply chain – farm production, assembly and transportation, crushing, and refining – are economically independent, since use of a particular farm, elevator, crusher, or refiner is not critical to the success of the system. Similarly, in the retailer-led private label chain, the retailer's investments have the greatest impact on added value in the system, but the value of these investments is affected by ability to control the quality assurance systems. Therefore, the retailers are chain leaders, and the quality assurance systems are retailer-owned. General Mills' chain leadership in the Wheaties supply chain can be explained in the same way.

In the Parmesan cheese supply chain, farmers whose land is located in the delimited production area are absolutely essential to the existence of the product. Milk processing plants and cheese aging facilities located in the region are also essential. The former are typically owned by producers through cooperatives. On the other hand, wholesalers control most of the aging facilities and distribution functions for the chain, and they are not owned by producer groups. Given the key role of producers in this chain, the fact that the chain leadership role is played by the Consorzio di Tutela del Parmigiano Reggiano, a cooperative organization that is democratically controlled by the producers, is consistent with the predictions of property rights theory. A similar line of reasoning supports the chain leadership role of third party certification agencies in the organic dry bean and the *filières* in the Label Rouge supply chains. In effect, these are jointly controlled organizations that allow a diverse group of agents to cede at least part of their residual control rights in the interest of achieving system-wide coordination that makes the chain possible.

Property rights theory contributes little to our understanding about other key aspects of supply chain design. Choice of quality assurance mechanisms and adaptability in the face of change are outside the scope of property rights models. Because patterns of asset ownership affect the distribution of net benefits among chain participants, property rights models do provide some insights on this important aspect of supply chain design. Ultimately, though, distribution of benefits is strongly influenced by bargaining, and most property rights models make fairly simplistic assumptions about bargaining outcomes.

4.4.   Resource and capability theory

While TCE provides a crucial theoretical framework for understanding and predicting how the need to reduce transaction costs affect the design of supply chains, it has also received several criticisms. In particular, TCE has been criticized for failing to recognize the importance of dynamic aspects, such as learning and innovation. Furthermore, because TCE focuses on a single transaction, it is not an appropriate framework for understanding learning and innovative processes when knowledge is broadly distributed and the locus of innovation is found in a network of interorganizational relationships. Finally, because of its static nature and its focus on a single transaction, TCE, as well as organizational economics in general, may not be helpful for analyzing the relationship between organization and competitive advantage.

Therefore, it is useful to integrate TCE with other perspectives provided by new approaches recently developed in the strategic management literature. In this literature concurrent emphasis is given to the role of intangible assets such as knowledge (Winter), core competencies (Prahalad and Hamel), learning (Senge), and brand image or corporate culture. Firms that accumulate complex resources effectively are capable of executing more complex strategies and sustaining competitive advantage. There are, of course, differences in emphasis among these approaches. However, their common characteristic is that they view the firm as an organization whose main purpose is to generate rents through creating and sustaining sources of competitive advantage. Finally, it is also important to note that this literature provides useful notions and analytical tools for examining the sources of competitive advantage not only at the level of a single firm but also at the more aggregate level of supply chains or networks of related firms.

In what follows, we provide a brief analysis of the most useful insights provided by this literature, focusing mainly on the resources-based view of the firm (RBV). While TCE is a static approach, concerned primarily with transactions that involve fixed, tangible assets, RBV offers a dynamic view focusing on rent-creating behavior through the ownership and exploitation of unique resources and capabilities. The RBV understands rents as being derived in large part from intangible assets such as organizational learning, brand equity, reputation and knowledge. The works of Shumpeter and Penrose provide the essential building blocks for this view, as developed more recently by Wernerfelt, by Barney, by Mahoney (1995), and by Grant.

Under the RBV perspective, resources (defined as tangible or intangible assets of the firm) are the basic unit of analysis. They can be given exogenously or created by activities within the firm. Examples include brand names, in-house knowledge, employee skills, trade contracts, physical assets, efficient procedures, and sophisticated computer systems (Wernerfelt). Capabilities (e.g., launching new products, product development, vertical differentiation and quality upgrading, delivering high-quality service, fast and flexible response to customer demands, etc.) emerge from the integration and combination of these resources. Capabilities are tangible or intangible assets which are firm-specific. They are created over time through complex interactions among the firm's resources and the exchange of information through the firm's human capital (Amit and Schoemaker). A firm's strategies are seen as a continuing search for above normal returns (rents), constrained by and dependent on its resources endowments.

While TCE focuses on how to integrate various activities in order to lower transaction costs, the RBV provides insights into the role resources and capabilities can play in providing a source of competitive advantage. It emphasizes that resources and capabilities must meet three criteria to confer competitive advantage. They must be valuable, rare, and imperfectly imitable. The RBV sees the firm's competitive advantage and competitiveness as determined by a process of resource accumulation and deployment. Resources and capabilities that are rare, difficult to imitate, and useful for creating value in a given industry or supply chain are often labeled "strategic resources."

A crucial prediction of the RBV is that firms are heterogeneous in terms of specific resources and capabilities. Asymmetries in the distribution of resources and capabilities across firms and supply chains and associated differences in performance do not depend only on luck, success in search, history, or inherent causal ambiguity. They also stem from the fact that some firms (and supply chains) are able to protect their unique resources and capabilities from expropriation or imitation more effectively than other firms and chains.

A further important insight from the RBV is that the ability of a firm to control the strategic resources needed to develop a competitive advantage depends on product or industry characteristics. For example, the necessary strategic resources might differ across products with different degrees of credence, experience, and search qualities. In this way, the RBV help to focus on the specific resources and capabilities that are most critical for a supply chain to develop sustainable competitive advantages in terms of quality attributes. For example, brand name reputation may be more valuable in industries that produce experience or credence goods than in industries where quality can be determined prior to purchase. Thus, in industries or supply chains where quality attributes are less observable, the development of brand names is more important and the presence of a chain leader endowed with adequate resources and capabilities is crucial.

Although the RBV is analytically less developed than TCE, the RBV can enrich the conceptual framework used to analyze supply chains. The reason is that the creation of economic rents through strategic initiative is essential to the choice of organizational structures. Avoiding opportunism and minimizing transaction costs may not be the only or the most relevant factors at work. In TCE, asset specificity is the basic concept for explaining the degree of vertical integration and coordination as well as predicting the chain leader. RBV provides a different explanation based on the idea that it is most unlikely that many firms have the ancillary capabilities necessary for all activities in the chain of production (Langlois and Robertson). According to this view, closely related activities are best undertaken under unified governance, since they require similar capabilities. When activities are dissimilar and require different capabilities, it may be more efficient for firms to rely on the capabilities owned by others firms accessed through linkages based on market transactions [12]. Thus, a crucial prediction of the RBV is that the boundaries of the firm are determined (at least in part) by the relative costs of developing ancillary capabilities internally or purchasing them externally from other firms. The RBV also helps focus attention on the importance of the chain leader's strategic resource endowments. For example, while TCE can explain General Mills' decision to own elevators, the RBV suggests that, independently from transaction cost considerations, the existence of a strategic resource such as a strong brand is crucial to explaining the leader's decisions and overall chain performance.

---

[12] These findings are consistent with propositions derived in the property rights framework (Hart and Moore; Hart).

Recent extensions of the RBV provide further insights on how food supply chains can accumulate resources and capabilities to ensure quality. For example, the dynamic capability paradigm (Lado and Wilson; Teece, Pisano, and Shuen) is an integrative approach that offers a better understanding of the mechanisms by which firms faced with significant changes in the competitive environment can accumulate and disseminate new skills and capabilities, despite conditions of path dependency and core rigidity in technological and organizational processes. Firms that have dynamic capabilities are flexible and able to adjust quickly.

The RBV also emphasizes the importance of capabilities that foster cooperation and trust. It has been argued that one weakness of TCE is its focus on a single firm and on single transactions. Less theoretical and empirical work has been dedicated to analyzing processes by which multiple firms, working collaboratively, develop individual and common capabilities. The capabilities arising from interorganizational relationships may be crucial for the creation of sustained competitive advantage through the development of collaboration and cooperation based on trust.

Williamson (1996) argues that "trust" is a much misused term, with apparent examples of transactions based on trust actually being examples of calculative relationships which can be rationalized as individual self-interest. On the other hand, trust can play a significant role. With complete trust there is no opportunism, and with no opportunism, the transaction costs decrease. If supply chains differ in terms of the trust embedded in them and if trust affects transaction costs, the issue is worth exploring.

In this regard, the RBV of networks developed by Gulati, Nohria, and Zaheer provides interesting insights. They show that one important implication of firms' embeddedness in networks is that enhanced trust between firms can mitigate moral hazards. Networks promote trust and reduce transaction costs in several ways. First, networks enable firms to gather superior information about each other's capabilities. They can also facilitate due diligence, since each partner has greater knowledge about the other's resources and capabilities and greater confidence in their mutual assessments. Thus, networks can greatly reduce the informational asymmetries that increase transaction costs. In addition to this knowledge-based trust, networks can also create deterrence-based trust, since it is easier to discover opportunistic behavior in a network and its costs are greater because of reputational effects. Gulati, Nohria, and Zaheer observe that the loss of reputation can influence not just the specific relationship in which one behaved opportunistically, but all other current or potential relationships.

A further insight from this work on strategic networks is provided by the notion of "network resources." While the RBV was originally developed to examine the relationships between the resources of a firm and its performance, it is quite possible to develop a resource-based view not only of a firm but of a strategic group of firms and of an industry as well. Even if this extension has not been well developed up to now, it is possible to apply the RBV approach to a supply chain and examine the key questions of what resources generate a sustainable competitive advantage at the level of the supply chain and by what mechanisms they achieve this. Since the key concept in the RBV is that a resource can produce a rent if it is valuable, imperfectly imitable, and substitutable by all other firms, it is natural to extend this notion to the resources that can be located and created within a given chain and that are difficult for firms in other chains to imitate or substitute.

Interfirm networks may provide an effective way to organize knowledge transfer or access in dynamically competitive domains, as well as in contexts where complex knowledge is scattered or specialized. An intensive exchange of knowledge, deliberately delivered, may help reinforce strategic positioning. In this sense, suppliers may be regarded as resources enabling firms to consolidate in-house competencies. The source of value-creating resources and capabilities can go beyond the boundaries of the firm. A network can be seen as an inimitable resource, by itself, and a means to access inimitable resources and capabilities (Gulati, Nohria, and Zaheer). A corollary to the proposition that networks (and supply chains) provide an environment through which the firm can access key strategic resources is notion that it being embedded in a successful chain can become almost a precondition for firms to develop resources and capabilities. The idea is that strategic networks at least potentially provide a firm with access to information, resources, markets and technologies.

Empirical support for the proposition that strategic networks give firms access to information, resources, markets and technologies is clearly provided by the case of retailer-driven supply chains in UK. Most notably, the beef supply chains established by Marks & Spencer and Tesco use information feedback from the retailer to farmers as a source of competitive advantage for the entire chain and as an incentive for farmers to remain loyal to the chain.

Analysis based on the RBV typically views networks and supply chain structures as the result of path dependent processes. For example, the institutions and trading partner relationships that are essential features of the Parmesan cheese supply chain are possible, at least in part, because of the centuries-old tradition of producing this product and the passage of laws in Italy and the EU that establish the PDO designation. Recognizing path dependence helps explain why successful chains and networks are difficult for competitors to imitate.

Finally, it is equally important to note that the firm's network structure may lock it into undesirable strategic situations. This is a real possibility for supply chains based on geographical designations and for the European chains producing artisanal products in general. History and tradition can lead to a chain characterized by poor structure and inappropriate behaviors. The firms locked in these chains do not receive resources and capabilities, nor do they transfer them to other firms in the chain. The result is an inefficient equilibrium. The Italian experience with producers' associations shows that, more often than not, they are rather slow in evolving towards the more sophisticated and strategic functions. As we have seen, these institutions are critical for (re)design of the supply chain to ensure quality. The performance of these supply chains is strongly related to the strategic role played by their producers' associations.

5.   CONCLUDING REMARKS

Increased emphasis on food quality and safety and increased reliance on interorganizational cooperation in integrated supply chains are key aspects of structural change in the food system. In this paper we have examined supply chain design strategies for ensuring food quality. Our tools for this task included descriptions of supply chains for six diverse product categories and application of four conceptual frameworks to the challenge of explaining similarities and differences in those chains.

The case examples we chose for this paper are generally representative of supply chains for six broad categories of food products: branded products, genetics-based products, products defined by special production practices, retailer private-label products, protected denomination of origin products, and products produced under a national quality assurance system. These supply chains differ significantly with respect to choice of mechanisms for ensuring food quality and the locus and strength of chain leadership.

While mechanisms for ensuring food quality are numerous, our empirical descriptions clearly show that there are two main types of chain. In the first, quality controls are based primarily on direct inspection and measurement. In these chains (Wheaties and LoSatSoy™ cooking oil) laboratory testing and control technologies are adequate for ensuring product quality. In the second type, key product attributes are not directly measurable or measurements are very costly. In these chains (organic dry beans, private label beef in the U.K., Parmesan cheese, and Label Rouge poultry) there is increasing reliance on trace-back systems and quality assurance mechanisms based on process controls designed and implemented through ISO quality systems and HACCP. To a large extent, then the selection of mechanisms is explained by the presence of hidden characteristics and by the need to minimize the costs for quality control. Given current trends in consumer demand and the increasing complexity of food product attributes, we expect reliance process controls to accelerate and become increasingly common in the years ahead.

With respect to chain leadership, our empirical descriptions once again suggest that chains can be grouped into two broad categories. In the first, the chain is established and organized by a well-defined chain leader whose activities are based in a key segment of the supply chain. This is the case for Wheaties, LoSatSoy™ cooking oil, and private label beef in the U.K. The leadership structure of chains in the second category is quite different. They are established and managed, at least in part, by overarching institutions established through collective action, legislation, or both. Supply chains for organic dry beans, Parmesan cheese, and Label Rouge poultry fall in this category.

In the second half of this paper, we briefly introduced and assessed the applicability of conceptual and analytical tools for supply chain analysis and design based on: transaction cost economics, agency theory, property rights theory, and the resource-based theory of the firm. There are important links and complementarities among these approaches. Transaction cost economics, with its emphasis on the governance and mechanisms for reducing transaction costs, provides crucial insights, predictions and explanations. Agency theory offers rigorous analytical tools for analyzing the choice of quality assurance mechanisms and for designing new mechanisms. Property rights theory, with its focus on optimal asset ownership patterns, helps explain chain leadership roles. However, these theories are not well suited for the analysis of competitive advantage and firm heterogeneity. They typically treat technology and institutions as fixed and so are not important tools for understanding how chains can be designed to assure dynamic performance in terms of adaptability, innovation, and sustained competitive advantages.

We find that useful insights for the design and management of supply chains to ensure food quality are also provided by strategic management literature. The resource-based view, though less fully developed analytically and methodologically than other theoretical frameworks, has the merit of adopting a more holistic and dynamic approach. It contributes to our understanding of the strategies firms and supply chains can use to identify and acquire resources and capabilities.

Ultimately, the paper highlights the significance and value of examining supply chains through an extended conceptual framework. Thus, our paper adds to recent contributions concerned with synthesizing insights for different analytical perspectives (Langlois and Robertson; Foss). We believe concepts and tools based on such a synthesis will be important for future work oriented toward explaining similarities and differences in existing supply chains and toward designing new supply chain structures that are well adapted to the changing food system.

6. REFERENCES

Akerlof G.A. (1970), The Market for 'Lemons': Qualitative Uncertainty and the Market Mechanism, *Quarterly Journal of Economics*, **84**: 488-500.

Alchian A.A. – Demsetz H. (1972), Production, Information Costs, and Economic Organization, *American Economic Review*, **62**: 777-795.

Amit R. – Schoemaker P.J.H. (1993), Strategic Assets and Organizational Rent, *Strategic Management Journal*, **14**: 33-46.

Barney J.B. (1991), Firm Resources and Sustained Competitive Advantage, *Journal of Management*, **17**: 99-120.

Canali G. (1996), Evolution of consumers' preferences and innovation: the case of Italian foods with denominations of origin, in: Galizzi G. – Venturini L., *Economics of Innovation: The Case of Food Industry*, Phisica-Verlag, Heidelberg, pp. 319-327.

Chambers W.B. (1999), *Changes in the Structure of the U.S. Food System: Evidence from the Dry Bean Industry*, Ph.D. Dissertation, Department of Applied Economics, University of Minnesota.

Coase R (1937), The Nature of the Firm, *Economica*, **4**: 386-405.

Darby M.R. – Karni E. (1973), Free Competition and the Optimal Amount of Fraud, *Journal of Law and Economics*, **16**: 67-88.

Demsetz H. (1991), The theory of the firm revisited, in: Williamson O.E. – Winter S.G., *The Nature of the Firm: Origins, Evolution, and Development*, Oxford University Press, New York, pp. 159-178.

Fearne, A. (1998), The Evolution of Partnerships in the Meat Supply Chain: Insights from the British Beef Industry, *Supply Chain Management: An International Journal*, **3**: 214-231.

Fearne A. – Hughes D. (1999), Success Factors in the Fresh Produce Supply Chain: Insights from the UK, *Supply Chain Management: An International Journal*, **4**: 120-128.

Foss N.J. (1999), Research in the Strategic Theory of the Firm, *Journal of Management Studies*, **36**: 725-755.

Galizzi G. – Venturini L. (eds., 1999), *Vertical Relationships and Coordination in the Food System*, Physica-Verlag, Heidelberg.

Grant R.M. (1996), Prospering in Dynamically-Competitive Environments: Organizational Capability as Knowledge Integration, *Organization Science*, **7**: 375-387.

Grossman S.J. – Hart O.D. (1986), The Costs and Benefits of Ownership: A Theory of Vertical and Lateral Integration, *Journal of Political Economy*, **94**: 691-719.

Gulati R. – Nohria N. – Zaheer A. (2000), Strategic Networks, *Strategic Management Journal*, **21**: 203-215.

Hart O. (1995), *Firms, Contracts, and Financial Structure*, Clarendon Press, Oxford.

Hart O. – Moore J. (1998), Property Rights and the Nature of the Firm, *Journal of Political Economy*, **98**: 1119-1158.

Harvey M. (2000), Innovation and Competition in UK Supermarkets, *Supply Chain Management: An International Journal*, **5**: 15-21.

Henson S. – Northen J. (1998), Economic Determinants of Food Safety Controls in Supply of Retailer Own - Branded Products in United Kingdom, *Agribusiness: An International Journal*, **14**: 113-126.

Hobbs J.E. (1996), A Transaction Cost Analysis of Quality, Traceability and Animal Welfare Issues in UK Beef Retailing, *British Food Journal*, **98**: 16-26.

Hobbs J.E. – Kerr W.A. (1991), The 1990 Food Safety Act and Scottish Agribusiness, *The Scottish Agricultural Economics Review*, **6**: 51-59.

Holleran E. – Bredahl M.E. – Zaibet L. (1999), Private Incentives for Adopting Food Safety and Quality Assurance, *Food Policy*, **24**: 669-683.

Holmström B. (1979), Moral Hazard and Observability, *Bell Journal of Economics*, **10**: 74-91.

Hughes D. (ed., 1994), *Breaking with Tradition: Building Partnerships & Alliances in the European Food Industry*, Wye College Press, Wye, U.K.

Iowa Sate University Office of Biotechnology (1997), *ISU and Iowa Companies Launch First Low-Saturated Soybean Oil*, Biotechnology Update XI.
URL: http://biotech.iastate.edu/biotech_update/Oct_97.html#low_sat

Kalaitzandonakes N. – Maltsbarger R. (1998), Biotechnology, identity preserved crop systems, and economic value, in: Ziggers G.W. – Trienekens J.H. – Zuurbier P.J.P., eds., *Proceedings of the Third International Conference on Chain Management in Agribusiness and the Food Industry*,. Management Studies Group, Wageningen Agricultural University, The Netherlands, pp. 649-658.

King R.P. (2000), *Supply Chain Design for Identity Preserved Agricultural Products*, seminar paper presented at the Mansholt Institute, Wageningen Agricultural University, The Netherlands.

Lado A.A. – Wilson M.C. (1994), Human Resource Systems and Sustained Competitive Advantage: A Competency-Based Perspective, *Academy of Management Review*, **19**: 699-727.

Langlois R.N. – Robertson P.L. (1995), A dynamic theory of the boundaries of the firm, in: *Firms, Markets, and Economic Change*, Routledge Press, New York, pp. 20-45.

Lohr L. (1998), Implications of Organic Certification for Market Structure and Trade, *American Journal of Agricultural Economics*, **80**: 1125-1129.

Mahoney J. (1992), The Choice of Organizational Form: Vertical Financial Ownership Versus Other Methods of Vertical Integration, *Strategic Management Journal*, **13**: 559-584.

Mahoney J. (1995), The Management of Resources and the Resource of Management, *Journal of Business Research*, **33**: 91-101.

Nelson P. (1970), Information and Consumer Research, *Journal of Political Economy*, **78**: 311-329.

Penrose E.T. (1959), *The Theory of the Growth of the Firm*, Wiley, New York.

Prahalad C.K. – Hamel G. (1990), The Core Competence of the Corporation, *Harvard Business Review*, **68**: 79-91.

Ross S. (1973), The Economic Theory of Agency: The Principal's Problem, *American Economic Review*, **63**:134-139.

Royer J.S. – Rogers R.T. (1988), *The Industrialization of Agriculture*, Ashgate, Aldershot.

Segerson K. (1999), Mandatory Versus Voluntary Approaches to Food Safety, *Agribusiness: An International Journal*, **15**: 53-70.

Senge P.M. (1990), *The Fifth Discipline*, Doubleday, New York.

Stiglitz J. (1975), Incentives, Risk, and Information: Notes Towards a Theory of Hierarchy, *Bell Journal of Economics*, **6**: 552-579.

Stiglitz J. (1974), Risk Sharing and Incentives in Sharecropping, *Review of Economic Studies*, **41**: 219-256.

Sylvander B. (1996), Normalisation et Concurrence Internationale: La Politique de Qualité Alimentaire en Europe, *Economie Rurale*, **231**: 56-61.

Teece D.J. – Pisano G. – Shuen A. (1997), Dynamic Capabilities and Strategic Management, *Strategic Management Journal*, **18**: 509-533.

Verrini L. (1999), La Cultura della Qualità nel Parmigiano, *Qualità*, **29**: 28-32.

Wernerfelt B. (1984), A Resource-Based View of the Firm, *Strategic Management Journal*, **5**: 171-180.

Westgren R. E. (1994), Case Studies of Market Coordination in the Poultry Industries, *Canadian Journal of Agricultural Economics*, **42**: 565-575.

Westgren R. E. (1999), Delivering Food Safety, Food Quality, and Sustainable Production Practices: The Label Rouge Poultry System in France, *American Journal of Agricultural Economics*, **81**: 1107-1111.

Williamson O.E. (1979), Transaction-Cost Economics: The Governance of Contractual Relations, *Journal of Law and Economics*, **22**: 233-261.

Williamson O.E. (1985), *The Economic Institutions of Capitalism*. Free Press, New York, 1985.

Williamson O.E. (1991), Strategizing, Economizing, and Economic Organization, *Strategic Management Journal*, **12**: 75-94.

Williamson O.E. (1993a), Comparative Economic Organization: The Analysis of Discrete Structural Alternatives, *Administrative Science Quarterly*, **36**: 269-296.

Williamson O.E. (1993b), Transaction Cost Economics and Organization Theory, *Industrial and Corporate Change*, **2**: 107-156.

Williamson O.E. (1996), *The Mechanisms of Governance*, Oxford University Press, New York.

Winter S.G. (1987), Knowledge and competence as strategic assets, in: Teece D., ed., *The Competitive Challenge*, Ballinger, Cambridge, MA, pp.159-84.

Ziggers G.W. – Trienekens J.H. – Zuurbier P.J.P. (eds., 1998), *Proceedings of the Third International Conference on Chain Management in Agribusiness and the Food Industry*, Management Studies Group, Wageningen Agricultural University, The Netherlands.

LAND & RESOURCE ASSESSMENT

WATER SCARCITY: INSTITUTIONAL CHANGE, WATER MARKETS, AND PRIVATIZATION

Cesare Dosi and K. William Easter [*]

SUMMARY

A number of countries face water shortages because they need to make some basic changes in their water management. Policy options do exist. Most of them share the objective of treating water and water services as an economic good, by regulating private inefficient appropriation of open-access resources, and by making the demand for water less independent of users' willingness to pay for it. The aim of this paper is to provide an overview of these policy options by illustrating their rationale and possible caveats. We begin by stressing the importance of improving countries' social capital (i.e., institutional arrangements and management rules for allocating water between competitive uses). We then concentrate on some economic approaches to improving water management, i.e., the establishment of water markets and the privatization of water utilities, by focussing on experiences and on-going developments in the United States and the European Union.

1. INTRODUCTION

Increasingly, water scarcity is described as a major challenge facing the world (Postel, 1999). However, its definition remains a controversial issue. For instance, water scarcity is not an absolute concept, and is dependent on a number of factors besides water's annual availability. A number of countries face water shortages because they need to make some basic changes in their water management including, in some cases, infrastructural and institutional changes to facilitate management. Another group of countries have water shortages because they need to make management changes and develop new

---

[*] Cesare Dosi, Department of Economic Sciences, University of Padua. William K. Easter, Department of Applied Economics, University of Minnesota. An earlier version of the paper was presented at the 2nd International Symposium of Water UNITWIN-UNESCO, Cannes (France), May 20-31, 2000. The authors would like to thank Tracy Boyer for her help in putting together material for the U.S. section of the paper, and Alberto Garrido for providing information about the recent Spanish provisions about water markets.

Economic Studies on Food, Agriculture, and the Environment
Edited by Canavari *et al.*, Kluwer Academic/Plenum Publishers, 2002

91

sources of water supply. A third group needs to change institutional arrangements for managing and allocating water.

Policy options do exist. Most of them share the objective of treating water and water services as an economic good, by making water demand less independent of users' willingness to pay for it. Policy options also include regulations and other instruments designed to help prevent the inefficient use of open access water resources.

The aim of this paper is to provide an overview of some of these management policy options by illustrating their rationale and possible caveats, both in terms of the available theoretical literature and real world applications. We begin in section 2 by stressing the importance of improving countries' social capital (i.e., institutional arrangements and management rules for allocating water between competitive uses).

In section 3 we describe a number of institutional issues that have arisen in the establishment of water markets and the privatization of water utilities. We also discuss some legal and regulatory changes that may be required if water is to be used more efficiently.

In section 4 we focus on water policies in the United States and European Union. The aim is to provide an overview of some ongoing developments, especially in regards to privatization and the use of water markets. National statistics in both areas seem to show a relative abundance of fresh water. Yet the persistent geographical water imbalances, the increasing importance of quality issues, the competition between traditional water uses and instream uses, and a reconsideration of the role of the public sector, raise serious questions regarding the traditional approaches to the operation and management of water services.

Finally, section 5 provides some general concluding remarks and policy suggestions.

2.   WATER SCARCITY AND SOCIAL CAPITAL

The term scarcity is commonly used to describe smallness of supply compared with demand. In the case of freshwater, both supply and demand are open to different qualifications and definitions. As far as supply is concerned, a frequently used indicator of countries' freshwater availability is the total amount of internal renewable flows, generally evaluated according to annual and long-term average precipitation and evapotranspiration rates.

Although (per capita) internal renewable water flows may prove useful to highlight situations of relative abundance, or potential scarcity [1], they may mask significant geographical unbalances, and inter-annual or seasonal fluctuations: *ceteris paribus*, the higher these unbalances and fluctuations are, the more a country is prone to water shortages.

Providing adequate supply has several components. It involves providing an adequate quantity of water, of the required quality, where and when is needed. Fulfilling these

---

[1]   Following Brouwer and Falkenmark (1989), renewable freshwater available in a country (TR) is defined as the total amount moving in rivers or aquifers; TR may be divided into the amount originating from «domestic» rainfall or by water received from neighboring countries in transboundary rivers and aquifers. According to the Falkenmark's water stress index, while a country with more than 1,700 cubic meters/year/person is expected to experience only intermittent and localized water shortages, the threshold of 1,000 cubic meters has been proposed as an approximate benchmark below which a country is likely to experience widespread and chronic shortfalls; at less than 500 cubic meters per capita, water availability becomes a primary constraint on socio-economic development.

supply conditions depends on a country's capital endowments, namely on man-made capital — infrastructures able to (re)distribute water over space and time and water treatment facilities — and on a country's social capital. The latter concerns the relationships between individuals, between institutions, and between individuals and institutions (Pearce and Atkinson, 1998). The empirical evidence suggests that communities relatively poorly endowed with freshwater, but which have been able to develop appropriate formal or informal resources management rules, perform better than societies which, despite their theoretical global abundance, are unable to properly allocate water resources between different users: ceteris paribus, the more binding are the economic and political constraints facing a society in developing appropriate man-made or social infrastructures, the more water scarcity is likely to emerge.

For water demand, a distinction should be made between water needs and the much larger set of wants for water to provide additional goods and services (Lundqvist and Gleick, 1997). Tracing this borderline is not easy, in that the concept of needs reflects individual and social value judgements, which vary over time and space. But if the term need is intended to mean "basic human requirements," then it is possible to draw a distinction, within water demand, between what is independent of economic and social conditions, and what can be manipulated through prices, or other social and legal rules [2]. In this respect, it must be emphasized that in many countries, the lack of institutional arrangements aimed at regulating private appropriation of common (open access) resources, and the subsidization of water services, have helped transform the concept of water demand into one that should be more properly described as "water requirements under the expectation of quasi-null cost" (Arrojo, 1999).

The failure to treat water (and water services) as an economic good is also generally responsible for a circularity between rising demand, inadequate supply, and increasing (perceived) scarcity. When water is demanded at prices below supply costs, consumers do not provide enough revenue to expand water supply systems (Garrido, 1999). Consequently, users feel deprived of what they perceive as a need, simply because water demand has been derived, de facto, substantially independent of their willingness-to-pay for it.

The number of water scarce communities becomes large only when you add those who lack the economic, administrative, and political ability to reduce the water resource constraint. Many areas, around the world, fall within this latter category. But in many of these countries the current unbalances between water demand (wants for water) and supply could be bridged through improving their "social capital," i.e., through appropriate institutional arrangements (demand-side management reforms).

3. IMPROVING WATER MANAGEMENT THROUGH INSTITUTIONAL CHANGES

For most of the U.S. and Europe the lack of water is not likely to be a major constraint to future increases in food production or cause people to change how they use water to meet their basic needs. However, future water availability may well depend on how fast both regions pursue policies to improve water management, including the intro-

---

[2] Water for basic human requirements would have almost a completely inelastic demand.

duction of water markets and the privatization of water utilities. Cheap water has weakened the link between any pending scarcity and innovation. Thus, while we work to conserve and substitute for scarce resources, we continue to use "cheap" water lavishly.

The rapid growth in water demands between 1940 and 1980 has forced a number of countries to rethink their approach to water management. The shift in thinking has been slow in coming but seems to have taken hold in some countries. The new approach has involved changing the role of government in managing water. More emphasis is placed on water as an economic good and less on the idea that since people have a basic right to water it should be provided no matter what the cost.

There is a growing realization that water management provides a bundle of services that can be divided up, with some of the services better (more efficiently) provided by the private sector (Easter and Feder, 1997). By unbundling services, the public sector can maintain its role where it is most important, i.e., protect against monopoly power, negative externalities, the under-provision of "public goods," and the overuse of "open access" water. The private sector and market forces can then be used to help better manage and allocate water services.

The western United States, Chile, and Mexico have used the water markets approach fairly extensively over the past two decades. Water markets have also been developed for groundwater in a number of countries (Easter, et al., 1998). The privatization of public water entities has occurred more extensively in Europe and Latin America. England was one of the leaders when it sold off its public water assets to large private companies in 1989.

3.1. Institutional changes: objectives and constraints

Clearly, as a country takes stock of its water resources and considers the need for institutional change, they have to spell out their objectives and potential constraints. How important are efficiency considerations relatively to equity concerns? For example, how important are concerns about potential monopoly rents, or the ability of the poor to pay for a reliable and clean water supply? If the decision is to move toward an efficient allocation and management of water over time, a number of institutional arrangements need to be in place.

After examining their objectives, many states in the United States and countries in Europe would probably find that in the last two decades a higher priority has been given to cleaner water for recreation and ecological uses. This is illustrated by the decision to reallocate some of the water going to irrigation in the central valley of California to improving water quality for fish habitat.

Another objective that is important for some countries is the desire to reduce public spending, especially for large projects involving big physical changes in natural water systems. This is important in the United States as illustrated by the fact that the last major United States federal water project (the Central Arizona Project) was approved for construction over 25 years ago. European countries also want to reduce public spending on the delivery of water services and are focusing on water quality for domestic services.

In designing their water institutions, European countries are more constrained because of the international nature of most of their river systems; most European rivers cross three or more countries. In the United States, only a few rivers cross more than one country, and those that do, cross only two countries. For example, the Mississippi River basin is almost totally within the United States' boundaries, while the Great Lakes involves only Canada and the United States. The United States, along with its two neighbours, has generally developed institutional arrangements (such as the International Commission on the Great Lakes) to coordinate the management of much of its international water resources. In contrast, if the Mississippi or the Great Lakes were in Europe, water management would involve at least six or more countries. The February 2000 cyanide spill from a Romanian gold mine illustrates this co-ordination problem. The spill first polluted the Tisa River (which flows into Hungary and Yugoslavia) before it reached the Danube River (Associated Press, 2000).

3.2.   Water markets

Establishing the institutional and organization arrangements for effective water markets may not be worth the cost, until the value of water is high enough at the margin or soon will be. The eastern United States and northern Europe have not reached this level of water scarcity and are not likely to in the near future (these areas are primarily concerned with water quality problems). In contrast, the western United States and southern Europe have sizable areas that face periodic water deficits (during years with low rainfall) and could benefit from establishing such arrangements.

To establish an active water market, water laws may need to be changed. Rights-to-water use must be authorized separate from land, and should be granted for a long enough time period and be secure enough to provide users an incentive to invest in water conserving practices. For surface water, this may require establishing priorities in water use, such as the seniority system for appropriative rights as used in the western United States. In the case of groundwater, where recharge is limited or nonexistent, volumetric rights need to be established. If rights are not clearly established, groundwater users will ignore the "user cost" or "scarcity rent" attributable to the groundwater stock. Consequently, the groundwater will be used more rapidly than is economically optimal. If recharge is adequate to replace the water pumped, then rights need to be established so that pumping does not interfere with neighbouring wells or with surface water rights. Where surface water and groundwater are closely linked, then water rights must be jointly established on some noninterference or volumetric basis.

While establishing a system of water rights is essential, it is also important to develop a procedure for registering, monitoring, and enforcing the rights. This can be done nationally, regionally, or locally depending on how the region's water management is structured. Many times it will be cheaper to do it locally, particularly if water-user organizations are active. Where return-flows or economic third-party effects are an important concern, some state or national involvement may be necessary to assure that the interdependency in basin's water use is taken into account. In some cases, sales of water rights have been limited to consumptive use or not allowed between basins because of possible third-party effects.

Once users can buy and sell water, institutional and organizational changes may be needed to broaden the market and make it more competitive. For example, improvements in canal infrastructure may be needed so that trading can take place over a larger area, e.g., canals connecting different systems. In addition, management may have to be improved so that orders to buy and sell water can be easily implemented. The managers may also need improved control structures so that they can easily increase the flow in one canal and decrease it in another.

Another important institutional arrangement for effective water markets is some mechanism for resolving conflicts over water rights and water delivery, such as a committee of respected water users elected by the community. The role of such a group would be closely tied to how well the water rights are specified and how they were established and distributed to users. If water rights are unclear, and the distribution contentious, then there will be numerous difficult conflicts that may have to be resolved, probably in court rather than by a committee of users. One of the contentious issues is likely to be how much water is allocated to ecological and environmental uses. As suggested above, this issue will be of growing importance in the future. Still, if water rights are clearly specified and their distribution is based on past use, then most of the disputes can be resolved by an elected body of respected users.

Since the assignment and allocation of water rights generally involves the distribution of significant economic rents, the allocation rules must be carefully developed in a transparent manner. In many cases this can be done based on past use, much like it was done in Chile, where water-use associations played an active role in verifying past use. However, this will not work when past water allocations significantly exceeded actual supplies. In cases where you have overlapping claims for water use, such as in Peru, then you may need a water court or tribunal to determine who has priority. Another possibility is to uniformly reduce all claims enough to balance supply and demand during a normal year's water supply. Some combination of setting priorities and reducing claims could also work.

3.3.   Privatization of water services

Reforms addressing the water sector have, particularly in Europe, concentrated on reformulating public sector's functions and activities, and on involving new actors in the supply of water services [3]. For instance there is a tendency towards decoupling or stressing distinctions between water resource-planning and operation of water utilities. In addition, there is a trend towards introducing more competitive pressures in the water industry, an objective often, although not exclusively, pursued through involving the private sector in the operation and management of water utilities.

The latter process is generally broadly referred to as privatization of water services, though the term privatization is sometimes also used to describe restructuring processes not necessarily involving the private sector. These alternative processes may take on the form of decentralization, where local authorities take over control functions which were previously state-owned, or the form of corporatization of public firms. On the other hand, privatization, *stricto sensu*, occurs through sales of previously state or municipally owned firms, contracting-out of water services, or through granting a concession for a private company to run a water utility (Hall, 1997a).

Broadly speaking, the main theoretical justification for the involvement of private companies in the provision of water services is to increase the water industry's perform-

---

[3] The term water services includes water abstraction, treatment and distribution, as well as interception and treatment of waste waters (sewage and sanitation systems). Although many considerations also apply to the latter, in this paragraph we will mainly consider freshwater supply.

ance, i.e., to get better and/or cheaper services (Cowen, 1997). In fact, the private sector is potentially capable of injecting technological, financial and managerial resources which the public sector may be unable to obtain, because of fiscal and bureaucratic constraints, and the lack of adequate incentives (Spulber and Sabbaghi, 1994). This is also, in general, the prime justification advocated by governments when embarking on a privatization process, although other economic and political considerations may influence such decision. These considerations may include the desire to obtain receipts from privatization to reduce public sector debt and/or finance tax reductions and shifting the cost of providing previously subsidized services from taxpayers to water users. For instance, using privatization receipts as a way of "laundering" taxation is not uncommon. [4]

There are four different means of introducing more competitive pressure in an utility market which include: (a) product-market competition, (b) yardstick competition, (c) competition for the market, (d) competition to supply inputs. Product-market competition can be further divided into the three different ways it can be promoted: (i) the installment of competing networks, (ii) private supply and retail competition, or (ii) common carriage competition (Webb and Ehrhardt, 1998).

In general, installing competing networks is not a serious option in the water supply and distribution industry (Cowan, 1998; Webb and Ehrhardt, 1998). Although it should not be always ruled out [5], duplication of networks is generally undesirable, given the current state of technology, in that it would imply losing scale economies and higher average water distribution costs [6]. However, the fact that existing main networks are naturally monopolistic does not mean that additions should be owned by incumbent operators (Cowan, 1998). For instance, reform of water services could in principle allow groups of consumers to supply themselves (private supply), or allow consumers to buy water from new operators (retail competition) (Webb and Ehrhard, 1998).

Shared use of existing networks by different firms (common carriage competition) is another option for injecting competitive pressure in a utility market. For instance, competition in telecommunications, public transport (railways), gas and energy have been, and are being mostly pursued by regarding transmission systems as a separate enterprise, and by allowing more firms to operate on the same system. However, opening up water distribution networks to common carriage may pose specific problems that need to be carefully addressed. One specific issue is that of water quality and whether the standards set can be safeguarded (Byatt, 1998). For instance, quality from different companies sharing the same distribution network may vary, and once contaminated water (water not satisfying existing standards) flows in the system, it mixes with and affects water quality from other sources. This is a serious and challenging issue, but not an insurmountable problem (Byatt, 1998). Other network industries (e.g., electricity) face similar (though generally

---

[4] For example, in France, local governments have imposed an «entry fee» for water concessions, and successful bidders have then been able to increase water charge to cover the fee; in Spain, the successful company pays an annual rent which can be transferred to water charges (Hall, 1997a); the water concession in Budapest was awarded to the private bidder offering the greatest financial benefit for the municipality, although it involved a higher price for consumers (Hall, 1997b).

[5] In some areas the available water distribution network may be so inefficient or provide such a poor service that the construction of a competing network could be economically feasible (Webb and Ehrhardt, 1998).

[6] The only economically feasible competitive distribution networks are bottled water distribution systems, which are a minor (typically, a complementary) component of the whole water distribution system.

less severe) problems, and must ensure minimum quality standard and technical compatibility in order to avoid damages to the network (Webb and Ehrhard, 1998).

Although theoretically appealing, opening up existing networks is not a common mean for injecting competition pressure in the water industry. Even in countries which have adopted a strong privatization process, common-carriage is either prohibited by current legislation, or has not occurred to a large extent. In one of the few examples, OFWAT (the Office of Water Services in the UK which regulates the water industry) in 1995 relaxed the definition of a qualifying site for "inset appointments," [7] and allowed new utilities to supply large consumers in an incumbent's interconnected system by paying a fee for using the system. Only a few cases of inset competition have occurred so far, but the interest manifested by various companies and customers to take advantage of the possibility of common-carriage has had an effect on incumbent firms. In response to the threat of competition, the vast majority of UK water companies have lowered tariffs for large users with cuts ranging from 1 percent to about 25 percent (Webb and Ehrhardt, 1998).

Yardstick competition is a potentially powerful means for regulating and putting pressure on incumbent firms when the intrinsic nature of the privatized utility does not allow the entry of an adequate number of competing firms. In particular, as far as the water industry is concerned, because of its naturally monopolistic features, a single company is likely to emerge in each location. However, comparative competition between geographically separated suppliers may be injected (either through setting an appropriate regulatory regime or through providing comparative information on charges and levels of service) in order to stimulate firms' efficiency and pass back efficiency gains to customers (Byatt, 1998). To be effective, comparative competition requires an adequate number of comparable agents, diversity of ownership, and consequent variety of management styles. It follows that a low degree of competition within the whole water industry, and/or mergers between firms may significantly undermine the potential advantages of yardstick competition.

Large initial investment requirements and economies of scale inevitably provide the incumbent firms with market power and reduce market contestability. However, the sector's performances could be improved through institutional arrangements designed to introduce competition for the market. In other words, competitive pressure can be injected through awarding the right to operate, in a given area, to the company able to submit the best offer in terms of overall quality and cost of the water service.

In recent years, many countries have used competition for water concessions, and the approach has often delivered substantial benefits to consumers (Webb and Ehrhard, 1998). While in developing countries the involvement of private companies, through awarding the concession to run a water utility, is often regarded as the only sustainable way to acquire an adequate infrastructure (World Bank, 1994), in developed countries the prevailing approach seems to be improvement of the operation and management of existing assets. For instance, in developed countries, franchising of water services is often combined with public ownership of the main infrastructures (Cowan, 1998). One of the

---

[7] The concept of «inset appointment» was introduced by the Water Act 1989. A new company can apply for an appointment to supply water (or sewerage) to customers located in a defined area. Originally (until the Act was amended in 1992), the only customers eligible for inset supply were new customers, i.e. sites that were not already connected and that were more than 30 meters from the local water utility's distribution main (or sewer).

best known example is the French *gestion deleguée*. Thus, competitive bidding for water concessions has become one of the most common forms of competition (World Bank, 1993). [8]

The benefits, in terms of private companies' performance, stemming from competitive bidding for running water utilities, obviously depends on a number of conditions. These conditions include the degree of competition in the water industry and the long-term credibility of initial contractual arrangements. In the former case, similarly to yardstick competition, the effectiveness of the competition for the market can be undermined by the shortage of competitive bidders. In this respect, it is worth noting that the water industry is dominated by a small number of companies (Hall, 1999). The majority of privatized water concessions in large and small cities on every continent are run by two French companies: Vivendi (previously known as Generale des Eaux) and Suez-Lyonnaise des Eaux. Where present, national companies often operate through joint-ventures with these multinationals.

As far as the contractual arrangements is concerned, a critical issue is the appointing authority's ability to maintain regulatory pressure on the agent during the life of the contract. For instance, while there may be competition in awarding the concession, maintaining competitive pressure may prove to be difficult, because of the lack of credible sanctions. Attracting private companies into the market may be relatively easy, but expelling them may be much more difficult (Clark and Mondello, 1998; Moretto and Valbonesi, 1998). This, in principle, makes short-term contracts the preferable solution, but they could prove to be unattractive for firms due to the large sunk costs involved in acquiring the water concession.

Competition to supply inputs is another option for improving the overall performances of a water utility. If there is sufficient competition between input suppliers (e.g. for billing and revenue collection, and maintenance of infrastructures), competitive procurement may allow for significant cost-reductions which, in principle, could be passed on to consumers (or taxpayers). In the public sector, competitive contracting out has become commonplace. For example, EU legislation imposes competitive procurement on member states' public sectors. However, paradoxically, the privatization of a water utility may reduce the potential benefits of competitive contracting out of specific activities. For instance, if a water concession is a concession in the strict sense, then the company running the business is not subject to requirements of EU legislation, and the company can contract out a specific activity (e.g., construction of new infrastructures), to a fellow-subsidiary, without advertising it for tender (Hall, 1997a).

3.4.   Regulatory issues

Important regulatory issues have arisen as countries and states move to use water markets and privatize various aspects of their water supply and waste water treatment

---

[8] In France there are three types of "concessions". For instance, a private company may: (i) either acquire the complete responsibility for operating the water system, making the necessary investments in the infrastructure, and take responsibility for financing them at its own risk (concession in the strict sense, «concession» in French), or (ii) operate the business and carry on maintenance at its own risk, but a public authority is responsible for investments (operating concession, *"affirmage"* in French), or (iii) receive a flat fee to manage the system, without taking any responsibility for investments (management contracts, *"gérance"* in French) (Hall, 1997b).

systems. In some cases laws and regulations have had to be changed to allow water markets to develop and management of public water and treatment systems to be privatized. In a number of countries this has involved changes in tax codes. For example, the U.S. federal tax code was changed so that contract terms for private firms managing public water and treatment facilities could be for more than 5 years. The removal of these limits in 1997 eliminated a key constraint to private management of water and treatment services. Under the old rules a facility that was managed under a contract of more than five years was not eligible for tax-exempt financing. The new limits have been set at 20 years which allows private firms more time over which to spread investment costs. However, private ownership of infrastructure-intensive water and treatment systems is still discouraged by the U.S. federal tax code. Municipal debt is tax-exempt while private debt is not. Changes may also be required in U.S. state and local property tax codes where municipal-owned facilities receive tax breaks but privately-owned facilities do not.

Another major problem facing privatization efforts are the regulations that have been introduced at different levels of government to regulate water prices. There has been a general concern that private firms or even public utilities may use their monopoly power in setting water prices. Yet it is not clear that monopoly pricing is a legitimate concern for domestic water users. Given our need to encourage users to conserve water and the fact that the water bill is no longer a significant share of most families' income in the U.S. and western Europe, higher water prices are in most cases socially desirable. If current price regulation is resulting in low prices and the overuse of water, then maybe it is time for a general deregulation of water pricing. This would also remove a barrier to privatization of water system management.

Modifications in environmental laws may also be necessary. In most cases these laws were not written with a concern for efficient water use or the need to have water reallocated among different uses. Consequently, many environmental laws and regulations tend to be inflexible when questions arise concerning the trade-off among different uses, including environmental uses. For example, the U.S. Endangered Species Act may be used to prohibit water sales and transfers since the transfers may affect the habitats of species that are listed as endangered.

There are a number of other regulations and laws that are likely to limit the future operations of water markets in the U.S. and Europe. One which has received particular attention in the U.S. west is the "public trust doctrine." It has been used in U.S. state courts to limit private water uses that infringe on certain public water uses such as water for fish habitat. The basic idea is that these public rights have existed since the U.S. and individual states were founded and they must be recognized (Huffman, 1997). In some cases this means existing uses such as irrigation are faced with reduced water allocations.

A second set of issues arises because of the lack of consistency in country and state laws which can act as barriers to market transactions across borders or even within borders. The difficulty occurs when rivers cross country or state borders and no sharing arrangements have been established among the states or countries along the river. Even when appropriative rights have been established with priorities set for each state or country, there are no priorities regarding the whole river. Thus downstream users will find that what they receive for their water right is held hostage to how water is used upstream, which will change over time. The uncertainty created by these transborder problems will lead to inefficiencies in water use and poor investment decisions. A similar uncertainty is created by laws that reserve rights for potential future uses or users. In the U.S., the reserve rights for Native American tribes are an example where a certain quantity or share of water is allocated to a tribe whether or not it is or will be used now or in the future (Colby, 1998).

A third area of concern involves specific national (federal) or state laws that restrict water use and limit water transfers outside their borders. Many U.S. federal and state projects disallow sales to users outside the districts that have water contracts (Howe, 1997). These institutional restrictions on water transfers are primarily the product of national and state resource protectionism (Huffman, 1997). In the U.S. the commerce clause prevents the states from limiting interstate commerce while NAFTA may have a similar impact on future water trading in North America (Frerichs and Easter, 1990).

A fourth institution that will limit water trading and increase transaction costs is the no-injury rule. This is found in a number of institutional arrangements designed to protect third parties against damages from water sales or transfers. These damages might be in terms of reduced return flows for downstream users or reduced economic activity (fewer input sales, or fewer products to process and/or market) in communities serving the irrigators selling water. "The big issue facing U.S. western states [and other areas introducing water markets] [9] is how to address third-party effects associated with the reallocation of [...] water without making transaction costs too high" (Howe, 1997, p. 83).

Fifth, the "beneficial use doctrine" connected with many water rights may no longer be appropriate in a time of growing water scarcity and the development of water markets. The beneficial use requirement may cause owners to misuse or waste water so that they can maintain ownership. If water sales are judged to show that the original owner is not making beneficial use of the water, then it will essentially prevent water trading.

3.5. The intersection of markets and privatization

Although the water markets and privatizing water utilities have been discussed separately, there is no conceptual reason why both could not be used to improve water use and allocation in a given water system. A private firm could manage and operate a water system and water rights' holds (users) would pay the firm a fee for the service. The sale of water rights among users could be done through the water firm or some other coordinating body. The water company would just have to have an up-to-date record of who owns how much water. Whether or not a market or privatization of the water system or both should be pursued depends on where the big inefficiencies are in the system. If the problem is inefficiencies in the over all management of the water utility, then some form of privatization would be good medicine. In contrast, if the problem is inefficient water use by consumers, then water markets could be the solution. Markets would provide consumers a sense of the value of their water and an incentive to use it efficiently. Effective water pricing by the water company could also have a similar effect.

Of course, water markets have been found mostly in areas with a significant amount of irrigation. They are generally used to transfer water among irrigators and from irrigators to water companies or water utilities. There are also cases of individual households in rural areas buying water from irrigators (Easter et al., 1998) as well as trade among utilities. In principle, transfers could also take place among urban households, although no good examples currently exist. For urban households the problem of allocating the water

---

[9] Added by the authors.

rights and keeping track of relatively small transactions may mean high transaction costs relative to the value of the water saved. Thus, for most urban settings the preferred option is likely to be some form of privatization of water utilities. Most efficiencies gained from a water market could then be achieved if the water company is allowed to raise prices enough to cover all costs plus some charge for the scarcity value of the water.

4.   US AND EU: EXPERIENCE AND ON-GOING DEVELOPMENTS

The U.S. and European experience with privatization of the water sector and establishing water markets has been partly guided by their respective water policies. Their policies regarding water quantity are limited at the federal level in the U.S. and at the European Union level in Europe. Thus, the guiding water policies in the U.S. have been those from individual states while in Europe they are from individual countries. This means that the privatization and water market experience will vary from country to country in Europe and from state to state in the U.S. The institutional arrangements developed to either promote or limit water markets and privatization are also likely to be quite different. This situation could be changing in Europe as the Union is increasing the uniformity of policies particularly regarding water quality. However, in the U.S. states' policies are likely to continue to play a central role.

4.1.   The U.S. experience with privatization and markets in the public water sector

Part of the reason why U.S. federal water policies regarding quantity are limited, is the fact that water development is viewed primarily as a state or local responsibility. For example, almost all cities and towns are supplied by local or state entities. The major exception to this, for consumptive water use, has been irrigation development which has in the past received large federal subsidies in the western U.S. However, even for irrigation the federal role has changed as more and more responsibility is shifting to local user entities. In terms of regulating water use the federal government has had an even more limited role with regard to water markets and privatization. The major exception has been federal court decisions regarding the sale of water across state boundaries (states cannot restrict such sales) (Frerich and Easter, 1990). For privatization the main impact of the federal government has been the tax laws that favor public entities relative to private firms and federal regulations for drinking water quality.

4.1.1.   U.S. water policy

Currently, the United States does not have a comprehensive water resources policy. Since the Water Resources Planning Act of 1965 was eliminated in the early 1980s, along with the river basin commissions and funding for state planning assistance, federal water policy has been limited to water quality concerns (Muckleston, 1990). Questions of water allocation and pricing have been left to state and local entities as well as to a few regional entities such as the Great Lakes Commission that remained active after the 1980s debacle.

Even with the federal policy concerning water quality, the responsibility for improving quality is shared with the states. The water pollution control act was first enacted in 1948 and became the Clean Water Act in 1972 with additional amendments in 1977, 1987 and 1996 (Davis and Mazurek, 1998). The primary focus of federal water pollution control efforts has been on surface water and point sources of pollution (end of pipe). Most of the federal funds have gone for construction grants for municipal sewage treatment facilities, with over $700 billion spent from 1972-1996. This has added to the capital intensity of the industry which is 3 to 4 times more capital intense than telephone and electric utilities. Because of the focus on capital investment in municipal waste treatment, nonpoint sources of pollution have been almost ignored and the efforts to protect groundwater have been, at best, fragmented. Thus the major challenge for future water pollution control efforts in the U.S. will be to devise a system that effectively addresses nonpoint pollution control issues and develops a more unified and effective approach to groundwater protection.

4.1.2.   *Privatization*

The United States in the past two decades has experienced a shift towards deregulation and privatization in the public sector. Much of the shift was seen as a way to increase competition and reduce costs through improved incentives. However, the water sector has only recently seen a move towards privatization [10]. Two factors seem to have stimulated this new interest. First is the financial pressure, particularly for smaller systems, due to the general need to upgrade or expand water and sewer systems. In some areas, such as the eastern part of the United States, the water and sewer systems are old and in need of repair or replacement. The second factor, which is closely related to the first, is the increasing cost of complying with new health and environmental standards for water quality. Under the Safe Drinking Water Act Amendments (SDWA) of 1996, water utilities must meet stricter requirements for removal of contaminants. The SDWA imposes a tougher standard on bacterial and microbial contaminants and reduces the acceptable levels of harmful byproducts from disinfection. In 1998 the EPA estimated that compliance with these new requirements will cost over $1 billion annually nationwide above and beyond the existing need to replace aging distribution and treatment infrastructure (Seidenstat et al., 2000). The American Water Works Association, the nation's largest drinking water industry association, estimates 20-year costs of $325 billion for infrastructure (Seidenstat et al., 2000).

Interest in involving the private sector in managing water systems has been complemented by an expansion in the activities of U.S. investor-owned water utilities. In 1994, Price Waterhouse found that 90% of the investor-owned water utilities surveyed had either closed transactions with or were considering proposals to provide services to other cities (Beecher, et al., 1995). These private sector activities in the U.S. have taken a number of different forms, ranging from outright private ownership to public utilities contracting with private firms to do their water billing. With the exception of outright acquisition or ownership, the different options for privatization (listed in Table 1), involve

---

[10] Historically, private provision of water services was the norm through the early to mid 1800s, but by the 20th century, municipal monopolies comprised roughly half of the water works in the U.S. (Baker, 1899). Today, ownership of water services varies by size of the community served, with private ownership predominating only in communities serving less than 500 individuals. The EPA estimates that while 33% of community water systems are privately owned, only 15% of all people are served by privately-owned companies (USEPA, 1999). Most large water systems are municipally owned and provide safe drinking water at affordable cost.

Table 1.   Privatization Options Found in the United States

Options for Private Sector	Description
• Acquisition	• Public utility sells the facility to private entity resulting in private ownership and operation.
• Joint venture	• Private entity owns facility in conjunction with public utility
• Design, build, own, operate and transfer	• Private entity builds, owns, and operates the facility. At the end of the specified period, such as 30 years, the facility may be transferred to a public utility.
• Concessions to design, build and operate	• Private entity designs, constructs, and operates the facility. The public utility retains ownership and financing risk, while the private entity assumes the performance risk for minimum levels of service and/or compliance.
• O & M concessions	• Public utility contracts with private entity for a fee to operate and maintain the facility. The public utility owns the facility.
• Concessions to design and /or build	• Private entity designs and/or constructs the facilities and turns it over to the public utility to operate.
• Contract for specific services	• Private entity contracts to provide public utility with specific services such as meter reading or billing and collection.
• Management concessions	• Private entity manages and supervises the public utilities personnel.

Source: Adapted from Beecher et al., 1995

either contracting to supply inputs or concessions to provide management, or construction and design services. Whether these arrangements provide adequate competition in the bidding process is not clear and probably varies a lot across the U.S. Yet there is evidence to suggest that the design, build, operate, and transfer option has significantly reduced construction and operating costs (Siedenstat et al., 2000).

In a 1997 survey of 261 U.S. cities, 40% currently had some form of private/public partnership and another 14% were considering proposals. The most common arrangement was for private design and construction, particularly for water treatment facilities (71%). Meter reading (33%), billing and collection (31%), as well as operation and maintenance (44%) were the other private sector activities reported (Callahan, 2000). As this survey illustrates, public asset sales to private firms is very limited. Asset sales usually take the form of transfers of small water systems to investor-owned utilities that are located in neighboring areas. Until the early 90's, the federal tax code discouraged private ownership of infrastructure- intensive industries by making federal construction grants available only to public utilities. If a public utility had received such a government grant, the municipality was required to pay back 100% upon sale of the facility. In 1992 a Presidential executive order on infrastructure privatization required that only the remaining "undepreciated" portion of government grants be paid upon sale of the facility. Despite this change, other disparities in the tax law, whereby municipal debt is tax-exempt while private debt is not, have discouraged private purchase of water facilities. In addition, political concerns about maintaining control over water resource facilities have stymied efforts to establish private ownership.

Most water facilities in the U.S. are constructed by private firms and managed by municipalities after completion. This approach has been criticized for its lack of integration and inferior product delivery. The Design-Build-Operate and Transfer Model has been used to give the private firm control over the project in its entirety, creating a more integrated and better performing facility which has been estimated to have saved roughly 25% of the construction costs and 20-40% of the operating cost (Siedenstat et al., 2000). For example, Seattle's Tolt River Project is estimated to have saved 40% compared to the conventional model of private construction and public management of the facility.

Concessions for operations and maintenance are largely the result of tax code changes in 1997 as discussed above in the section on regulatory issues. Contracts for operations and management of facilities have varied significantly by provisions and length. For example, Buffalo, New York held a contract in 1997 with American Anglian Environment Technologies for operation of its system with a guarantee of no layoffs (Seidenstat et al., 2000). Under agreements in Cranston, Rhode Island, the private contractor paid cash up-front for significant capital improvements when the existing short-term lease was renegotiated for 25 years.

Privatization has brought about increased entrance of foreign, highly-specialized water companies and fostered consolidation among American water companies. The merger of French-owned firms Aqua Alliance with Metcalf and Eddy in 1998 resulted in their management of approximately 200 wastewater and 170 water utilities in the U.S. (Seidenstat et al., 2000). The British have also entered into the American water utilities market, i.e., Anglian Water PLS has joined in partnership with American Water Works, the largest U.S. investor-owned water company to compete for O&M contracts as American Anglian Environment Technologies.

The potential for inter-utility cooperation between the energy and water sectors provides another interesting potential for new competitors in the water industry. Deregulated, newly competitive energy utilities could choose to ally themselves with large, captive customer bases of water utilities in order to achieve efficiencies in billing, metering, and office staff across utilities. Although energy utilities are restricted by the U.S. Public Utility Holding Company Act to refrain from expansion into other sectors, this Act may be repealed. If this happens, cross-utility endeavors are clearly possible. In 1998 New Jersey's Public Service Electric and Gas Company announced that its nonregulated subsidiary, Energis Resources, would partner with United Water Resource (Seidenstat et al., 2000).

The majority of private water utilities are subject to regulation by state public utility commissions and a traditional rate-of-return regulation. As of 1995, state commissions regulated roughly 8,537 water utilities (Beecher et al., 1995). Four states and the District of Columbia do not oversee water utilities due to the small number of privately-owned utilities in these areas. Many larger, publicly-owned utilities are not regulated except for certain issues such as selling water outside of their system.

The scope of regulation by public utility commissions varies and could include any of the following: approval of loans, mergers, service areas, acquisitions; oversight of management; review or audit of drought management practices and accounting; and setting of the rate of return. State regulatory commissions and local agencies have traditionally engaged water utilities in protracted rate-setting cases and have favored average cost-pricing schemes. Tight controls over rate-setting originally arose because of concerns about monopoly pricing power but, as pointed out above, these concerns may no longer be appropriate. The result has been a lack of flexibility in pricing water and a general decline in water systems. "Traditional rate-setting methods, employed by state regulatory commissions as well as local government agencies, appear to have produced a situation of rapidly deteriorating water systems, both rural and urban, characterized by aging capital facilities and under-maintained water systems" (Mann, 1981, p. 101). An interest in reducing these price subsidies and additional unfunded mandates under the SDWA have rendered these regulatory issues more serious and increased interest in privatization.

4.1.3. Water markets

Given the dry conditions and rapid growth in the western half of the United States, it is not surprising to find water market activity concentrated in the U.S. west. During dry years, most of the river basins in the southwestern U.S. have water demands that exceed supply, e.g., the lower Colorado and the Rio Grande. In the drier areas water markets have played an increasing role in balancing water supply and demand over the past 25 years. The California State Water Bank is a good example of how water trading has helped prevent serious economic damage that could have occurred during the 1990s drought (Easter et al., 1998).

Water markets can be used to deal with supply and demand variabilities as well as with overall growth or shifts in demand. Short-term or temporary water transfers are best suited for managing supply and demand variability while permanent or long-term transfers are more suitable for handling increases or shifts in demand. California's water trading has been best suited to deal with supply and demand variability, because most of the water trading has been short-term in nature.

Hundreds of thousands of acre-feet (1 ac. ft. = $1,233m^3$) of water are transferred annually in California but most of the transfers are for less than one year. These temporary trades occur mostly in the same irrigation district or system. Long-term trades have been limited to about 25 in the last 20 years and are mostly in response to urban demands. For example, in 1989 the Los Angeles Metropolitan District (MWD) obtained 106,000 ac. ft. from the Imperial Irrigation District (IID) while in 1998 the San Diego County Water Authority obtained 200,000 ac. ft. from IID. Most of this water will come from water conservation within IID's irrigation system (e.g., canal lining) and conservation by farmers.

In 1998 California's Department of Water Resource reviewed 23 transfer proposals that would use State Water Project facilities (Table 2). This, of course, leaves out the transfers that occur within districts or would be transferred in other facilities. Still the total proposed transfers in State Water Project facilities amounted to 362,102 ac. ft. in a year of relatively good rainfall. Over 44 percent of the transfers were from urban to urban while another 27 percent were from agriculture to agriculture.

The emphasis on temporary or short-term transfers in California is somewhat different from trades in Colorado, New Mexico, and Utah where many more permanent trades took place. MacDonnell (1990) reported that these states had 5,844 permanent water transfers during 1975-1984. Not surprisingly, the direction of these trades has been primarily from agriculture to urban users since Western agriculture receives nearly 85 percent of all water diversions (Howe, 1998). There have also been numerous temporary trades within the Western states but a trend towards more permanent trades is likely to occur because of shifts in demand to urban uses which require a more assured water supply than irrigation. Agricultural water use peaked in the 1980s while urban use has continued to expand (Gleick, 1998). The shift to more permanent trades would be facilitated

Table 2.   Water Transfers Using State Water Project Facilities 1998

Type of Transfers	Number of Transfers	Amount of Water Conveyed (acre-feet)
For Agricultural Use		
Agriculture to Agriculture	11	99,026
Agriculture/Urban to Agriculture	1	85,700
Urban to Agriculture	1	100
Subtotals	13	184,826
For Urban Use		
Urban to Urban	4	159,701
Agriculture to Urban	4	2,494
Subtotals	8	162,195
For Environmental Use	2	15,081
Totals	23	362,102

Source: Department of Water Resources, 1999.

by institutional arrangements that help clarify water rights and streamline the procedures for determining if there will be adverse affects on other water users. Howe (1998) points out that one reason Utah and New Mexico had the most permanent water transfers is because they both have a clear, nonlegalistic means of determining if there will be any adverse affects on other water users caused by proposed transfers. Once the assessment of potential adverse effects has been completed the proposal is approved or modified by the Office of the State Engineer. Court appeals occurred in only about one percent of the cases.

The shift towards more permanent transfers is evident in the 1998 data. Thirteen western states reported 140 separate water transactions in 1998. They included 104 transfers for municipal purposes, 24 for irrigation and 12 for public trust purposes. Most of the sales were by agricultural users. Of the total, 102 were purchases of water rights while the remaining 38 were short-term leases or exchanges. Because of good rainfall in the western U.S., 1998 was a low year for water trades. The number of unit transactions involving the Colorado-Big Thompson illustrates this point. In 1998, 3,187 units were traded as compared to 6,426 units (a record) in 1997 and 5,167 units in 1996 (one unit equals an average annual yield of 0.7 acre-feet).

California is also likely to experience an increase in longer-term trading. Many farmers are beginning to find that they can make more money selling water to urban areas than growing crops. Cities are willing to buy the water but they will usually want a long-term commitment and even permanent trades. If this trend accelerates, it could have major implications for agriculture and communities in the southern half of California's large San Joaquin Valley. This is the area where some of the world's largest corporate farms are located and sales of water could reduce agricultural activity in this part of the valley.

4.2.   The European Union

Like U.S. state policies, country policies in the European Union have provided the guidelines for their privatization of the water sector and development of water markets. An important question is how will this change in the future as the concerns for water quality grow and more areas find it difficult to obtain low cost clean water supplies? The growing concern about water quality combined with the fact that most of the rivers in Europe are international, seems to suggest that European Union directives will become increasingly important in future privatization and water markets in Europe. Yet there is still a strong sentiment in favor of local control over resources particularly in Northern Europe. This creates a growing dilemma in water resources management: the desire for local management and control of natural resources whose uses have many international repercussions that require a broad management prospective.

4.2.1.   The Community water policy

European water policy began in the 1970's, with environmental standards for surface waters used for potable water supplies, and later on, by setting binding quality targets for drinking water and by addressing other water resource uses (fish and shellfish habitat, recreation, and groundwater). While some directives have proved to be quite effective, other European legislation — such as the Nitrate Directive addressing pollution from agricultural sources — have been poorly implemented by member countries. According to a recent EC Commission's communication on the state of Europe's environment, while there have been substantial improvements in surface water quality due to reductions in point source discharges; pollutant emissions from diffuse or nonpoint sources have shown little change. EU maximum groundwater concentrations of nitrate and certain pesticides are frequently exceeded (European Commission, 1999).

By the mid 1990s, pressure emerged for a substantial redirection towards adopting a more global approach to the Community water policy, by integrating disperse pieces of legislation as well as by adding new objectives. In response to a request by the European Parliament's Environment Committee and from the Council of Environment Ministers, in 1997 the European Commission produced its proposal for a Water Framework Directive (COM(97)49), subsequently amended (COM(97)614, COM(98)76, COM(99)271). Currently it is being negotiated by the European Parliament and the Council of Ministers and the final adoption is foreseen for 2000.

The draft directive which, *inter alia*, would be the first piece of European legislation to address the issue of water quantity, is aimed at: (a) streamlining Community water legislation, (b) expanding the scope of water protection and achieving a "good status" [11] for all waters by a certain deadline (2010), (c) promoting in all member countries the adoption of single system of water management, namely management by river basin, instead of according to administrative or political boundaries; (d) increasing awareness of citizens and getting them more closely involved; (e) ensuring that the price charged to water users is based on the full costs of water.

The introduction of pricing oriented to full-cost recovery is undoubtedly one of the most important European water policy innovations (Blöck, 1999) and has been one of the most controversial component of the Commission's proposal. Whereas this principle has a long tradition in some countries, this is not the case in many countries, where, in general, the recovery rates vary considerably between water use(r)s and economic sectors. According to the draft Directive, "by 2010 Member states shall ensure full cost recovery for

---

[11] The term "good status" come directly from the EC directive and appear to mean water quality consistent with EC water quality standards.

all costs for services provided for water uses overall and by economic sectors." However, while full-cost recovery appears to convey a clear-cut criterion, when one considers how the concept is actually practiced, significant differences are found (OECD, 1999a). For instance, there are countries where the principle is translated into recovery of operation and maintenance (O&M) costs, in others into recovery of O&M and capital costs, while in other countries there is an effort to include scarcity values and negative externalities caused by water users.

In this respect, the draft directive does not provide an unambiguous and clear-cut operational translation of the full-cost principle and tends to leave some freedom to member states. However, "while the European Commission has set up modest objectives with regards to cost recovery, it has not ruled out the incorporation of scarcity values and environmental externalities in full cost recovery" (OECD, 1999a, p. 35). According to the Commission's proposal, the principle of full-cost recovery must be applied to all sectors, and cross-subsidization between sectors should be avoided. Member countries are allowed to grant exemptions, e.g., for providing basic services to households at an affordable price or for regions entitled to structural fund support, but all deviations from full-cost recovery prices should be explicit.

Besides European legislation directly addressing water issues, member states' management and operation of water services have also been indirectly affected by other policy developments at the European level. These developments include the legal requirement to integrate environmental protection into other EC policies, a principle which was established by the Single European Act and was given a more comprehensive legal basis in the Maastricht Treaty. A second indirect driving force has been the process of creating the European Monetary Union which has imposed a more rigorous budget discipline on member countries. A third force is the process of introducing more competitive pressure in the European utility markets traditionally dominated by publicly owned monopolistic companies.

4.2.2.   Member Countries

European legislation has exerted an external pressure and has helped slightly reduce the heterogeneity of national environmental policies. Nevertheless, partly because of the lack of a global and comprehensive community water policy, significant differences remain between member countries regarding their legal and institutional arrangements, organization and management practices, and pricing principles. Yet a comprehensive overview of these national water policies goes beyond the scope of this paper. Instead we will focus on some significant features and developments involving privatization, water markets, and water pricing.

Increasing regulation of water uses. In Member Countries there has been an increase in the regulatory power of public authorities over water use. Generally speaking, increasing state regulation and centralized water management has been a common feature in southern countries, while involvement of local authorities has remained a typical feature of the institutional design in northern ones (Barraqué, 1998).

Contrary to countries which have embarked on a process of privatization of water rights, and establishing water markets, in Europe, although water markets are gaining interest, only Spain has recently taken steps in this direction [12]. In fact, there has been a tendency towards increasing the share of public waters, by subjecting water uses (abstractions and wastewater disposal) to licensing. Even groundwater is being included in this evolution to public waters. Recent reforms adopted in member states such as Spain and Italy, where groundwater abstractions were de facto traditionally considered as part of landowner rights, all water resources, including groundwater, have been formally placed in the public domain.

Water pricing. European water users face significantly different prices. Observable variations in service charges are only partly traceable back to differences in natural resources endowments, accessibility to sources, patterns of urbanization, or efficiency levels in the provision of water services. In fact, inter and intra-country price heterogeneity is mostly explained by differences in national and local water policies, and patterns of operation of water services, involving heterogeneous pricing structures and cost recovery rates.

Agricultural water pricing principles vary considerably within the European Union. While there are countries where farmers do not receive any special treatment, such as The Netherlands, England and Wales (OECD, 1999a), in general European farmers enjoy a privileged status, and are charged less than other water users. Capital costs for supplying irrigation water are usually covered through public budgets. In countries like Greece and Italy, water charges for agriculture are often insufficient to generate enough revenue to cover operation and maintenance costs.

In other countries which have experienced increasing competition between agricultural and other water uses, a tendency towards increasing farmers' contributions is gradually emerging. For example, France and Portugal, although maintaining a special status for agricultural water users, have made progress towards matching water charges and supply costs. Also in Spain, despite the administrative problems encountered in implementing the 1985 Water Law (which *inter alia* has introduced a levy intended to cover capital investments), there are many examples of collective and private efforts to use irrigation water more efficiently. In Italy, an innovative program is being implemented in the southern Capitanata region (one of the country's most important irrigated areas) with the aim of increasing cost recovery, introducing water-saving farming practices, and penalizing excessive consumption (OECD, 1999a).

Even household water pricing varies considerably among countries (Table 3), and often within the same country. In Italy, where in 1998 the average price was less than IT 800 lire per cubic meter as compared to 3,600 lire in Berlin (Lobina and Hall, 1999), the range of price variation for water supply and sewerage services is 1 to 10 or more (Massarutto, 1999). In general, although not exclusively, in Southern countries, households have benefitted from various measures aimed at tempering the industrial cost recovery rates for water services. This objective has been pursued through different mechanisms, such as public financing for infrastructural development, free use or "political" pricing for the operation of public water networks, or cross-subsidization among different utilities at the municipal or intermunicipal level.

---

[12] The Spanish Government submitted in May 1999 Bill to the Parliament with the intention to reform the 1985 Water Law. Among the bill's breakthrough changes is the possibility of water rights holders to sell or lease-out their rights. The bill attempts to promote water markets, thus facilitating trading among right-holders or water claimants, although the priority allocation rules established by the 1985 Law will still remain in force (Garrido et al., 1999).

Table 3.   Average   Pricing   (m^3)   for   Household   Water   Services   (water   supply+wastewater services) in European Countries (Germany = 100)

Country	index	Country	index
Austria	19	Italy	20
Belgium (Flanders)	67	Luxembourg	24
Belgium (Brussels)	51	The Netherlands	74
Danemark	74	Spain	25
Finland	64	Sweden	68
France	73	England and Wales	73
Germany	100	Scotland	34
Greece	27		

Source: Massarutto, 1999.

However, the general trend observable in Europe is towards an increase of cost recovery rates. This process will become mandatory in all Member states if the EC Water Framework Directive is approved. Even without the directive, cost recoveries are gradually increasing even in countries such as Italy where support for subsidizing water users has been traditionally stronger than in other countries (Massarutto, 1999). The 1994 Italian Water Law (36/1994) provides a new framework for water supply and sewerage through vertical integration of the water cycle (abstraction+ sewerage + treatment + discharge) into single territorial units (*"ambiti territoriali ottimali"*). In principle, the units should be defined according to river-basin criteria. The national act foresees a single system of water charges for the entire water cycle where prices have to be set so as to generate enough revenue to cover investment and operating costs. Tuscany, one of the most advanced regions in implementing Law 36/1999, has identified investment plans and proposed new charges in each territorial unit. In general, the process has been slow and on average, an increase of water charge of 35% is foreseen over the next 20 years in Tuscany (Il Sole 24 Ore).

Other European countries, either at the national or local levels, have conducted or are conducting household pricing reforms, aimed at integrating traditional social considerations with the objectives of using water more efficiently and treating it as an economic good. In Spain, the city of Barcelona has pioneered the use of prices designed to combine strong incentives to save water with equity considerations. In France, the 1992 Water Law prohibited the use of "flat fees" thereby ruling out both nonvolumetric schemes and prices combining a fixed charge with volumetric charges. For Portugal, Decree-Laws 379/93 and 319/94 prescribed that charges for privatized water services must be fixed at economic levels. Finally, Denmark's 1996 government declaration requires water utilities to ensure that all properties connected to public water supply have meters installed (OECD, 1999b).

Restructuring and privatization of water services. Private companies' contribution to the provision of water services varies considerably between member states. For water supply, in 1996 the private contribution ranged between about 90% in England and Wales to almost nothing in Denmark, Greece, Ireland, Luxembourg, The Netherlands, and Austria (Hall, 1997a). Generally speaking, there is an increasing involvement of private companies in service provision, although this takes different forms. Full privatization has only occurred in England and Wales, while delegation (concessions), traditionally dominant in France, is increasing rapidly in Spain and Portugal (OECD, 1999b).

Primarily water utilities in France are publicly owned, but 75% of the population is served by private companies working under operation and maintenance contracts (Seidenstat et al., 2000). Privatization in France has principally involved either concessions to construct and operate facilities or *affirmage* whereby the municipality bears the initial expense of construction and the private firm takes over operation. *Affirmage* is often chosen when the municipality can receive preferential interest rates. The price of water is often set by a contracted formula which may include a surcharge which the operator returns to the city as debt service. As a result of this system, the French water utility industry is characterized by multiple large and influential companies known for technological innovation and high-quality service delivery.

Private concessions have also appeared in Italy (the municipality of Arezzo has recently awarded a concession to a consortium headed by Lyonnaise des Eaux). However, the dominant process appears to be corporatization (sometimes associated with partial privatization through sale of equity) of municipally owned firms (*aziende municipalizzate*) which are starting to operate beyond their own traditional territories. In Germany, where like Italy, the water sector has been dominated by municipally owned firms (*Stadtwerke*), in 1999 the sale of 49.9% equity in Berlin's water company was announced. In addition, the city of Bremen has partially privatized its sewerage company (Hall, 1999).

5.   FINAL REMARKS

One of the important reasons why countries are exploring economic approaches to water resources management is the potential economic and environmental benefits from doing so. Privatization and water markets both have the potential for establishing user incentives for efficient water use. This can reduce the need for expensive future water development and reduce government capital expenditures.

To realize these benefits some important institutional changes will be needed. In some cases it may be a change in tax laws and in others it may involve amending the basic laws regarding the sale of public assets. For water markets it means changing the rules so that private tradable water rights or water-use rights are established.

However, in developing the condition for privatization of water service or for establishing water markets it is important to guard against possible market failures. In water markets it is important to develop a nonlegalistic means for making sure that water trades don't have significant third-party effects. For privatization, regulation may be needed to prevent entities from taking advantage of their monopoly control over either the water supply or the distribution network. Clearly second-best issues arise if significant third-party impacts occur because of water trades or monopoly control over the distribution system are ignored.

There are no unique answers to dealing with problems of water scarcity. What works in a community facing "real" water scarcity will be different from one where the problem is inadequate institution's and/or management. Thus problem identification and designing solutions in relationship to existing institutions are important in dealing with water issues.

The comparison of U.S. and Europe illustrates some interesting differences in their approach to water scarcity concerns. What explains these differences and what have been the results? The first question is not easy to answer but it may be associated with the relative importance of the domestic water supply relative to irrigation. Where irrigation is important, it is the largest user of water. Consequently, when water scarcity and the need for change hit Chile, Mexico, and California in the United States — where irrigation is important — they made use of markets to improve water-use efficiency and save water. In contrast, when England decided to privatize public water entities, it was in response not so much to water scarcity but to a government budget scarcity and the need for future water infrastructure investments. Privatizing water entities took them out of the public sector budget. In addition, it allowed the government to make the monopoly provision of water contestable where different firms could compete periodically to operate or manage a water system under a long-term contract.

In Europe, only Spain has shown some interest in water markets. One reason for this may be an emphasis on the basic right-to-water as opposed to water as an economic good. Another reason for the difference in approach could be the difference in past institutional arrangements. For some countries, past institutional arrangements may have lowered the transaction cost of making new institutional changes. For example, the western United States and Chile both had some experience with establishing tradable water rights.

There are some similarities between Europe and the U.S. in that both have a relative abundance of fresh water from a global perspective. Yet at the regional level there are areas that are facing serious shortages, at least, in some years. Many times the shortage is due to inadequate institutions or water management rather than an absolute scarcity.

A further similarity is their approach to water policy. Unfortunately, both lack a comprehensive federal water policy, especially for water quantity. The one exception to this is that both Europe and the U.S. seem to be moving towards a comprehensive policy for water quality at least for point sources of surface water pollution.

Another interesting similarity between the European and U.S. approach to water resources issues is their failure to effectively address groundwater and nonpoint pollution problems. These are clearly difficult problems to solve and none of the countries has devised an innovative means to approach either issue. For example, Italy's and Spain's approach of putting groundwater in the public domain could be a step backward because without effective enforcement, it will likely provide the current users an incentive to pump the groundwater even more rapidly. Passing laws is one thing, enforcing them is quite another.

Nonpoint pollution has some of the same monitoring and control problems as groundwater. Since pumpers as well as the nonpoint polluters are widespread and numerous, they are difficult to monitor and control. This means that institutions need to be created that change the incentives and allow more local monitoring and control. Can we create local pumping districts that set pumping rates for their aquifer? For nonpoint pollution a targeting of conservation practices and monitoring of critical areas might work. Another possibility for some areas may be to develop markets for pollution permits that will allow point sources to buy permits from nonpoint sources. Both U.S. and European countries need to try new approaches, such as those mentioned above, to address their critical water policy issues.

6. REFERENCES

Arrojo P. (1999), *The Impact of Water Prices on Agricultural Water in Spain*, paper presented at the Conference on Pricing Water -Economics, Environment and Society, Istituto da Agua, Lisbon, September 7-9.

Associated Press (2000), *Cyanide Spill in Europe Turning into a Disaster*, Star Tribune, Minneapolis, MN, Feb. 13: A19.

Baker M.N. (1899), *Water-Works*, in: Bemis, E.W., *Municipal Monopolies*, Thomas Crowell & Co, New York, NY.

Barraqué B. (1998), *Groundwater Management in Europe*, in: Hilding-Rydevik-I.Johansson, *How to Cope with Degrading Groundwater Quality in Europe*, Swedish Council for Planning and Coordination of Research, Stockolm, pp.74-101.

Beecher J. – Richard G.D. – Stanford J.P. (1995), *Regulatory Implications of Water and Wastewater Utility Privatization*, National Regulatory Research Institute, Columbus OH.

Blöck H. (1999), *The EU Water Framework Directive: Taking European Water Policy into the Next Millennium*, paper presented at the Conference on Pricing Water - Economics, Environment and Society, Istituto da Agua, Lisbon, September 7-9.

Boland J.J. – Whittington D. (2000), The Political Economy of Water Tariff Design in Developing Countries: Increasing Block Tariffs Versus Uniform Price with Rebate, in: Dinar D., *The Political Economy of Water Pricing Reform*, Oxford University Press, New York NY, pp.215-235.

Brouwer F. – Falkenmark M. (1989), Climate-Induced Water Availability Changes in Europe, *Environmental Monitoring and Assessment*, **13**: 75-98.

Byatt I.C.R. (1998), Competition in the Water and Sewerage Industry, in: Helm, D.-Jenkinson, T., *Competition in Regulated Industries*, Oxford University Press. Oxford, pp.234-246.

Callahan N.V. (2000), *Once Monopolistic Water Utilities Are Becoming Competitive; Is Yours?*, Water Online, www.wateronline.com.

Clark E. – Mondello G. (2000), *Water Management in France: Delegation and Irreversibility*, Journal of Applied Economics, **III**: 325-352.

Colby B.G. (1998), Negotiated Transactions as Conflict Resolution Mechanisms: Water Bargaining in the U.S. West, in: Easter.K.W. - Rosegrant M.W. - Dinar A., *Markets for Water Potential and Performance*, Kluwer Academic Publishers, Boston MA, pp.77-94.

Cowan S. (1998), The Water Industry, in: Helm, D. - Jenkinson, T., *Competition in Regulated Industries*, Oxford University Press, Oxford, pp.160-174.

Cowen P.J.B. (1997), *Getting the Private Sector Involved in Water - What to Do in the Poorest of Countries*, The World Bank Group, Private Sector Development Department, Note n.102. www.worldbank.org/html/fdp/notes.

Davis J.C. – Mazurek J. (1998), *Pollution Control in the United States*, Resources for the Future, Washington, D.C.

Department of Water Resources (1999), *The Role of Water Transfer in Meeting California's Water Needs*, http://www.lao.ca.gov/090899_water_transfers.html, Legislative Analyst's Office, September 8.

Easter K.W. – Feder G. (1997), Water Institutions, Incentives, and Markets, in: Parker D.D. – Tsur Y., *Decentralization and Coordination of Water Resource Management*, Kluwer Academic Pub., Boston MA, pp.261-282.

Easter K.W. – Rosegrant M.W. – Dinar A. (1998), *Markets for Water Potential and Performance*, Kluwer Academic Pub., Boston MA.

European Commission (1997), *Proposal for a Council Directive Establishing a Framework for Community Action in the Field of Water Policy*, COM (1997) 49 Final, Bruxelles.

European Commission (1999), *Europe's Environment. What Directions for the Future?*, COM (1999) 543 Final, Bruxelles.

Frerich S. – Easter K.W. (1990), Regulation of Interbasin Transfers and Consumptive Uses from the Great Lakes, *Natural Resources Journal* , **30** (Summer 1990): 561-579.

Garrido A. (1999), *Pricing for Water Use in the Agricultural Sector*, paper presented at the conference on Pricing Water – Economics, Environment and Society, Istituto da Agua, Lisbon, 7-9.

Garrido A. – Iglesias E. – Kujal P. (1999), *Testing Some Provisions of the Recently Created Water Markets in Spain*. Awarded proposal by IFREE/ESL Madrid Post-Conference Prize, Tinker Foundation, New York.

Gleick H.P. (1998), *The World's Water 1998-1999 (The Biennial Report on Freshwater Resources)*, Island Press, Covdo, CA.

Hall D. (1997a), *Restructuring and Privatization in the Public Utilities—Europe*, University of Greenwich - Public Services International Research Unit, Report n. 9707-WE-Eur-emp. www.psiru.org.

Hall D. (1997b), *Public Partnership and Private Control – Ownership, Control and Regulation in Water Concessions in Central Europe*, University of Greenwich – Public Services International Research Unit, Report n. 9705-W-Eur-JV. www.psiru.org.

Hall D. (1999), *The Water Multinationals* , University of Greenwich - Public Services International Research Unit, Report n. 9909-W-U-MNC. www.psiru.org.

Hall D. (2000), Public Choice and Water Rate Design, in: Dinar, A., *The Political Economy of Water Pricing Reform*, Oxford University Press, New York NY, pp. 189-212.

Howe C.W. (1997), Increasing Efficiency in Water Markets: Examples from the Western United States, in: Anderson T.L. - Hill P.J., *Water Markets - The Next Generation*, Rowman Littlefield Pub., New York NY, pp.79-99.

Howe C.W. (1998), Water Markets in Colorado: Past Performance and Needed Changes, in: Easter K.W., *Minnesota Water: Potential and Performance*, Kluwer Academic Publishers, Boston MA.

Huffman J.L. (1997), Institutional Constraints on Transboundary Water Markets, in: Anderson T.L. - Hill P.J., *Water Markets -The Next Generation*, Rowman Littlefield Pub., New York NY, pp. 31-42.

Il Sole 24 Ore (2000), *Acqua, rischio-rincari dalla 'Galli'*, Il Sole 24 Ore, 5 February.

Lobina E. – Hall D. (1999), *Italian Water Industry 1999: A Profile*, University of Greenwich - Public Services International Research Unit, Report n. 9902-W-Italy-Leg. www.psiru.org.

Lundqvist J. – Gleick P. (1997), *Comprehensive Assessment of the Freshwater Resources of the World*, Stockholm Environment Institute, Stockolm.

MacDonnell L.J. (1990), *The Water Transfer Process and a Management Option for Meeting Changing Water Demands*, Natural Resource Law Center, School of Law, University of Colorado, Boulder, CO.

Mann P.C. (1981), *Water Service: Regulation and Rate Reform*, National Regulatory Research Institute, Columbus, OH.

Massarutto A. (1999), *Comparing Water Pricing Policies in the EU: A Positive Analysis*, paper presented at the Conference on Pricing Water - Economics, Environment and Society, Istituto da Agua, Lisbon, September 7-9.

Moretto M. – Valbonesi P. (1998), *Option to Revoke and Regulation of Local Utilities*, mimeo, Department of Economics, University of Padova.

Muckleston K.W. (1990), Integrated Water Management in the United States, in: Mitchell, B., *Integrated Water Management*, Belhaven Press, New York NY, pp. 22-44.

OECD (1999a), *Agricultural Pricing in OECD Countries*, OECD, Environment Directorate, ENV/EPOC/GEEI (98) 11/Final. www.oecd.org/env/docs.

OECD (1999a), *Household Water Pricing in OECD Countries*, OECD, Environment Directorate, ENV/EPOC/GEEI (98)12/Final. www.oecd.org/env/docs.

Pearce D. – Atkinson G. (1998), The Concept of Sustainable Development: An Evaluation of its Usefulness Ten Years After Brundtland, *Swiss Journal of Economics and Statistics*, **134**: 251-269.

Postel S. (1999), *Pillar of Sand (Can the Irrigation Miracle Last?)*. W. W. Norton & Co., New Yok, NY.

Price Waterhouse (1994), *The Public Utility Industry: 1994* Survey of Industry Developments and Financial Reporting, Price Waterhouse.

Siedenstat P. – Nadol M. – Hakim S. (2000), *America's Water and Wastewater Industries: Competition and Privatization*, Public Utilities Reports.

Spulber N. – Sabbaghi A. (1994)*, Economics of Water Resources: From Regulation to Privatization*, Kluwer Academic Publishers, Boston, MA.

United States Environmental Protection Agency (1999), *National Characteristics of Water Systems Serving Populations under 10,000*, USEPA 816-R-99-010, July 1999, Washington D.C.

Webb M. – Ehrhard D. (1998), *Improving Water Services through Competition*, The World Bank Group - Finance, Private Sector, and Infrastructure Network, Note n.164. www.worldbank.org/html/fdp/notes.

World Bank (1993), *Water Resources Management: A World Bank Policy Paper*, The World Bank, Washington D.C.

World Bank (1994), *World Development Report 1994: Infrastructures for Development*, Oxford University Press, New York, NY.

EVALUATION OF THE RECREATIONAL USES OF RURAL LAND: A CASE STUDY

Guido Maria Bazzani, Davide Viaggi, and Giacomo Zanni [*]

SUMMARY

The recreational use of agricultural land is emerging as one of the main opportunities for income production and rural development in many rural areas of the world. Many agri-environmental measures are directly aimed at improving the recreational value of rural land through the provision of direct services or landscape improvement.

The aim of this study is to assess the value for the consumers of such recreational improvements. In particular, an assessment was carried out in an area of Bologna Hills in order to evaluate the willingness to pay (WTP) for the provision of a set of given recreational services. The collection of information was achieved through a questionnaire addressed to households living in the Province of Bologna. The information collected includes recreational behaviour of the households, as well as the WTP for the services proposed. Results are discussed in relation to actual policy implementation, comparing recreational benefits with the public expenditure necessary to produce such benefits.

The results show that the provision of recreational goods in the countryside is valued by households as a relevant good, the quality of which significantly affects their behaviour in relation to the use of the countryside. Nevertheless, the WTP is probably not sufficient to fully cover the public expenditure necessary to carry out the proposed intervention. Fine tuning is thus necessary to target intervention, to improve policy efficiency and to create mechanisms for direct payment the services provided.

1. INTRODUCTION

The recreational use of agricultural land is emerging as one of the main opportunities for income production and rural development in many rural areas of the world. Many

---

[*] Guido Maria Bazzani, CNR Land and Agri-System Management Research Centre (Ge-STA), Bologna. Davide Viaggi, Department of Agricultural Economics and Engineering, University of Bologna. Giacomo Zanni, Department of Economics, University of Foggia.

agri-environmental measures are directly aimed at improving the recreational value of rural land through the provision of direct services or landscape improvement.

Nevertheless, doubts exist about the capacity of such policies to target problems relevant to consumers of landscape and recreational services and, as a consequence, about the social benefit/cost ratio of such policies.

The aim of this study is to assess the value for the consumers of such recreational improvements. In particular an assessment was carried out in an area of Bologna Hills in order to evaluate the willingness to pay (WTP) for the provision of a set of given recreational services. The collection of information was achieved through a questionnaire addressed to families living in the Province of Bologna. The information collected includes recreational behaviour of the households as well as the WTP for the services proposed. Results are discussed in relation to actual policy implementation and to public expenditure necessary to produce the services.

2. DEMAND, SUPPLY, AND POLICIES FOR RECREATIONAL SERVICES PRODUCED BY AGRICULTURE

Recreation in the countryside is a growing activity in both private use and public attention. For example, the surface area of protected areas in Italy has shown a sharp increase in the last ten years, reaching about 10% of the national land surface (ISTAT, 1997a; Marinelli et al., 1990). This is associated with a related increase in the number of visits to these areas.

Policy intervention providing direct incentives for the provision of recreational services is growing as well; the surface encompassed by regulation EEC 2078/92 is rapidly increasing, reaching over 1 million hectares in 1998. About half of this area concerns measures directly affecting the landscape and the recreational quality of the countryside. About 50 percent of the Italian population takes regular trips to woodlands or the countryside, and over 20 percent practice outdoor sports (ISTAT, 1997b).

The subject of recreation in rural areas is multiform, complex and characterised by high fragmentation of demand. First of all, these recreational services are often positive externalities produced by farmers, though they usually do not receive compensate for such service. This is because such services are characterised by a low level of excludability and rivalry of consumption and are, as a consequence, close to being public goods.

The complexity of the demand for recreation in the countryside is due not only to the number of functions of agricultural territory and the variety of possible recreational activities, but also to the incertitude of consumers' preferences about recreation in rural areas.

Individual objectives are affected by personal education, income, household structure, as well as a variety of psychological aspects of recreational choices. These aspects have to be cast with in the more complex issue of the perception of rurality (Walsh, 1986; Blanc, 1997). Further, the collective demand of recreation cannot be reduced to the sum of individual demands, but must represent a product of interaction among consumers. Such interaction creates "bodies organising the demand", which mainly focus on very narrow typologies of services (for example, one single sport). Consequently these bodies find difficulties in interacting with the wide variety of environmental management functions (Beuret, 1997).

The structure of the objectives of the public actors is also complex. In particular, the preservation of the non-market functions of the territory overlaps sector objectives, such as the support of the agricultural income. Within the set of public objectives, those aiming at the development of recreational services have only recently been considered deserving of attention.

As for the willingness to pay (WTP) for consumption of outdoor recreation, it is necessary to distinguish that expressed through the behaviour of the public administration from that expressed by the private members of the society (single or associated citizens, etc.).

Public funds for recreationincluded those regulated by reg. EEC 2078/92, namely those elements directly or indirectly promoting the recreational usability of the territory. The engagements aimed at the production of landscape-recreational services in Italy in the last five years have involved an average yearly expenditure of about 30 billion ITL. This figure is insignificant when compared to the value of national agricultural production (under 0.1%).

There is increasing literature about the evaluation of citizens' WTP for the safeguard of landscape and recreational use of protected areas. They are mainly location specific applications, that do not help to understand the total dimensions of the phenomenon. A synthetic evaluation indicates that the Italian national WTP for the production of landscape-recreational services in the rural zones of Italy could reasonably stand between 0.3% and 3% of the agricultural gross production. They are not negligible values, but they are insufficient to cover the costs of policy intervention and to represent a significant income alternative for farmers (Viaggi and Zanni, 1998).

These figures can be better understood though a more precise analysis of the behaviour of society in relation to farming. In order to interpret the attitudes of the collectivity towards the farmers producing recreational services, it is necessary to distinguish the activities that are part of the traditional mission of the farmers from those that are not. In the former category, we find practices aimed at the reduction of the chemical impact from pesticides and the mitigation of mechanical operations on the soil, with the aim of minimising erosion; in the latter category we find set aside constraints, or the commitment to plant non-remunerative crops of landscape value in place of cash crops.

The first attitude is based on the defence of a natural right to not be polluted. Instead, the second admits the existence of a cost to farming, and therefore accepts the idea of offering compensation in order to promote the production of the services. However, the border between the rights and the duties of the various actors is quite fuzzy, and this reduces the total amount of the services effectively produced.

The ability to transform the WTP for incentives to farmers depends on the way in which the consumers relate themselves to the agricultural world. The users can influence the producers through direct payments, indirectly through informative pressures (through the associative agencies), or through the ballot (through the public administration). The public administration can intervene with the payment of incentives or through command and control legislation. It tends to intervene when a certain threshold of gravity of the environmental problem is exceeded. This gravity may be difficult to assess because often the requirements of the demand cannot be directly linked to the supply. These mechanisms of collective action favour delays and divergences between the demand of environmental services and the relative supply.

In order to understand the supply of recreational services in the countryside, the main elements to be analysed are the following: i) the features of the production processes; ii) the typologies of the goods produced and iii) the production costs.

Recreational services in agriculture are not very tangible and of a varied quality. The supply includes a mix of natural resources, facilities and personal services that the consumers combine according to their preferences in order to produce recreational services (Walsh, 1986).

The subjects that produce this kind of recreation are typical multiple goalers. The public bodies (for example, parks) associate with recreational aims also embrace goals for conservation of natural resources and environmental protection. Such bodies interact both with a structure fully devoted to the production of recreational commercial services and with agricultural production realities that incorporate mixed strategies and finalities (profit, environmental protection, landscape protections, etc.).

From a quantitative point of view, the supply appears insufficient to cover the demand by consumers. Farmers are mainly devoted to a modest maintenance of the emergencies already present inside the farms and the parallel development of business activities of tourist or agritourist type. The first activity maintains the landscape or vegetable and animal elements having cultural or historical importance, and is performed mostly as the result of public incentives or legislative constraints. This are, however, practices linked to the level of activation of normal business production processes. The second activity responds more directly to the demand of specific recreational services. In this situation, there is correspondence between the demand of the user and the supply of the agricultural firm.

In order to complete the description of the supply function, it is necessary to consider the shape of the marginal costs of production from which this is determined. The following elements of cost are important (Walsh, 1986): i) construction costs; ii) maintenance costs; iii) opportunity cost of land; iv) transaction costs. The costs incurred for the building, the restoration and the maintenance of shapes of environmental and landscape agriculture are often so high that the application of such practices would reduce the income of the farm byone half (Tempesta, 1990). Transaction costs play an important role. In fact, the plurality of the customers and the producers, constrained by the particular nature of the recreational service and the specificity of the agricultural structures, makes the coordination necessities particularly relevant.

In the determination of the subsidies for recreational services, it is necessary to take into account an incentive margin to add to the coverage of the costs. For example, in order to promote participation in the engagements of landscaped-recreational character of the reg. EEC 2078/92, in the Emilia-Romagna Region, an average margin of 50% of the cost to support farmers has been admitted (Viaggi, 1998).

The whole cost tends to assume excessive dimensions that are unlikely to be completely supported by public agencies. On the other side, if left to the farmers, such burdens could become indefensible, increasing confrontation and encouraging the abandonment of the agricultural activity. For actions of public participation in developing the production of landscaped-recreational services by farmers, it is necessary to recall two general principles: i) the beneficiary pays principle; ii) the provider gets principle.

The "beneficiary pays" principle states that the one who benefits from a service or a good has to pay for it. The problem is to define how much of such sums must be paid by the beneficiaries, considering that the benefit is composed of use and non-use values.

Tariffs can be used to implement this principle and it is well recognized that tariffs can have various operating modalities (depending on the market, the logistic problems, the costs of collection, etc). Such tariffs can be aligned to the marginal cost of production and maintenance of the services, or, according to a more "Coase-oriented" formulation, to the benefits that the customers effectively enjoy.

Territorial resources supply direct and indirect recreational benefits. Based on the beneficiary pays approach, those who make direct use of the resources have to pay first, while for indirect uses the collectivity has to pay through taxation. In the cases in which it is possible to obtain a direct payment for the service, the legislation would need to establish property rights, in order to promotemarkets, using direct incentives only in the starting phase. In the cases in which the distributed services are, and are likely to remain, pure externalities it is more effective to just use incentives.

The "provider gets" principle asserts that the producer of a service must be compensated for it. It is complementary to the previous principle, indicating who must receive the payments. To this purpose, the OECD (1996) emphasises that such a principle cannot be used indiscriminately in order to justify payments to farmers. The payment possibility is dependent on the active role of the farmers in connection with the task of other entities (parks, local bodies) that take part in the production of the services. This role could be measured, for example, as a function of the production costs sustained.

Various modalities of classification for policy instruments are available for the public administration to improve landscape and recreational services. A general classification distinguishes: i) persuasion; ii) expropriation and management by public agencies; iii) command - and - control instruments; iv) direct economic incentives; v) definition of property rights; vi) development of co-ordination and negotiation between actors.

Persuasion, even though it is considered little intrusive by economists, seems to assume a pre-eminent role for improving communication between customers of the agricultural and rural territory, and farmers considering alternative income opportunities. Participation by local communities in administrative solutions, and institutional procedures for co-ordinating recreational plans, can reduce conflicts. The expropriations of areas by public agencies face the problem of high costs and frequent inefficiencies of management. Therefore, they are solely of interest for the acquisition of areas of limited dimensions and highest landscaped or naturalistic value (Colman, 1991; Hanley, 1991; Hodge, 1991).

The use of command-and-control instruments introduces problems related to economic efficiency and creates conflicts between the farmers and the public agencies. The imposition of constraints without some payment can be interpreted as a re-appropriation of rights by the collectivity, especially when the constraints have a modest impact on farming activities. Public subsidies represent the more direct form of public participation in order to promote the supply of public goods in the countryside. Their wide use, however, faces the budgetary problems of the public agencies. The specific forms of incentives would have to be adapted to the business typologies, sincethe attitudes of the single farmers vary in relation to recreational problems (Castello et al., 1996).

The creation of markets for assets that currently have no market can partly solve the problem of the remuneration of the producers and the attainment of a suitable level of supply.

It is easier to identify the price paid for the assets acquired during the visits to the countryside (for example the lunch), as opposed the benefits obtained from the assets for which some direct payment does not take place (for example the landscape). In addition, the problem of redistributing this surplus between all the producers arises. This is straightforward operation if both activities belong to the same actor (for example the agritourist operator), but it becomes rather complex if it is necessary to redistribute the sum between various actors (for example between a "conventional" restaurant and a farmer).

A development of coordination can take place through public support of organisations that represent the users of the territory, or mediation efforts of the public administration in the direct negotiation between private agents. To this purpose, it is necessary to set up innovative forms of organisation between farmers, consumers and the public. The essential element of innovation could emerge through appraisal of mixed forms of participation, composed of the integration of several available instruments. In particular, public participation would have to place high priority on the translation of economic signals into economic incentives for the farmers, in conformity with the effective social value of the activities that they implement and to the WTP of the households. In this context, new theoretical approaches could be usefully employed so as to better formulate the conditions of well-being of the consumers as a function of a number of services (Casini et al., 1997). They could also be used to interpret, in a more articulated way, the operation of the markets and the nearly-markets, in order to supply answers to economic questions on the border between the public and the private sector (Bromley, 1991; Bromley, 1997).

3. METHODOLOGY

The methodology adopted is Contingent Valuation (CV) based on questionnaires sent to a number of representative households in the province of Bologna. The questionnaire was submitted to households in the spring 1998, after an analysis of the problem based on focus groups and a testing.

The outline of the questionnaire is the following:

- Information about recreation activities;
- Interaction between recreation and agriculture;
- Presentation of the study area: the hill and mountain area of Bologna Province;
- Evaluation of intervention proposals;
- Socio-economic data about the interviewed;
- Self-evaluation of the questionnaire.

The survey's focus is the hill and mountain area of the province of Bologna. The territory is divided into a number of almost parallel valleys that take their name from the main rivers of each valley: Samoggia, Savena, Idice, Sillaro and Santerno, and Reno. The whole area isabout 210 thousand hectares, divided into 60 municipalities. Economic activity is more developed close to the plain. A number of historical and archaeological sites are spread over the territory (Blum, 1982). The interesting features of the landscape make the area a centre of tourist activity, in particular during spring, summer and autumn. Agriculture is the main activity of the area, using 75% of the land. The main crops are wheat and barley. Woodland is 25% of the surface, while 10% is covered by permanent grassland and pasture.

Table 1. General statistics of the survey

	No. questionnaires	Percent
Sent	414	100%
Returned	243	58.7% of those sent
Usable	197	81.1% of those returned
Favourable	153	77.7% of those usable

Source: Authors' processing of own data

At the beginning of 1998, in the area, 2300 hectares and 950 animal heads partici-pated in reg. 2078/92. The scenario submitted for evaluation proposes a widening of the measures to obtain a major change of landscape and of recreation usability of the terri-tory. These results would be obtained in two years time, through the implementation of 4 actions:

- safeguard of all animal heads present in the area;
- upkeep of about 30% of grassland and woodland of the area;
- conversion of about 25% of arable land into grassland and other landscape ele-ments;
- free access to about 50% of private agricultural and wooded land of the hill and mountain area of Bologna.

The actions would be implemented by the farmers of the area.

The knowledge of the area among the interviewed was shown to be quite good, as in-dicated by pre-tests. In effect, about 86% of the sample go periodically to the area, guar-anteeing an informed answer to many inquires of the questionnaire. Table 1 shows the main statistics concerning the survey.

Of 414 questionnaires envoyed, only 243 were returned. This is common in this type of study as is shown by the literature. Of the returned questionnaires, after discarding incomplete and protest questionnaires, 197 were usable, 81% of those returned. One hun-dred fifty-three, or 77.7% of the respondents, were in favour of the project implementa-tion that was suggested in the CV exercise.

4. RESULTS OF THE SURVEY

The survey included an open question about the willingness to pay, in the form of a yearly payment. Interviewees were reminded that their actual WTP would be constrained by their income and that the sum, collected as a tax, would be managed by the public administration. This is currently the real situation in Italy.

The average WTP is equal to 97 thousand ITL per household (SD 188 thousand ITL)[1]. The distribution of values is rather skewed (3.8). Values are concentrated around multiples of ten thousand ITL. A few extreme values are located over 200 thousand ITL, up to 1 million. The critical analysis of such questionnaires (10) has cast doubts about their reliability, so they have been discarded from further analysis. The final sample con-sists of 143 questionnaires, i.e. 35% of the questionnaires sent. After censoring the WTP

[1] Variable	Mean	Std. Dev.	Skew.	Kurt.	Minimum	Maximum	Cases
WTP	97.3856	188.2066	3.8	17.9	0.0000	1000.0000	153

and its distribution, the results were altered [2] and the new average value is 53.85 thousand ITL (SD 54.4).

It should be noted that the majority of the people who favored project implementation, revealed an active frequenting of the territory. This occurence and observed high income among the respondents, with a distribution skewed to right, makes the sample not perfectly representative of the population. For this reason caution should be adopted in extrapolating the WTP estimated on the whole population. In this situation, assuming a very prudent approach, it is possible to attribute WTP=0 to all non respondents (171 households), as well as to people opposing the project implementation (44 households). As a consequence, the average WTP over the whole population drops to 21.5 thousand ITL. This value represents the average yearly benefit obtained by households.

The motivations for the WTP have been further explored by identifying the distribution of preference among the four actions included in the intervention proposal. The answers can be interpreted as a vote in favour each action.

The result shows how the protection of flora and fauna has the higher priority (36%), followed by upkeep of non-cultivated land (33%). Landscape (20%) and free access (11%) show lower importance (Table 2).

Another interesting way of testing the coherence of the questionnaire is to ask about the impact of agriculture on the recreational experiences of each person interviewed.

Sixty percent judge the impact to be positive, while only 1% consider it to be negative. The remaining percentage show indifference. The correlation of positive responses with the WTP is very high (about 0.30). The WTP is strongly stratified in relation to the perception of a positive impact of agriculture, being 67 thousand ITL for positive impact, 33 thousand for indifference and 15 thousand for negative impact (Table 3).

The recreational use of the territory by respondents represents a particularly important dimension of the problem. Questions were posed about both usual fruition of the area and the likely change in behaviour if the proposed intervention was implemented. As for the former, households claims to have made an average of 5.3 visits each to the area with a minimum of 0 and a maximum of 10 (SD 3.65 and Skew 0.1). This confirms direct knowledge of the area. Fifty-one percent of the sample further answered that they would go more frequently if the intervention was carried out, while 49% answered that they would go with the same frequency. The WTP is, respectively, 84 thousand ITL and 23 thousand ITL (Table 4).

As for the household's location, the WTP is slightly higher for residents in town (about 55 thousand ITL) compared with the countryside (50 thousand ITL).

A positive correlation (0.3) is found between income and WTP. This is in accordance with economic theory and previous studies (Table 5).

The aggregate analysis of covariates has been done through the estimation of a linear regression model:

$$WTP = f(I, X)$$

where I represents income and X other covariates.

---

[2]

Variable	Mean	Std. Dev.	Skew.	Kurt.	Minimum	Maximum	Cases
WTP	53.8462	54.4198	1.0	3.5	0.0000	200.0000	143

Table 2.   Relationship between WTP and expenditure distribution

Favourable to	WTP	SD	N	%
Landscape	63.0	46.3	45	20
Non cultivated areas	66.4	49.0	76	33
Flora and fauna	67.1	55.7	83	36
Free access	52.5	58.3	25	11

Source: Authors' processing of own data

Table 3.   Relationship between opinion about agricultural impacts and WTP

Impact of agriculture	WTP	SD	N	%
Positive	67.9	59.2	86	60
Indifferent	33.3	38.2	55	38
Negative	15.0	7.0	2	1

Source: Authors' processing of own data

Table 4.   Change of behaviour in case of program implementation

Visit the area	WTP	SD	N	%
More often	83.7	51.7	73	51
The same	22.7	37.2	70	49
Less often	-	-		0

Source: Authors' processing of own data

Table 5.   Relationship between WTP and income (million ITL)

Income class	WTP	SD	N	%
0-10	-	-	5	3
10-30	36.4	52.9	32	22
30-50	53.7	50.9	61	43
50-100	69.0	53.0	40	28
100-150	100.0	79.0	5	3
>150			0	0

Source: Authors' processing of own data

The model assumes a constant proportionality between WTP and R^2. Among different solutions explored, the one that best fits shows an adjusted $R^2 = 0.443$ and F= 23.64 with 137 d.o.f.. Together with income, 4 covariates (3 of which are dummy variables) and the intercept enter the model (Table 6).

The coefficients and Standard Deviation (SD) are as follows: income (I) 0.561, (4.1), for million lit; the behaviour after the intervention (AFT) 46.94, (6.4); the evaluation of the effects of agriculture on recreation (IMP) 21.78, (3.4); the frequency of visits to the area in the past (VIS) 28.9, (2.8); and the number of households components (FAM) – 7.96, (2.3). Only the constant (-10.8) is not significant. All signs are considered to be correct. The coefficient on the family variable canbe interpreted in the sense that the larger the family, the lower is the WTP at a given income. These results confirm the hypothesis that the WTP is linked to the active use of the territory as shown by the

Table 6.   Results of the regression model

Dependent variable is WTP   Mean = 53.84615,   S.D. = 54.4198
Model size: Observations = 143,   Parameters = 6,Deg.Fr. = 137
Residuals: Sum of squares= 225776. Std.Dev. = 40.59560
Fit: R-squared = 0.46312,   Adjusted R-squared = 0.44353
Model test: F[5,137] =23.64,   Prob value = 0.00000

Source	Degr. of fr.	Sum of squares	Mean square	F
Regression	5	194758.20734	38951.64147	23.64
Residual	137	225776.40805	1648.00298	
Total	1	420534.61538	2961.51138	

Variable	Coefficient	Stand. Error	t-ratio	P[t]	Mean of X
I	0.56100	0.13681	4.100	0.00007	47.06
AFT	46.943	7.3090	6.423	0.00000	0.5105
IMP	21.782	6.7181	3.242	0.00149	0.5874
VIS	28.907	10.146	2.849	0.00506	0.8601
FAM	-7.9607	3.4395	-2.315	0.02213	2.937
Constant	-10.797	66.987	-0.161	0.87218	

Source: Authors' processing of own data

the WTP is linked to the active use of the territory as shown by the positive and high coefficient of the two variables which take into account past and future visits.

5.   RELATIONSHIPS WITH AGRO-ENVIRONMENTAL POLICIES

The relationship between the demand for recreational services in the countryside and the policies to be put into action for their provision can be analysed on two strictly linked dimensions. The first is the comparison between WTP and the costs of the policy; the second is the discussion of the best way to organise the production of services. The close relationship between these two aspects stems from the fact, that, as confirmed by the questionnaire, the WTP is a function of the modalities of payment from households and of management of the funds collected. These contribute to the determination of the costs of the policy.

Given the limits of the methodologies adopted, considerable caution is needed in comparing policy costs and WTP. Taking into account such limits, the comparison can be made both at the level of budgetary costs and at the level of actual social costs.

The comparison of budgetary costs and benefits can be interpreted as the likely input and output of money from the accounts of the public administration, hence representing the financial interpretation of the decision problem by the competent public body. Under this perspective, comparing WTP and policy cost in the case study carried out, it is possible to affirm that the WTP is not able to cover the full cost of the intervention under evaluation (Figure 1).

In effect, the cost of the intervention proposed by the questionnaire and submitted to the judgement of surveyed households can be estimated around 15 million euro per year. On the other side, applying the average WTP obtained from the surveyed households to whole population of the Province of Bologna, the total WTP is lower than 5 million euro

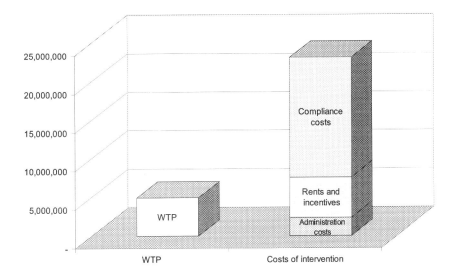

Figure 1. Comparison between WTP and cost of proposed measures (euro)

per year. This amount is a significant sum, that allows us to cover about one third of the costs of the policy. It is clear that an actual interest for this intervention by the population does exist, but the household's WTP is insufficient to justify the implementation of this policy.

Given the high degree of approximation of these estimates, it is necessary to reflect on possible elements of correction. On one hand, the WTP could be overestimated due to *embedding*[3], even if the results obtained on the problem are quite reassuring. On the other hand, non-recreational impacts (such as possible environmental, hydro-geological, eco-system and cultural benefits) have been excluded from the evaluation. They could be linked to the proposed intervention, being in some way "included in the cost", and would without doubt increase the WTP of consumers if included in the evaluation.

On the side of expenditure, it is necessary to point out that the total cost of intervention has been evaluated on the basis of the actual payments by farmers for analogous measures implemented under reg. EEC 2078/92. An estimation of the likely public administration cost was summed up using these values. Such payments, on the basis of the outcome of monitoring studies, are much higher than compliance costs. Figure 1 proposes a division of the total cost into three cost categories: compliance costs, rents and incentives for farmers, and administrative cost. The evaluation of this distribution has been obtained on the basis of a hypothesis constructed on average monitoring data about the application of reg. EEC 2078/92 in Emilia Romagna.

While the WTP is not able to cover the full cost of the intervention, however, it allows coverage of a significant part of compliance costs. Basically, compliance costs are

---

[3] Embedding is the anomalous result by which the WTP for a good is the same as WTP for a more inclusive good (i.e., one in which it is "embedded").

costs for technological change, calculated as the difference between the ex ante technology and the ex post technology, after the adoption of the proposed measures. This suggests that the relationship between the demand and supply of landscape-recreational services could benefit considerably from an increase in policy efficiency, through a minimisation of administration costs and a reduction of rents for farmers.

From a social viewpoint, the measures proposed appear clearly not profitable, as the WTP (4.5 million euro) is lower than the pure compliance costs (15.5 million euro). The difference represents actual profits foregone. Even assuming that a relevant part of the profits forgone depend on the reduction of current transfers, it appears reasonable to estimate that, while the project under evaluation is surely valuable from a social point of view, the benefit/cost balance is still negative. This means that, in order to make the intervention socially profitable, it is necessary to reduce compliance costs and/or to increase the social value of the intervention proposed. Compliance costs can be reduced, for example, through targeting and improved technical solutions. On the other hand, an increase in the value of recreational services for consumers would be needed, through an improvement of the services provided, in order to improve the WTP. Of course, the desiderability of producing more services should be evaluated comparing their cost with the increase in WTP.

Altogether, while the measures proposed appear not currently feasible from both a budgetary and a social perspective, there is lots of room for improvement in making the intervention more valuable and acceptable. This is necessary in order to allow for a better exploitation of the opportunities of new markets in the field of recreational services. Such results can be obtained through an improvement of the organisation, targeting and incentive mechanisms of agri-environmental measures. The same can be said about a reduction of administration costs through a better organisation of the intervention.

These considerations bring our attention to the field of organisational aspects of agri-environmental policies. The work has confirmed the literature with evidence of how the demand for recreational services is split into a variety of objectives, often linked to specific sports, fragmented into a variety of organisations, and frequently having a conflicting relationship with the rural environment. The supply is based on a number of actors, distributed through out the territory, often having objectives different from the production of recreational services and oriented by specific production or environmental objectives.

Among the many options linking demand and supply (see paragraph 2), direct intervention through subsidies shows all of the problems pointed out by the previous analysis, both in terms of implementation costs and of acceptability by tax payers. In some cases, markets can be improved, but a large part of services demanded by the population do not fit into markets and are expected to be free. These services include upkeep of non-cultivated areas, picnic structures, etc.. An important role could be played by services normally paid for by users, such as rural tourism, that cover only a comparatively small part of all countryside activities.

In order to produce those services, the trend showed by the questionnaires is to ask for innovative forms of participation instruments, able to involve all of the interested categories in the collection of funds and in the decisions of the actions to be taken. While the use of simple public subsidies is not appreciated, very much, by the population interviewed, respondents recognise the need of a relevant degree of public intervention, tempered by participation of recreational organisation in the decision making process. This is reasonable also because not all services demanded can actually be sold on a market.

In a sense, the issue can be defined as a problem relating to an increase in communication and to giving the right signals for activating market transactions.

This can be translated into policies related to communication *tout court*. In fact, an issue related to how to make countryside tourism markets function better is the degree of information owned by both partners in the possible transaction. This is true both for countryside users, who often do not know how the services they use are provided, and for farmers, who are not aware of the income opportunities given by recreational activities in the countryside. Improvements in communication can have effects over the long term. While this is now being done, there is still very much to do.

One second way in which communication would be made more effective is by regulating market signals in order to allow them to better reach all the actors involved. In particular, while a number of services in the countryside are actually paid for, the profit is not distributed among all actors producing the services that are appreciated by users. For example, restaurants benefit from services produced by farmers and other actors collaborating in landscape maintenance. A more co-operative relationship between all actors producing services could allow a higher response to the needs of the consumer, and better profits for farmers.

If a need for policy exists, it is important that it comes from an actual collaboration of agricultural entities with all local stakeholders in particular, recreational organisations. This could feed knowledge about consumer appreciation of countryside features into policies related to agriculture. There is an experience in the management of subsidies that should be exploited by improving existing mechanisms before rejecting them.

In addition, recreation organisations could be involved in the creation of local plans, together with farmers' organisations, for accessing financial aid under more organised and finalised conditions, rather than by simply submitting farm-size plans.

Finally, in some cases, the production of services by recreational organisations themselves could be necessary. This implies the purchase of rights for land use from actual owners.

6. DISCUSSION

Agriculture has made a relevant contribution to the quality of recreational experiences in the area of the Bologna Hills. In accordance with this general appreciation, households show a positive WTP for the funding of intervention aimed at improving the landscape quality of the countryside. The WTP is positively correlated to income.

However, households' WTP is not able to cover the costs of the agri-environmental intervention proposed, nor is it able to cover the actual social costs. Such incomplete ability is based, among other reasons, on the limited capacity of consumers to fully recognise the role played by agriculture in the production of recreational services, and in the difficulties in targeting intervention on the services directly used by the consumers.

The main policy implication is the need to find innovative policy instruments, based on negotiation and coordination, aimed at not substituting incomplete markets, but at completing and promoting existing ones through better informed policy making and better redistribution of profits (environmental agreements, single contracts). Such instruments would involve both farmers and recreational organisations in the creation of local agri-environmental plans.

It is necessary to create markets in which connections exist between producers of marketable services and all the producers of the services, such as the landscape, that make them appreciated by consumers.

7.   REFERENCES

Basile E. – Cecchi C. (1997), Differenziazione e integrazione nell'economia rurale, *Rivista di Economia Agraria*, **1-2**: 3-27.
Baumol W.J. – Oates W.E. (1988), *The theory of environmental policy*, 2nd ed., Cambridge University Press, Cambridge.
Beuret J.E. (1997), L'agriculture dans l'espace rural. Quelles demandes pour quelle fonctions?, *Economie Rurale*, **242**: 45-52.
Blanc M. (1997*)*, La ruralité: diversité des approches, *Economie Rurale*, **242**: 5-12.
Bromley D.W. (1991), *Environment and economy - Property rights and public policy*, Blackwell, Oxford.
Bromley D.W. (1997), Rethinking markets, *American Journal of Agricultural Economics*, **79**(5): 1383-1393.
Casini L. – Bernetti I – Menghini S. (1997), Teoria delle "libertà" e metodi multicriterio per l'analisi delle condizioni di sviluppo territoriale, *Rivista di Economia Agraria*, **1-2**: 29-59.
Castello L. – Viaggi D. – Zanni G. (1997), *Agri-environmental policies and protected areas: a case study in the "Parco del Taro"*, Parma (Italy), Proceedings of the 52nd Seminar of the European Association of Agricultural Economists, "EU Typical and Traditional Productions: Rural Effect and Agro-industrial Program", Parma, June 19-21, 1997, pp. 453-462.
Cavailhes J. – Dessendre C. – Goffette-Nagot F. – Schmitt B. (1994), Analyses des evolutions récentes de l'espace rural, *Economie rurale*, **223**: 13-19.
Choe C. – Fraser I. (1998), A note on imperfect monitoring of agri-environmental policy, *Journal of agricultural economics*, **49**(2): 250-258.
Clawson M. – Burnell Held R. – Stoddard C.H. (1960), *Land for the future*, The Johns Hopkins Press, Baltimore.
Clawson M. – Knetsch J.L. (1971*)*, *Economics of Outdoor Recreation*, Johns Hopkins University Press, London.
Colman (1991), Land purchase as a means of providing external benefits from agriculture, in: *Farming and the countryside*, Hanley N., ed., CAB International, Wallingford, pp. 215-229.
Dillman B.L. – Bergstrom J.C. (1991), Measuring environmental amenity benefits of agricultural land, in: *Farming and the countryside*, Hanley N., ed., CAB International, Wallingford, pp. 250-271
Dowson B. – Hill T. (1998), Community Forest Recreation: a Dynamic model for our Future Countryside?, *Managing Leisure*, **3**(1): 26-36.
Drake L. (1992), The non-market value of the Swedish agricultural landscape, *European Review of Agricultural Eonomics*, **19**(3): 351-364.
Gallerani V. – Zanni G. (1998), L'inserimento dei campi da golf nel territorio italiano. Problemi e criteri di valutazione dell'impatto, *Agribusiness, Paesaggio & Ambiente*, **2**(2): 164-177.
Guglielmi M. (1995), Vers de nouvelles fonctions de l'agriculture dans l'espace?, *Economie rurale*, **223**: 17-21.
Hanley N. (1989), Valuing rural recreation benefits: an empirical comparison of two approaches, *Journal of agricultural economics*, **40**(3): 361-374.
Hanley N. (ed., 1991), *Farming and the countryside*, CAB International, Wallingford.
Hodge I. (1991), The provision of public goods in the countryside: how should it be arranged?, in: *Farming and the countryside*, Hanley N., ed., CAB International, Wallingford, pp. 179-196.
ISTAT (1997a), *Annuario statistico italiano – 1997*, Roma.
ISTAT (1997b), *Musica, sport, computer e altre attività del tempo liber*o *– Year 1995*, Roma.
Marinelli A. – Casini L. – Romano D. (1990), Valutazione economica dell'impatto aggregato e dei benefici diretti della ricreazione all'aperto di un parco naturale della Toscana, *Genio Rurale*, **9**: 51-58.
OECD (1996), *Amenities for rural development. Policy examples*, Parigi.
Pearce D.W. – Turner R.K. (1989), *Economics of natural resources and the environment*, Harvester and Wheatsheaf.
Pruckner G.J. (1995), Agricultural landscape cultivation in Austria: an application of the CVM, *European Review of Agricultural Economics*, (22) **2**: 173-190.

Romano D. (1990), Tempo e domanda di ricreazione all'aperto, *Studi di economia e diritto*, **1**: 159-186.

Signorello G. (1990), La stima dei benefici di tutela di un'area naturale: un'applicazione della "contingent valuation", *Genio Rurale*, **9**: 59-66.

Signorello G. (1994), Valutazione contingente della "WTP" per la fruizione di un bene ambientale: approcci parametrici e non parametrici, *Rivista di Economia Agraria*, (XLIX) **2**: 219-238.

Slangen L.H.G. (1992), Policies for nature and landscape conservation in Dutch agriculture: an evaluation of objectives, means, effects and programme costs, *European Review of Agricultural Eonomics*, (19) **3**: 331-350.

Stellin G. – Rosato P. (1990), Un approccio multicriteriale alla gestione del territorio: il caso del Parco Naturale Regionale delle Valli di Caorle e Bibione, *Genio Rurale*, **9**: 67-75.

Tempesta T. (1990), Una stima delle variazioni di prodotto netto aziendale conseguenti al ripristino dei caratteri del pesaggio agrario tradizionale in un parco del Veneto, *Genio Rurale*, **9**: 86-93.

Tempesta T. (1995), La stima del valore ricreativo del territorio: un'analisi comparata delle principali metodologie, *Genio Rurale*, **12**: 15-34.

Tempesta T. (ed., 1997a), *Paesaggio rurale e agro-tecnologie innovative*, FrancoAngeli, Milano.

Tempesta T. (1997b), La valutazione del paesaggio nell'area tra Isonzo e Tagliamento: un approccio di tipo monetario, in: *Paesaggio rurale e agro-tecnologie innovative*, Tempesta T., ed., FrancoAngeli, Milano, pp. 195-263.

Viaggi D. (1998), L'applicazione del reg. CEE 2078/92 in Italia: una valutazione, *Agribusiness, paesaggio & ambiente*, (2) **1**: 58-66.

Walsh R.G. (1986), *Recreation economic decisions: comparing benefits and costs*, Venture Publishing, State College, Pennsylvania.

Whitby M. (1991), *The changing nature of rural land use*, in: *Farming and the countryside*, Hanley N., ed., CAB International, Wallingford, pp. 12-25.

Zanni G. (1997), Non profit e sport: proposta per la definizione di criteri di utilità sociale, in: *Enti non profit: quale futuro per lo sport?*, Zanni G. – Bertolotti A., eds., Maggioli, Rimini, pp.73-88.

THE VALUE OF LICENSES FOR RECREATIONAL USE
OF NATURAL RESOURCES

Michele Moretto and Paolo Rosato *

SUMMARY

The paper provides a behavior model of a consumer considering the purchase of a license that allows him to benefit from a natural resource in a certain place, for a certain period of time, and according to pre-established rules and procedures. We developed a model starting from the assumption that the opportunity to purchase a license can be likened to a call option, given that the uncertainty concerning the future benefits and the irreversibility of the expenditure may make it expedient to wait before purchasing. Furthermore, the model includes the main factors conditioning the perceived utility of the recreational activity (i.e. the stock of resources, the number of rivals, and operations designed to enrich/deplete stock levels) with particular attention to the uncertainty that characterizes activities pertaining to biological resources. The first result concerns the effect of uncertainty. It acts as a deterrent to the purchase of the license and lengthens the optimal duration. Generally speaking, therefore, the first tool available to an agency appointed to regulate the purchase of licenses for recreational activities is the quality of the information to be circulated.

A subsequent analysis enabled us to highlight the effect produced by an uncertain sharp change in the stock of resource on the optimal duration of the license. It emerged that, as the discount rate increases, shorter duration is not always favored, as might be expected with the net present value criterion.

Optimal license duration also depends on the elasticity of the supply function: as the elasticity decreases, optimal duration decreases. This highlights a second possibility of intervention for the agency: by modulating the supply function with respect to the duration, it can directly condition consumers' choices. Finally, irreversibility and uncertainty lengthen the optimal duration of the license, as longer duration reduces the effect of the variability of future benefits.

* Michele Moretto, Department of Economic Sciences, University of Padua. Paolo Rosato, Department of Civil Engineering, University of Trieste. The authors are grateful to Valentina Zanatta for the patient editing support.

1. INTRODUCTION

The management of recreational use of natural resources, such as fishing and hunting, is a topic of great theoretical and practical interest. This interest is basically due to a continuously increasing demand for recreation and to rapidly diminishing stocks of recreational amenities. The growth in demand is linked to improvement of the social and economic conditions of consumers in many countries, accompanied by increasing pressure from the economic interest groups operating in the recreation sector (e.g. tour operators, equipment manufacturers and retailers, television entertainment, etc.). Resource stocks, on the other hand, are rapidly diminishing due to exploitation for production purposes, environmental pollution, and reduction of reproduction habitat.

As a result of these changes in supply and demand, various tools to regulate recreation have been conceived, studied and used. The most common one is the license. The license is a right to exercise, in accordance with pre-established rules and procedures, a certain activity, for a certain period of time, in a given area. This right can be acquired in various ways - it can be inherited and/or related to citizenship, residence and land ownership - but in the majority of cases it has to be purchased from a public agency. The definition of the right varies and is closely linked to the objectives of the regulation. Generally speaking, the aim of the regulation is to guarantee a sustainable use of the resource in the long term (i.e. to ensure an extraction rate compatible with the reproduction capacity of the resource). In addition, the regulation system should boost the efficiency of the activities that use the resource. The license is usually nominal, meaning it can be used by one person only, and it cannot be transferred or refunded. Licenses can be granted for different periods (daily, weekly, seasonal, annual etc.) and can permit the user to keep a certain amount of fish, mushrooms, game, etc. It is evident, therefore, that setting up a license system to regulate recreational use of natural resources is complicated as it requires a number of presuppositions. In this regard, extensive literature exists concerning the analysis of professional license systems (Anderson, 1958; Campbell and Linder, 1990; Karpoff, 1989), but the consequences of a license system for the recreational use of resources is still relatively unexplored.

This paper provides a behavior model of a consumer considering the purchase of a license that allows him to benefit from a natural resource in a certain place, for a certain period of time, and according to pre-established rules and procedures. As the purchase of a license has some characteristics similar to an investment decision (e.g. it is irreversible), the future benefits are uncertain and, to a certain extent, the purchase can be postponed. The focus of this analysis is to establish if and when the consumer decides it is optimal to purchase the license [1]. Due to the above characteristics associated with natural resource license systems, it is reasonable to assume that the optimal purchase decision must satisfy a slightly more restrictive rule than the one suggested by the usual Net Present Value. The uncertainty surrounding the availability of natural resources implies that in the future the consumer may wish to abandon the recreational activity. Irreversibility means that if the

---

[1] This aspect is of particular interest as it coincides with the initiation of the recreational activity. It is reasonable to assume that the consumer can decide to purchase the license at any time even though the seasonal climatic trends, variability of the quality of the resource and the characteristics of the license can considerably limit the possibility of choice. For example, in many countries annual fishing or hunting licenses must be purchased before the beginning of the hunting or fishing season. Daily or weekly permits, on the other hand, can be purchased at any time except, of course, during the reproduction period.

consumer purchases a license now, he cannot re-sell it in the future. Finally, the expediency of postponing the purchase gives him the chance to gather further information before committing himself to an irreversible decision. The possibility of purchasing a license is therefore similar to a perpetual call option, or a right, not an obligation, to make an investment at any time in a certain financial activity at a pre-established operating price. As a financial option, the uncertainty of future benefits generates a value related to the opportunity to invest. [2]

The plan of the paper is as follows. Section 2 sets out the basic model and examines the consumer's value of the option to buy the license and the optimal purchasing time. The optimal duration of the license and the trade-off between cost of the license and its duration are discussed in section 3, followed by the conclusions in section 4.

2.   THE MODEL

The expediency of purchasing a license is assessed by weighing its cost against the satisfaction expected from the future recreational activity. This assessment can be made at any time and therefore it is not sufficient for the value of the expected satisfaction to be higher than the expenditure - it must also be higher than the satisfaction value that can be obtained, currently, by exercising the purchase at any time in the future. This aspect is very important when dealing with resources, like stocks of biological resources, the level of which varies due to natural causes like seasonal variation or due to management such as restocking, environmental rehabilitation operations or over-exploitation of the resource by professional activities.

In this paper, the basic assumptions are the following: a) the license cannot be transferred and therefore the decision is irreversible and generates a sunk cost; b) the license can be purchased at any time, i.e. the purchase can be postponed; c) the actions authorized by the license can be undertaken immediately after purchase; d) the duration of the license can vary: a day, a whole season (annual), several seasons or even for life. Having said this, let's consider a consumer who, at a given time, assesses the expediency of purchasing a license. He will obviously assess the expenditure with respect to the expected utility of the recreation. The latter will reasonably depend on the quantity of natural resources he expects to acquire. Therefore, the instantaneous net benefit expected by the consumer from the recreational activity is given by the following function $B(x, n)$ $(0 < x < X \leq \infty, 0 \leq n < N \leq \infty)$ where x is the stock of resources and n indicates the number of rival consumers drawing from the same stock. The marginal willingness to pay for the recreational activity is positive and decreasing with respect to the amount of stock, i.e. $B_x > 0, B_{xx} \leq 0$.

The number of rival consumers is included in the benefit function as we assume that the instantaneous quantity of resources available for the consumer is negatively correlated with the number of consumers simultaneously present due to a competition effect [3]. An

---

[2] Dixit (1992), Pindyck (1991) and a recent manual by Dixit and Pindyck (1994) provide a thorough review of this approach applied to various economic questions. For example, with respect to fishing, the only work that uses this approach, as far as we know, is that of Karpoff (1989), in which the author highlights how the option component of each license triggers a process of regulation of the number of licenses.

[3] It should be noted that the problem faced by a single consumer sharing a resource with others differs from that of the policy maker who considers the aggregate benefit of all the consumers. In this context, a rational

increase in rival agents produces a decline in the benefit and a reduction in the marginal willingness to pay, i.e. $B_n < 0$, $B_{nn} > 0$ and $B_{xn} < 0$.

The amount of the resource is measured by a variable of the biomass x_t which is assumed to satisfy the following stochastic differential equation. [4]

$$dx_t = \mu^x (x_t) \, dt + \sigma^x x_t \, dW_t^1 + x_t \, dQ_t \tag{1}$$

with $x_0 = x << X,\ \sigma^x > 0$

$$\mu^x (x_t) = \gamma(x_t) x_t - h (x_t, n_t) \, n_t$$

In (1) $\gamma (x_t)$ represents the expected growth rate of stock and is decreasing in the amount of the stock x_t [5]; $h (x_t, n_t) \, n_t$ is the expected individual harvest rate multiplied by the number of consumers present at time t and represents the reduction of the stock produced by recreation activity. This reduction is generally positively related to the stock level. The expected individual harvest rate is positively related to the stock and negatively related to the number of rivals [6]. In addition, it is assumed that certain random factors, such as weather, affect the stock and the growth rate of the biomass. The term $\sigma^x x_t$ measures the instantaneous volatility and dW_t^1 is the increment of a Brownian motion with mean $E (dW_t^1) = 0$ and variance $E [(dW_t^1)^2] = dt$. Finally, dQ_t is the positive or negative variation of a Poisson process, independent of W_t, with mean arrival time λ. This process represents sudden variations in x due to "exceptional" events such as pollution, poaching or illegal harvesting, or restocking. This variation in stock is assumed to be a fixed percentage $\phi (-1 \leq \phi \leq 1)$. The process dQ_t has the following probability distribution: $dQ_t = \phi$ with probability λdt; $dQ_t = 0$ with probability $1 - \lambda dt$. Therefore, (1) establishes that the stock of resources x fluctuates over time according to a Brownian motion, but that in each interval of time $(t, t + dt)$ there is a probability λdt of instantaneous variation occurring (positive or negative), bringing the stock to a level $(1 + \phi)$ times the initial level, subsequently beginning to fluctuate again until the next variation. [7]

As far as the number of rivals is concerned, it is driven by the following stochastic differential equation:

---

consumer will include an assessment of the stock of resources in its benefit function and this assessment will vary inversely to the number of rivals present (Arnason, 1990). This assumption is obviously slightly restrictive as the effect of congestion concerns not only competition between consumers in use of the stocks but, as we are considering recreational activities in the open air, also disturbance caused by the intrusion of non-rivalry users, such as trekkers and boaters. However, it is assumed that the main effect is that of competition.

[4] For a discussion of differential stochastic equations and Brownian motions, see Cox and Miller (1965) and Harrison (1985).

[5] The biological growth processes most widely used in models to optimize exploitation of sustainable resources (fish, wild animals etc.) incorporate the fact that when stocks increase, self-limitation processes are triggered with consequent reduction in the growth rate (see Clark, 1990).

[6] This assumption is in line with the classic approach to modeling of the use of sustainable resources, where the effort spent in harvesting (E), understood as the aggregate of capital energy, and work expended in a certain interval of time (Schaefer, 1954), necessary for the collection of a certain quantity (h) of resources, is inversely proportional to the stock level (x), i.e. $E = hx^{-1}$ (Pearce and Turner, 1989, p. 242). It can therefore be assumed, for the sake of simplicity, that collection is proportional to the stock level, $h(x)n = (\vartheta E x) n$, $0 \leq \vartheta < 1$, where $E = 1$.

[7] Without altering the results, we can assume that the size of the jump is uncertain.

$$dn_t = \mu^n(n_t)\,dt + \sigma^n n_t\,dW_t^2 \tag{2}$$

with $n_0 = n << N,\ \sigma^n > 0$

$\mu^n(n_t) = n_t u(x_t)p(n_t) - bn_t$

Equation (2) models the so-called "social dimension" of the recreational activity. We assume that the number of consumers basically depends on the knowledge they have, and/or think they have, of the stock of resources. This knowledge derives essentially from the circulation of information by people who consider themselves "informed" as they have recently benefited (or attempted to benefit) from the resource [8]. By indicating $u_t = u(x_t)$ as the contact rate between people who discuss stock levels x_t, at time t the "informed" consumers communicate with $n_t u_t$ people, only a fraction of whom $p(n_t)$ will be newly informed. In addition, the number of consumers drops in time at a constant rate b for reasons not directly connected with the recreational activity, such as saturation of the need to go out fishing or hunting. Finally, a level of uncertainty is considered with regard to the number and trend of the rivals: this is expressed by the term $\sigma^n nt$, where dW_t^2 is the usual increment of a Brownian motion with mean $E(dW_t^2) = 0$, and variance $E[(dW_t^2)^2] = dt$. This is possibly correlated with the process W_t^1 from model (1), i.e. $E(dW_t^1 dW_t^2) = \rho dt$, where $-1 < \rho < 1$ indicates the instantaneous correlation coefficient between the two processes.

2.1.   The value of the license

Assuming, in this section, that the license will run indefinitely, its value is given by the discounted flow of benefits:

$$V(x_T, n_T) = E_T\left\{ \int_T^\infty H(t)B(x_t, n_t)e^{-r(t-T)}\,dt \right\} \tag{3}$$

where T is the time at which the license is purchased, r is the discount rate and $H(t)$ represents the probability of instantaneously benefiting from the resource at time t. [9]

The consumer will purchase the license when its expected present value exceeds the cost K. This is a reasonable policy only if the purchase cannot be postponed; in reality, although the net present value $V(x_T, n_T) - K$ is positive, uncertainty of benefits, the seasonal trend and the sunkness of the cost may induce the purchaser to be cautious before committing himself to an irreversible decision. We therefore need to assess the value of waiting in order to model the expediency of postponing the purchase, in other words, to assess the option value to purchase the license.

---

[8] Here it should be noted that this information is often misleading. This possibility has not been included in the model.

[9] Generally, this probability varies according to the seasonal trend; it is zero during periods when the activity is prohibited and maximum when the availability of the resource, and/or the availability of time on the part of the consumer, is at a maximum.

2.2.   The value of the option to purchase the license

As the net benefit deriving from purchase of the license at time T is $V(x_T, n_T) - K$, the opportunity value at time zero, indicated by F, can be calculated by maximizing the following expression:

$$F(x,n) = \max_{T} E_0 \left[(V(x_T, n_T) - K)e^{-rT} \mid x_0 = x, n_0 = n \right] \qquad (4)$$

The maximization problem involves two state variables: x and n. However, as the option will probably be withheld if x is low or if n is high and, conversely, will be exercised when n is low enough for a given value of x, the rule that guides purchase can be represented by a curve $x = x^*(n)$, where x^* is always increasing with respect to n.

By a non-arbitrage argument, the *Bellman equation* of (4) simply requires the capital gain $E[dF(x,n)]$ to be equal at all times to the return $rF(x, n)dt$, i.e.:

$$rF(x, n)dt = E[dF(x, n)] \qquad (5)$$

which can be rewritten as a partial differential equation [10]:

$$1/2 \left[(\sigma^x)^2 x^2 F_{xx}(x,n) + (\sigma^n)^2 n^2 F_{nn}(x,n) + 2\rho\sigma^x\sigma^n xn F_{xn}(x.n) \right] +$$
$$+ \gamma(x)x - h(x,n)n)F_x(x,n) + (nu(x)p(n) - bn)F_n(x,n) + \qquad (6)$$
$$+ \lambda F((1+\phi)x,n) - (r+\lambda)F(x,n) = 0$$

Equation (6) must also satisfy the following boundary conditions:

$$\lim_{x \to 0} F(x,n) = 0 \qquad (7)$$

$$F[x^*(n),n] = V[x^*(n),n] - K \qquad (8)$$

$$F_x[x^*(n),n] = V_x[x^*(n),n] \qquad (9)$$

Condition (7) establishes that when the resource stocks x tend to zero, the option value also tends to zero. Condition (8) states that, within the purchase rule $(x^*(n))$, the option value must equal the current value of the license, net of the outlay for the purchase

---

[10] Expanding dF and applying Itô's lemma for combined Brownian and Poisson processes, we get the following partial differential equation (Dixit and Pindyck, 1994):

$$1/2 \left[(\sigma^x)^2 x^2 F_{xx}(x,n) + (\sigma^n)^2 n^2 F_{nn}(x,n) + 2\rho\sigma^x\sigma^n xn F_{xn}(x.n) \right]$$
$$+ \gamma(x)x - h(x,n)n)F_x(x,n) + (nu(x)p(n) - bn)F_n(x,n) +$$
$$+ \lambda \left[F((1+\phi)x,n) - F(x,n) \right] - rF(x,n) = 0$$

which can be reduced as in the text.

(matching value condition). As long as $x_t < x^*(n)$, we get $F[x(n), n] > V[x(n), n] - K$, so the consumer considers it convenient to postpone the purchase. In other words, $V[x(n), n] < K + F[x(n), n]$, or the value of the license, is lower than its total cost obtained by adding its direct cost K to the opportunity cost of the option expressed by $F[x(n), n]$. Finally, condition (9) excludes the possibility of the investment being made, for reasons of speculation, at a moment other than the one identified by (8) (smooth pasting condition). That is, the tangency point between the curve that represents the option value ($F[x(n), n]$) and the net present value ($V[x(n), n] - K$) identifies the critical value $x^*(n)$. Consequently, it is expedient to purchase the license if, for every given value of n, stock levels at that time are higher than the critical value, i.e. $x_t \geq x^*(n)$. [11]

2.3. Optimal purchase time

Although the model illustrated in section 2.2 fully represents the strategy of the license purchaser, to highlight the effect of context factors (like uncertainty) on consumer choices, a number of simplifying assumptions have been adopted to obtain a closed solution of the model. The assumptions are the following:

1. the instantaneous benefit function is $B(x,n) = x^\alpha n^{-\beta}$, with $0 < \alpha \leq 1$, $\beta > 0$.

2. the weighted sum of the growth rates of x and n is constant, i.e.

$$\mu(x,n) \equiv \frac{\mu''(x)}{x} - \frac{\mu''(x)}{n} = \hat{\mu} . \text{ [12]}$$

3. the probability of the consumer benefiting from the resource at time t has an exponential distribution $H(t) = e^{-ht}$. [13]

Making use of the above assumptions, the optimization problem is reduced to one dimension. In particular, by expanding dB_t and applying Itô's lemma it can be easily shown that B varies according to the following function [14]:

$$dB_t = \mu_B B_t dt + \sigma_B B_t dW_t + \alpha B_t dQ_t \text{ with } B_0 = B \tag{10}$$

where: $\mu_B = \hat{\mu} + \frac{1}{2}(\sigma^x)^2 \alpha(\alpha-1) + \frac{1}{2}(\sigma^n)^2 \beta(\beta+1) - \rho\sigma^x\sigma^n\alpha\beta$

$\sigma_B = \sqrt{(\sigma^x)^2 \alpha^2 + (\sigma^n)^2 \beta - 2\rho\sigma^x\sigma^n\alpha\beta}$

---

[11] In this case we have $(dx^* dn^{-1}) > 0$, i.e. if the number of rivals increases, the stock must increase more than proportionally for the purchase to remain viable.

[12] This assumption implies that: 1) the growth rate of the resource is constant, $\gamma(x) = \gamma$; 2) the harvesting rate is linear, $h(x,n) = a(xn^{-1})$ or $h(x,n) = ax$; 3) the contact rate is constant, $u(x) = u$; 4) the fraction of possible people contacted by the consumer is constant, $p(n) = p$. This set of conditions gives $\hat{\mu} = \alpha(\gamma - a) - \beta(up - b)$.

[13] This implies that the possibility of taking a trip between time t and time t + dt can be described by a constant rate h.

[14] This derives from the fact that the log-linear function of a log-normal random variable is distributed log-normally. The first and second moment of the process Bt are given by $E(dB \ B^{-1}) = (\mu_B + \lambda\phi)dt$ and $V(dB \ B^{-1}) = (\sigma^2_B + \lambda\phi^2)dt$ (see Dixit and Pindyck, 1994, p.169).

In short, we assume that the stock of resources, x_t, and the expected number of rivals, n_t, increase (or decrease) at a certain constant mean rate, but that the actual growth is stochastic, described by a normal distribution and independent over time [15]. The model has now been sufficiently simplified to provide a closed form solution for the purchasing rule.

PROPOSITION 1. [16]

(i) The value of the license is:

$$V(B_T) = E_T \left\{ \int_T^\infty B_t e^{-(r+\lambda+h)(t-T)} dt \mid B_0 = B \right\} = \frac{B_T}{\delta} \tag{11}$$

with: $\delta = r + \lambda + h - \mu_B > 0$

(ii) The option value is an increasing convex function of the instantaneous benefit B:

$$F(B) = AB^\xi \text{ for } x \in (0, B^*] \tag{12}$$

where: $A = \dfrac{(\xi-1)^{\xi-1} K^{1-\xi}}{(\delta \xi)^\xi} > 0$

and $\xi > 1$ is the positive root of the following non-linear expression:

$$\Phi(\xi) \equiv \frac{1}{2} \sigma^2{}_B \xi(\xi-1) + \mu_B \xi - (r+\lambda) + \lambda(1+\phi\alpha)^\xi = 0$$

(iii) The purchasing rule is given by:

$$B^* = \frac{\xi}{\xi-1} \delta K > 0 \tag{13}$$

The optimal threshold value B^* identifies the net benefit which makes it expedient for the user to purchase the license. The consumer will purchase the license the first time B_t exceeds the threshold defined by B^* [17]. It should also be noted that B^* is greater than the "common" opportunity cost of the license, expressed by δK. In fact, in the absence of

---

[15] In this specific case, the exogenous variables x and n are not stationary and do not possess long period distributions (non-conditioned). According to the information at time $t = 0$, the future values of the exogenous variables x and n possess a joint log-normal distribution, with variance proportional to the time horizon of the forecast. In addition, since it is a log-linear function (with constant elasticity) of a Brownian motion, the benefit B_t follows a Brownian motion as described by (10). The trend and the standard variation of the function B_t are a result of the linear combination of the corresponding parameters contained in the primitive variables x_t and n_t, with weights given by the elasticity α and β and by the correlation coefficient ρ.

[16] Proof: see Appendix.

[17] As $B(x,n) = x^\alpha n^\beta$, the purchase rule (13) can be easily represented by a strictly increasing function:
$x^* = [\xi(\xi-1)^{-1} \delta K]^\alpha n^{\beta/\alpha}$ (14)
with $dx^*/dn \geq 0$, and $d^2x^*/dn^2 \geq 0$ if $\alpha \geq \beta$ and $d^2x^*/dn^2 \leq 0$ if $\alpha \leq \beta$.

uncertainty and/or if the license cost was recoupable, it would be expedient to purchase it when the current value of the benefits exceeds the cost of the license, i.e. when B assumes a value such that:

$$B/\delta = K \tag{15}$$

In the presence of uncertainty, however, the critical level B^* is increased by the option factor $\xi(\xi-1)^{-1}>1$, which measures the additional benefit required by the consumer before making an irreversible outlay that provides uncertain benefits.

Using (11), (12) and (15) we can compare the opportunity cost of immediate purchase of the license with the corresponding benefit that can be obtained by purchasing at the optimal time. This can be assessed via the difference $F(B) - F_0(B)$, where we indicate the value of the license $F_0(B)= V(B) - K$ when purchased immediately ($t=0$). Since $F(B) = AB^{\xi}$, if the current value of B is below B^*, thus inducing the consumer to postpone the purchase, we get:

$$F(B) - F_0(B) = K + AB^{\xi} - (B\ \delta^{-1}) \tag{16}$$

The first term of the right-hand side of the equation (16) is the actual cost of the license. The second term is the value of the purchase option. Since purchasing implies eliminating this option, in the formula, it represents a cost. The third term is the present value of the future flow of benefits and therefore measures the value of immediate purchase of the license. In short, as long as $B < B^*$ and $F(B) - F_0(B) > 0$, the total cost of the license (given by the direct cost and the option value) will be higher than the present value of the benefits, and therefore the decision should be postponed. [18]

Via equations (10) and (13), it is possible to perform some comparative statics. For example, an increase in the discount rate, r, increases the value of the purchase option and therefore also increases B^*. In fact, as it has been assumed that the cost K is sustained when the license is purchased, an increase in r implies a greater reduction in the present value of the cost. Therefore, the option value increases but will tend to be exercised later (Dixit and Pindyck, 1994, p. 152-161).

Finally, since $\delta\xi(\delta\xi_B)^{-1}<0$, an increase in uncertainty of the future trend of B determines an increase of $\xi(\xi-1)^{-1}$ and therefore a greater gap between B^* and δK. Although the purchase option value increases when uncertainty as to future benefits increases, the expediency of postponing the purchase decision also increases (i.e. for purchase to be expedient, the expected benefit flow must increase).

3.   THE OPTIMAL DURATION OF THE LICENSE

The presented model is also useful for solving the problem of identifying the optimal duration of the license. If we assume that the consumer can purchase only one type of license allowing him to use the resource forever (or for such a long period that it can be

---

[18] Similar considerations apply also in the context of tax incentives for the adoption of eco-compatible environmental processes, see for example Dosi and Moretto (1996a,b) and Dosi and Moretto (1997, 1998).

considered indefinite), the expected value is equal to the discounted integral of the net benefit, as expressed in (3). In reality, the consumer can choose from a number of alternatives and can purchase licenses entitling him to benefit from the resource for one day, one week, one year etc. Obviously, the shorter the license period, the lower the total cost and the higher the mean cost per trip. In this case, the consumer's problem is not only when to exercise the purchase option, but also what duration to choose. Since the purchase of a license is always an irreversible decision, the possibility of choosing from licenses with different durations generates a further opportunity for postponing the purchase. In other words, there is an option value for every type (length) of license that can be purchased. [19]

In this paragraph it is assumed that the consumer can choose from a menu of N licenses of different duration identified by i, with cost K_i and expected present value, ranked in increasing license duration, given by:

$$V_i(B_T) = E_T \left\{ \int_T^{T+\tau_i} B_t e^{-(r+\lambda+h)(t-T)} dt \mid B_0 = B \right\} = \frac{B_T}{\sigma(\tau_i)} \qquad (17)$$

where T is the time at which the license is purchased, τ_i its duration, and $[\sigma(\tau_i)]^{-1} = (1-e^{-\delta\tau_i})\delta^{-1}$ the "real" discount factor that depends on the duration of the license. The shortest duration is the case in which the consumer purchases at time T a license for one single trip out between T and $T + dt$, so that $V_1(B_T) = B_T$ [20]. The longest duration is given by $\tau_N = \infty$ so that $V_N(B_T)$ is equal to (11).

A consumer who assesses the purchase with respect to the benefit, B_T, will choose the license that appears preferable at time T. The net purchase value is $\max_i [V_i(B_T) - K_i]$, for $i = 1, 2, ..., N$. In practice, we have to calculate when (T) and for what type of license (i) the option value is maximized.

The solution to the problem is shown in Figure 1 where three types of licenses are illustrated: V_1, V_2 and V_N. The first one permits only one trip, the second is of intermediate duration, $1 < \tau_2 < \infty$, and the third is "for life", $\tau_N = \infty$.

Since the inverse of the discount factor is a decreasing convex function with respect to the duration, i.e. $\delta'(\tau_i) < 0$ and $\delta''(\tau_i) > 0$, with $\delta(0) = +\infty$ and $\delta(\infty) = \delta$, the net value of each license is a straight line, growing in B, with a gradient that increases as the duration increases, while the greater value of the three determines the upper envelope curve.

Unlike (4), by the above arguments, the investment option value at the present time can be obtained from the following equation:

$$F(B) = \max E_0 \left\{ \left[\max_i V_i(B_T) - K_i \right] e^{-rT} \mid B_0 = B \right\} \qquad (18)$$

---

[19] Dixit (1993b) highlights that the choice between investment projects of different scales creates an option extravalue to wait. Moretto (1999) extends this result to the case of abandonment of a multiplant firm.

[20] By expanding in series the right-hand side of (17), we get $e^{-\delta\tau} = \sum_{n=0}^{\infty} n!^{-1} (-\delta)^n \tau^n$. By truncating the series expansion at the first term and placing $\tau_1 = 1$ we get $B\delta^1[1-(-\delta)^0 \tau_1^0-(-\delta)^1 \tau_1^1]=B$.

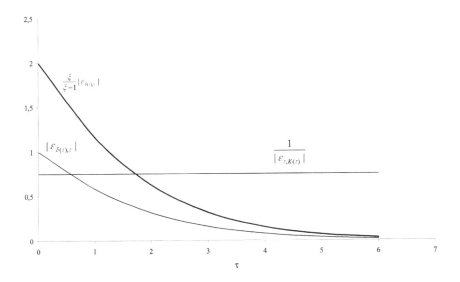

Figure 1. The case of the three licenses

As the differential equations (6) and (12) still apply, the solution to (17) is fairly easy. The set of option values, with respect to B and τ, is represented by a sheaf of exponential curves. The optimal solution can be identified by calculating, for each hypothetical τ, the tangent between the exponential curve (12) with the respective straight line indicating the net value $Vi(B) - K_i$, and choosing, from all the solutions, the one that provides the highest $F(B)$, or equally the one with the highest value of the constant A_i.

PROPOSITION 2. [21]

(i) A simple application of the procedure performed for a single license shows that:

$$A_i = \frac{(\xi - 1)^{\xi-1}(1 - e^{-\delta \tilde{\tau}_i})^{\xi} K_i^{1-\xi}}{(\delta \xi)^{\xi}} > 0 \qquad (19)$$

(ii) and the investment rule is given by:

$$B_i^* = \frac{\xi}{\xi - 1} \delta(\tau_i) K_i > 0 \qquad (20)$$

---

[21] Proof: straightforward from proposition 1.

Choosing the highest value of the constant A_i is equivalent to choosing the license with the highest term $(1-e^{-\delta\tau})^{\xi}K_i^{1-\xi}$. For example, from Figure 1, this condition occurs for a license with intermediate duration $1 < \tau_2 < \infty$. If the present value of the benefit, B, is below $B*_2$, the consumer will wait until it exceeds $B*_2$ and will then purchase the license with duration τ_2 at cost K_2. It can also be easily seen that it is not optimal to purchase the license with the longest duration (infinite) $\tau_N = \infty$ [22]. In other words, in the numerical given example, the intermediate license represents the best compromise between duration and cost (K, τ). It is not expedient for the consumer to wait and purchase the longest license; he will prefer to take advantage of the possibility offered by a shorter license which costs less. Neither is the daily license $\tau_1 = 1$ an optimal choice, as can be seen again from Figure 1. In this case, the consumer will by-pass this opportunity and waits for the expected benefit to grow sufficiently to make the intermediate license worthwhile.

3.1. The trade-off between cost and duration of the license

The trade-off between cost and duration may be analyzed by generalizing the maximization of the function, $(1-e^{-\delta\tau})^{\xi}K_i^{1-\xi}$, and then exploring the possibility of purchasing licenses with a continuum of durations. In order to do this, a fourth assumption is necessary.

4. the inverse supply function of the agency with respect to the duration is given by $K = K(\tau)$, with $K' > 0$, $K'' \leq 0$, and $K(0)=K_{min}\geq 0$, and $K(\infty) = K_{max} \leq \infty$. [23]

Choosing the license implying the highest value of the constant A is equivalent to solving the following problem:

$$\max_{\tau}(1 - e^{-\delta\tau})^{\xi} K^{1-\xi} \quad , \qquad \text{subject to } K = K(\tau) \qquad (21)$$

For an internal solution, $0 < \tau < \infty$, the first order condition is the following:

$$\frac{K'(\tau)\tau}{K(\tau)} = \frac{\xi}{\xi-1} \cdot \frac{\delta e^{-\delta\tau}\tau}{1-e^{-\delta\tau}} \qquad (22)$$

and given that $|\varepsilon_{\tau K(\tau)}| \equiv K(\tau)[K'(\tau)\tau]^{-1}$ and $|\varepsilon_{\delta(\tau),\tau}| \equiv \delta e^{-\delta\tau}\tau(1-e^{-\delta\tau})^{-1}$, equation (22) can be expressed as: $|\varepsilon_{\tau K(\tau)}||\varepsilon_{\delta(\tau),\tau}| = (\xi-1)\xi^{-1} < 1$. This infers the necessary condition that the product of the elasticity of the supply function and the elasticity of the discount factor must be below one. This means that as the discount rate grows, the duration for which the previous condition is valid increases. This is reasonable because as the rate increases, shorter, and therefore intermediate, license terms ($< \infty$) are favored. In addition, the fact that the product of the two elasticities is less than one also depends on the form assumed for the cost function. With a linear cost function $K(\tau)$, condition (22) reduces to $|\varepsilon_{\delta(\tau),\tau}| =$

---

[22] This license will be chosen when a higher minimum present benefit threshold is exceeded, identified by the tangent between $F_N(B)$ (upper curve) and $V_N(B)-K_N$.

[23] Alternatively we can consider $K(\tau)$ as a function that expresses the cost sustained by the consumer to obtain a license of duration τ.

$(\xi\text{-}1)\xi^{-1}$, and the most expedient license will always have an intermediate duration. If, on the other hand, $K(\tau)$ is constant, the most expedient license will always have unlimited duration. Finally, if $K''(\tau) < 0$, then the preferred duration will depend on how $K(\tau)$ grows with respect to τ, the quicker it reaches $\overline{K}$, the greater the weight of the constant cost conditions with respect to the duration. In other words, the intermediate solutions are expedient for the consumer when the discount rate is high and the elasticity of the cost function is high (i.e. the elasticity of the supply function is low).

After clarifying the effect of the discount rate and the cost function on the optimal solution, we shall now consider the effect of irreversibility and uncertainty. The first point to note is that if the purchase of the license is a reversible decision, or, if there was no uncertainty about future benefits of the resource, condition $|\varepsilon_{\tau K(\tau)}||\varepsilon_{\delta\tau}| = 1$ guarantees an intermediate solution.

Let's consider the option value implicit in (22) and assume a supply function with constant elasticity:

4a. the supply function of the agency in terms of duration, with respect to the cost of the license, is given by $\tau = K^{1/a}$ with $0 < a \leq 1$.

From the properties of $\delta(\tau)$, we get $|\varepsilon_{\delta\tau}| \to 1$ for $\tau \to 0$ and $|\varepsilon_{\delta\tau}| \to 0^+$ for $\tau \to +\infty$, with:

$$\frac{d\,|\varepsilon_{\delta(\tau),\tau}|}{d\tau} = \begin{vmatrix} -\tfrac{1}{2}\delta & con \ \tau \to 0 \\ 0^- & con \ \tau \to +\infty \end{vmatrix}$$

which guarantee that if an intermediate solution exists, it is unique. Having said this, the implications of the uncertainty on the choice of purchasing the license can be assessed with reference to the Figure 2.

The thin lower curve represents the elasticity of the discount factor with respect to the duration τ, i.e. $|\varepsilon_{\delta\tau}|$. The horizontal straight line is the inverse of the supply function elasticity, i.e. $|\varepsilon_{\tau K(\tau)}|^{-1} = a$. Finally, the upper curve represents $[\xi(\xi\text{-}1)^{-1}]\,|\varepsilon_{\delta\tau}|$. From condition (22) it emerges that the optimal duration under uncertainty is obtained when the straight line intersects the upper curve. If there were no uncertainty, the optimal condition would be represented by the intersection of the straight line with the lower curve.

The distinguishing element is therefore given by the coefficient $\xi(\xi\text{-}1)^{-1} > 1$. The consequence of introducing the factor $\xi(\xi\text{-}1)^{-1}$ is that the optimal duration of the license increases. By increasing the duration of the license the consumer reduces the risk of not achieving the expected benefits. This first consideration has a very important implication because, from (20), as τ increases, the threshold benefit that triggers the decision to purchase the license also increases. This can be stated in the following proposition:

PROPOSITION 3. The consumer is willing to purchase a license with longer duration only if there is an adequate expected benefit.

Furthermore, if uncertainty grows, the consumer will tend to postpone the decision to purchase, while awaiting an increase in the benefit due to a reduction in the number of rivals and/or an increase in the stock of resources, and then to purchase a longer-lasting license.

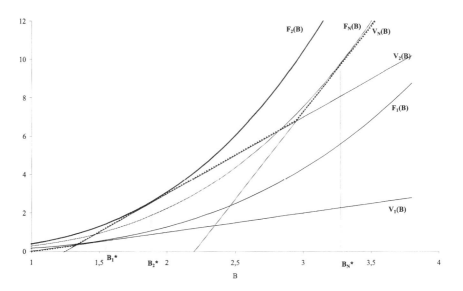

Figure 2. Optimal duration of the license with and without uncertainty

Now, expanding the discount factor, $\delta(\tau)$, in a Taylor series and ignoring the terms of second orders, condition (22) can be solved for the optimal duration τ as follows:

$$\tau = \frac{2}{\delta}\left(\frac{1-a}{a}\right) + \frac{2}{\delta a}\left(\frac{\xi}{\xi-1} - 1\right) > 0 \tag{23}$$

By replacing the linear approximation of $\delta(\tau)$ in (19) we get:

$$B^* = \frac{\xi}{\xi-1}\left(\frac{\delta}{2} + \frac{1}{\tau}\right)\tau^a \tag{24}$$

The first term on the right-hand side of (22) represents the optimal duration under certainty, whereas the second term incorporates the option value of postponing the purchase [24]. From (23), it can easily be seen that the optimal duration decreases as a increases, i.e. $\partial\tau/\partial a)^{-1} < 0$. It is also confirmed that an increase in the uncertainty of future benefits lengthens the optimal duration, i.e. $\partial\tau/\partial\sigma^2{}_B)^{-1} < 0$, and that an increase in the discount rate leads to a reduction in the optimal duration, i.e. $\partial\tau/\partial\sigma)^{-1} < 0$.

---

[24] It should be noted that an optimal duration can exist also in the presence of a concave supply function ($a > 1$), which would never be possible if $\sigma^2{}_B = 0$.

The negative effect of the discount rate deserves, nevertheless, further attention. First of all, as it is given by $\delta = r + \lambda + h - \mu_B$, the discount rate responds negatively to an increase of the growth rate of the expected benefits μ_B, thus increasing the optimal duration. In fact, with positive forecasts on the growth of stocks and negative forecasts on the trend of the number of rivals, the consumer will tend to purchase longer-lasting licenses. Secondly, an increase in the probability, λ, of a sudden change in the stock has an ambiguous effect on the optimal duration. The discount factor δ increases, with the effects already illustrated; in addition λ affects the option multiplier $\xi(\xi-1)^{-1}$ according to the sign of ϕ which acts both on the expected rate of increase of B and on its variance. If ϕ is positive, then μ_B will increase to $\mu_B + \lambda\phi$ and the variance will increase from σ^2_B to $(\sigma^2_B + \lambda\phi^2)$, involving an increase of $\xi(\xi-1)^{-1}$. The combined effect therefore depends on the prevalence of the "rate" effect, which reduces the duration of the license, or of the "growing stock" effect which, conversely, increases it.

If ϕ is negative, although the variance will still increase, there will be a drop in μ_B. The overall effect is negative, with a reduction of $\xi(\xi-1)^{-1}$ which, combined with the increase in δ, will reduce the optimal duration of the license.

4. CONCLUSIONS

This chapter has illustrated consumer behavior with respect to the purchase of a license for the recreational use of a natural resource. We developed a model starting from the assumption that the opportunity of purchasing a license can be likened to a call option, given that the uncertainty concerning the future benefits and the irreversibility of the expenditure may make it expedient to wait before purchasing. Furthermore, the model includes the main factors conditioning the perceived utility of the recreational activity (i.e. the stock of resources, the number of rivals, and operations designed to enrich/deplete stock levels) with particular attention to the uncertainty that characterizes activities pertaining to biological resources.

The first result of this analysis concerns the effect of uncertainty. It acts as a deterrent to the purchase of the license and it lengthens the optimal duration. Generally speaking, therefore, the first tool available to an agency appointed to regulate the purchase of licenses for recreational activities is the quality of the information to be circulated. From this it follows that operations designed to "program" future conditions by reducing certainty can accelerate purchases of the licenses.

A subsequent analysis enabled us to highlight the effect produced by an uncertain sharp change in the resource stock on the optimal duration of the license. It became clear that, as the discount rate increases, shorter durations are not always favored, as might be expected considering the criterion of the net present value. This is due to the fact that the various components of δ do not act univocally. As they are conditioned by the probability (λ) of sudden shocks in the stock (ϕ) and by a variance in benefits, they can produce effects contrary to those expected.

Optimal license duration also depends on the elasticity of the supply function: as the elasticity decreases, optimal duration decreases. This highlights a second possibility of intervention for the agency: by modulating the supply function with respect to the duration, it can directly condition consumers' choices. Finally, irreversibility and uncertainty lengthen the optimal duration of the license, as longer durations reduce the effect of the variability of future benefits.

In conclusion, this study has highlighted that any policy for the management of recreational activities via license must take account of the fact that the uncertainty of the benefits and the irreversibility of the expenditure generate an opportunity cost linked to the option value of the license, causing the consumer to behave very differently from what we would expect if we considered only the present value of the benefits.

5.   APPENDIX

First, from Fubini's theorem and from (10), it is possible to bring the factor $E_T(.)$ within the integral, thus obtaining, by simple integration (Harrison, 1985, p. 44):

$$V(B_T) = E_T \left\{ \int_T^\infty B_t e^{-(r-\lambda+h)(t-T)} dt \mid B_0 = B \right\} = \frac{B_T}{\delta} \tag{25}$$

where $\delta = r + \lambda + h - \mu_B$. We assume that $r - \mu_B > 0$, otherwise the maximization would be indefinite as an increasingly bigger moment T is chosen. It should be noted that $r - \mu_B > 0$ also implies $\delta > 0$. Yet, assuming $r - \mu_B > 0$, the option value can be estimated by maximizing $max\ E_0\ [(V(B_T) - K)e^{-rT} \mid B_0 = B]$, which permits reduction of the differential equation (6) to:

$$\frac{1}{2}\sigma^2{}_B B^2 F''(B) + \mu_B F'(B) - (r+\lambda)F(B) + \lambda F((1+\phi\alpha)B) = 0 \tag{26}$$

In addition, $F(B)$ must satisfy the boundary conditions (7), (8) and (9), purposely simplified.

$$\lim_{B \to 0} F(B) = 0 \tag{27}$$

$$F(B^*) = V(B^*) - K \tag{28}$$

$$F'(B^*) = V'(B^*) \tag{29}$$

At this point it is easy to verify that the solution of (26) can be expressed by a convex increasing function of B:

$$F(B) = AB^\xi \quad \text{for } x \in (0, B^*) \tag{30}$$

The parameters A and ξ, together with the critical value B^* can be obtained from the previous boundary conditions. In particular, by replacing (30) in equation (26), we obtain ξ as one of the roots of the following non-linear expression:

$$\Phi(\xi) \equiv \frac{1}{2}\sigma^2{}_B\xi(\xi-1) + \mu_B\xi - (r+\lambda) + \lambda(1+\phi\alpha)^\xi = 0 \qquad (31)$$

From condition (27) it is known that ξ must be positive and from (29) that it must be greater than one. The solution ξ can be represented more intuitively by re-writing (31) as $\Phi(\xi) = \Phi_1(\xi) + \Phi_2(\xi)$, with $\Phi_1(\xi) \equiv \frac{1}{2}\sigma^2{}_B\xi(\xi-1) + \mu_B\xi - r$ and $\Phi_2(\xi) \equiv \lambda + \lambda(1+\phi\alpha)^\xi$ and drawing the two functions as shown in Figures A1 and A2, with $\phi > 0$ and $\phi < 0$ respectively.

It should be noted that, if $\Phi_1(0) = -r < 0$, $\Phi_1(1) = \mu_B - r < 0$, $\Phi_2(0) = 0$ and $\Phi_2(1) = -\lambda\phi\alpha$, the solution of (31) is identified by the intersection of these two curves for $\xi > 1$. To guarantee $\xi > 1$ if $\phi < 0$, it is assumed that $|\phi\alpha\lambda| < |\mu_B - r|$. In addition, since (30) represents the option value of purchasing the license at the optimal time, the constant A must be positive, and the solution is valid in the interval of B within which it is preferable to keep the option alive $(0, B^*)$. By replacing (30) in (28) and (29), we get:

$$B^* = \frac{\xi}{\xi-1}\delta K > 0 \qquad (32)$$

$$A^* = \frac{(\xi-1)^{\xi-1}K^{1-\xi}}{(\delta\xi)^\xi} > 0 \qquad (33)$$

as shown in the text, and this completes the proof.

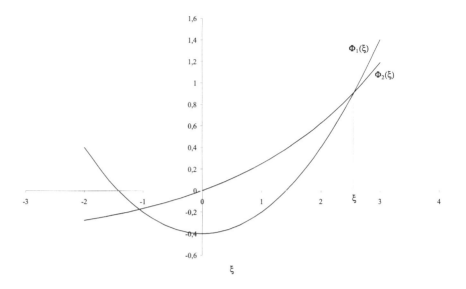

Figure A1.   Solution for ξ with $\phi > 0$

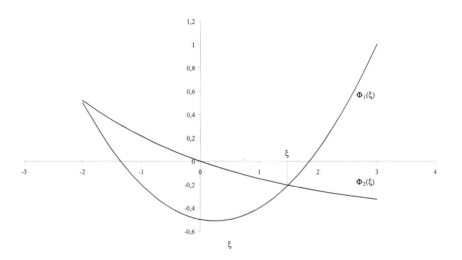

Figure A2.   Solution for ξ with $\phi < 0$ and $|\phi\alpha\lambda| < |\mu_B - r|$

6.   REFERENCES

Anderson L.G. (1985), Potential Economic Benefits from Gear Restrictions and Licence Limitation in Fisheries Regulation, *Land Economics*, **61**(4): 409-418.

Arnason R. (1990), Minimum Information Management in Fisheries, *Canadian Journal of Economics*, **23**: 630-653.

Campbell H.F. – Linder R.K. (1990), The Production of Fishing Effort and the Economic Performance of Licence Limitation Programs, *Land Economics*, **66**(1): 56-66.

Clark C.W. (1990), *Mathematical Bioeconomics* (2nd ed.), Wiley Interscience, Chichester.

Cox D.R. – Miller H.D. (1965), *The Theory of Stochastic Process*, Chapman and Hall, London.

Dixit A. (1992), Investment and Hysteresis, *Journal of Economic Perspectives*, **6**: 107-132.

Dixit A. (1993a), *The Art of Smooth Pasting*, Harwood Academic Publishers, Chur (CH).

Dixit A. (1993b), Choosing Among Alternative Discrete Investment Projects under Uncertainty, *Economic Letters*, **41**: 265-268.

Dixit A. – Pindyck R.S. (1994), *Investment under Uncertainty*, Princeton University Press, Princeton (NJ).

Dosi C. – Moretto M. (1996a), Toward Green Technologies: Switching Rules and Environmental Policies, *International Review of Economics and Business*, **43**: 13-30.

Dosi C. – Moretto M. (1996b), *Environmental Innovation and Public Subsidies under Asymmetry of Information and Network Externalities*, Nota di Lavoro 84.96, Fondazione ENI - Enrico Mattei, Milan.

Dosi C. – Moretto M. (1997), Pollution Accumulation and Firm Incentives to Promote Irreversible Technological Change under Uncertain Private Benefits, *Environmental and Resource Economics*, **10**: 285-300.

Dosi C. – Moretto M. (1998), Auctioning Green Investment Grants as a Means of Accelerating Environmental Innovation, *Revue d'Economie Industrielle*, **83**: 99-110.

Harrison J.M. (1985), *Brownian Motion and Stochastic Flow Systems*, John Wiley & Sons, New York.

He H. – Pindyck R.S. (1992), Investment in Flexible Production Capacity, *Journal of Economic Dynamic and Control*, **16**: 575-599.

Karpoff J.M. (1989), Characteristics of Limited Entry Fisheries and the Option Component of Entry Licences, *Land Economics*, **65**(4): 386-393.

Moretto M. (1999), *Optimal Capacity Adjustment by a Multiplant Firm*, Nota di Lavoro 28.99, Fondazione ENI - Enrico Mattei, Milan.

Pearce D.W. – Turner R.K. (1990), *Economics of Natural Resources and the Environment*, Harvester Wheatsheaf, New York

Pindyck R.S. (1991), Irreversibility, Uncertainty and Investment, *Journal of Economic Literature*, **29**: 1110-1152.

Scheafer M.B. (1954), Some Aspects of the Dynamics of Populations Important to the Management of Commercial Marine Fisheries, *Bulletin of the Inter-American Tropical Tuna Commission*, **1**: 25-56.

ANALYSIS OF LAND PRICES UNDER UNCERTAINTY: A REAL OPTION VALUATION APPROACH

Glenn D. Pederson and Tamar Khitarishvili [*]

SUMMARY

Recent empirical studies have found that traditional present value models do not adequately represent the process that underlies changes in farmland market prices. We briefly review the existing economic literature on land pricing and then develop a more specific analysis of land prices under uncertainty using a real option approach to valuation. Using the model, we illustrate the effects on farmland prices of factors that represent the major sources of investor uncertainty.

1. INTRODUCTION

Numerous studies have been published on land values in North America and Europe during the past 20 years. The authors of many of these studies have attempted to identify the causes and consequences of changing farmland prices. In the course of these investigations, various controversies and puzzles have emerged. In particular, the asset pricing puzzles challenge our traditional views of the processes by which farmland prices are determined. In this paper we focus on an asset-pricing puzzle that has its origin in the traditional present value approach to land pricing.

Present value models (PVM), either directly or indirectly, use the income capitalization formula where the price of land is equal to the discounted expected value of the stream of future annual returns from land. The constant discount rate PVM under rational expectations has been the most widely used approach in studies of farmland pricing due to its intuitive appeal and simplicity [1]. However, recent evaluations of the model using U.S. data have consistently rejected it as an adequate model of land price behavior. This is largely due to the inability of the model to explain the observed divergence of the returns

---

[*] Department of Applied Economics, University of Minnesota.
[1] In its simplest version the PVM assumes that land buyers are risk neutral, they consider only the economic return to land, and they discount future cash flows using a constant discount rate.

Economic Studies on Food, Agriculture, and the Environment
Edited by Canavari *et al.*, Kluwer Academic/Plenum Publishers, 2002

to land from land values since the 1970s. This is a puzzle [2]. For the PVM to hold, the relationship between land prices and the return to land should satisfy two criteria (Campbell and Shiller). First, land values and land rents should have the same time series representation (the stationarity test). Second, land prices and land rents should rise and fall at about the same rate (the co-integration test). Falk finds that the first criterion is met, but the second is not. Using a different data set, Clark et al. find that the first property is not met, and they conclude that the simple PVM is not appropriate.

The exact reason for failure of the PVM is not quite clear from the existing farmland pricing literature. It could be due to either the inadequacy of the PVM as a theoretical approach. Alternatively, it could be due to the fact that transaction costs in farmland markets are significant and violate the assumption of frictionless asset markets (Lence and Miller). If the PVM is fundamentally inadequate to represent farmland pricing, the implication is that an alternative model needs to be developed. For this purpose dynamic models have been suggested as a means for incorporating inflation, time preference, risk aversion and transaction costs in the pricing of farmland (Chavas and Thomas). Alternatively, a real option valuation approach, that considers rational investor behavior under uncertainty, would identify the conditions under which farmland prices could deviate from the standard PVM result and, thus, provide a better explanation for observed land price behavior (Turvey).

Our objective in this paper is to evaluate the role of uncertainty in the pricing of farmland from an investment perspective. We begin with a selective review of the recent research literature on farmland market prices. Second, we present a real option valuation model of land pricing under uncertainty. Using the real option model we analyze the effects of cash flow uncertainty on farmland prices. Finally, we draw some conclusions about the usefulness of these models for analyzing land markets and asset-pricing behavior under uncertainty.

1.1.   Land pricing literature

We find that the expanding body of applied research literature on agricultural land pricing suggests four general categories of models. Those include: structural models, present value models, capital asset (arbitrage) pricing models, and real option value models [3]. Each of these models is briefly considered in terms of the approach, the major findings, and the ability to explain the observed behavior of land prices.

---

[2]   This "present value puzzle" is not unique to land markets. Financial assets exhibit similar puzzles in standard asset pricing models. Some success in explaining these pricing anomalies in financial markets has been achieved by relaxing the assumption of perfect (frictionless) asset markets (see Aiyagari and Luttmer). Those models assume that transaction costs are strictly proportional to the value of the transaction and look like cost-adjusted versions of frictionless markets. The presence of transaction costs implies a degree of asset illiquidity.

[3]   Weersink et al. suggest that land value studies can be grouped into those that rely on the present value model (and use traditional time series analysis) and those that use co-integration analysis to determine the validity of the capitalization model and the assumption of a constant discount rate.

1.1.1. Structural models

Generally, structural models are attempts to explain land price behavior by specifying the determinants of the demand for, and supply of, land. These models have increasingly focused on the demand side of the land market (in recognition of the fact that land supply is relatively fixed). According to the early (ad hoc) structural models, land prices are driven by productivity increases, government programs, and urban pressures. Pope et al. find that the early models performed reasonably well in the period for which they were estimated, but they show that longer time series reveal that these models are ineffective for explaining the divergence between land rents and land values.

Contrary to the early ad hoc econometric models, Just and Miranowski formulate a theoretical model where land prices are motivated by changes in both the discounted value of returns and the desire for wealth accumulation. The model tests for the effects of: inflation; expectations about land prices, returns, government payments and risk; the opportunity cost of funds; and selected transaction costs. Using a naïve expectations formulation, they find that the dominant factors that explain large land price swings include both inflation rates and changes in the opportunity cost of funds. While government payments to farmers are a significant factor explaining the level of land prices, they account for only a small part of the volatility in land prices. Their analysis suggests that land prices tend to overreact to changes in the returns to land when there are inflation and opportunity cost of funds shocks. Thus, land prices are more volatile than the returns to land.

1.1.2. Present value models

Early present value studies assumed that net farm income (per acre) is the most appropriate measure of residual returns to land for determining its present value. Assuming appropriate discount rates, a buyer and a seller could theoretically establish their bid and ask prices, respectively, and then converge to a unique equilibrium price for land. Subsequently, it was pointed out that the returns to land include both current and capital gain return streams. Thus, the expected rate of growth in land prices (and, thus, the effects of both inflation and taxes) must be incorporated into the capitalization formula. Alston finds that inflation has a small effect on land prices and that most of the real growth in land prices occurs due to the real growth in rental income. Burt employs a dynamic framework that uses the capitalization formula with a constant discount rate and finds that rents are the main source of land price variation. [4]

Despite the apparent consistency of these findings, Weersink et al. find that there is no general consensus in the PVM literature due to the fact that researchers have not yet resolved the fundamental question of how cash flows relate to land bid prices. They conclude that the potential reasons for rejecting the present value model of land prices include: a) the presence of asset price bubbles, b) time-varying discount rates, and c) nonmonetary returns to farmland ownership. [5]

---

[4] The framework is dynamic in the sense that Burt uses a dynamic regression model with land prices determined by a distributed lag structure for rents.

[5] We add to that list the typical lack of accounting for transaction costs in PVM models.

For example, Featherstone and Baker test for the presence of an asset price "bubble" that signals the presence of speculative forces in the farmland market. By specifying both lagged land values and the expectations processes for rents and interest rates, they find that speculative forces may have played a major role in determining farmland prices in the U.S. during the 1980s. The presence of large and persistent asset price bubbles is inconsistent with the standard PVM formulation. Falk studies time series data using cointegration methods in order to determine if land prices and rents follow the same process. Although land prices and rents are highly correlated, Falk reports that land prices are more volatile than rents, and the series are not cointegrated. Falk suggests that nontraditional models, by which prices overreact to movements in current and recent returns due to speculative forces, may be useful approaches to the present value puzzle. In a subsequent analysis, Falk and Lee find that fads (speculative forces) and overreactions play a role in explaining short-run price behavior, but long-run price movements depend on fundamental forces (permanent fundamental shocks)[6]. Thus, by decomposition they find that land prices and rents are cointegrated series in the long run, but not in the short run.

Secondly, most studies of farmland prices assume a constant risk premium. Hanson and Myers test for time-varying risk premiums in the returns to U.S. farmland. They find evidence of significant time variation in the conditional variance of excess returns, suggesting that the risk from investing in farmland has varied over time. However, these changes in the variance of excess returns tend not to be strong predictors of mean farmland returns. They find also that the excess returns to farmland are strongly autocorrelated, a result that is inconsistent with present value models of farmland pricing.[7]

Lence and Miller analyze the rates of return from holding farmland in excess of transaction costs. Due to transaction costs, there is a range of inaction in which agents do not react to new information. This inaction occurs since the expected benefits from trading an asset (e.g., land) are not sufficient to offset the transaction costs. Thus, the excess return is bounded in equilibrium. These bounds vary between finite holding periods (a one-year interval where an agent buys and sells land and incurs both transaction costs) and infinite holding periods (where just the buying transaction cost is incurred). When the expected excess rate of return deviates from zero it suggests that the constant discount rate, frictionless PVM does not hold. However, in the presence of transaction costs it is possible for the expected excess rate of return to be nonzero and the constant discount rate PVM is still valid. Lence and Miller observe that the excess return series on farmland has a tendency to be positive during periods of rising land prices and negative during periods of declining land prices. They strongly reject the frictionless, constant discount rate PVM for both the finite and infinite holding period formulations. However, the evidence against the constant discount rate PVM in the presence of transaction costs is mixed. The results for the finite (one-year) holding period indicate that the constant discount rate PVM cannot be rejected, while the infinite holding period PVM model is clearly rejected.[8]

---

[6] Falk and Lee identify breakthroughs in crop genetics, global warming, trade agreements as possible permanent fundamental shocks (p. 697).

[7] This is inconsistent with the assumption that excess returns to farmland do not vary systematically.

[8] The implication is that it is not possible for speculators with short holding period horizons to obtain systematic excess returns, while speculators with infinite holding period horizons could obtain systematic excess rates of return by trading land.

1.1.3. *Asset pricing models*

In the context of the land pricing literature, capital asset and arbitrage-pricing models (CAPM and APT) can be viewed as extensions of the basic PVM. These models attempt to measure the premium associated with systematic risk when an asset is either held in a well-diversified investment portfolio or the asset is appropriately priced by a set of economic factors (Copeland and Weston). In asset market equilibrium there is no arbitrage opportunity and the risk premium is the additional rate of return that a risk-averse investor would require to invest in farmland. Theoretically, one can infer the price of farmland from the estimated risk premium if the stream of expected returns is known.

Bjornson and Innes apply these pricing models to U.S. agricultural data and conclude that farmland has relatively low systematic risk in a well-diversified portfolio. By comparing the returns on agricultural assets with those on assets with similar risk, they identify an anomaly where the mean rates of return on comparable risk farm assets and nonagricultural assets differ. This result refutes the no-arbitrage (equilibrium) condition. Hanson and Myers find a time-varying risk premium, which is also evidence against the capital asset pricing model.

Chavas and Thomas depart from traditional (static) asset-pricing models by developing a dynamic model of farmland prices. The model nests as special cases the static CAPM and the dynamic consumption-based CAPM models. They suggest that the reason for the failure of frictionless arbitrage pricing models is that the capital market is segmented due to costly arbitrage between farm and nonfarm equity markets. They specify an arbitrage pricing model that incorporates risk aversion, transaction costs, and dynamic investor preferences. The marginal transaction costs are allowed to vary for buyers and sellers, but these costs act as a disincentive for both types of agents to participate in land markets. Transaction costs are assumed to be proportionate to the aggregate volume (number of acres) of land transactions, not the value of those transactions [9]. Chavas and Thomas find evidence for the presence of transaction costs.

1.1.4. *Real option pricing models*

The real option value approach to farmland pricing is an application of the real option theory (Dixit and Pindyck; Dixit; Trigeorgis). The fundamental premise of real option theory is that flexibility has value. Real options "involve discretionary decisions or rights, with no obligation, to acquire or exchange an asset for a specified alternative price" (Trigeorgis, p. xi). As Trigeorgis points out, uncertainty in investment decisions is typically resolved gradually and the forecasting of cash flows is imperfect. Real option theory assumes that a capital investment decision resembles a call option on the future risky cash flows of a project [10]. In the context of this paper, the project involves the decision to sell farmland. The real option approach focuses more on the option value to the seller, instead of the rational bid price of the buyer. Thus, the maintained hypothesis is

---

[9] The assumption that costs are proportionate to the number of acres sold is contrary to the usual assumption that transaction costs such as commissions, fees and taxes are proportionate to the value of farm real estate sold.

[10] Trigeorgis identifies several categories of real options: option to defer, time-to-build option, option to alter, option to abandon, option to switch, option to grow (pp. 2-3).

that under uncertainty the economic value of farmland deviates from the bid price (as determined by the conventional PVM) due to the existence of real option value.

Contrary to standard real option models (where the option is held by the purchaser), Turvey suggests that the real option value of farmland rests with the seller. When the value of farmland is equal to its (zero net present value) bid price there is no advantage for the buyer of farmland to postpone the purchase. However, due to uncertainty about future cash flows, a seller may choose to postpone the sale in anticipation of an increase in the cash flows and a higher sale price, hence greater capital gains. If the decision is to sell, the seller may receive the current bid price plus a portion of the option value. It is this characteristic of the real option value model that makes it an attractive alternative to the standard PVM.

2.   REAL OPTION VALUE OF LAND

In a real option value framework we define the value of land to the seller as the sum of its fundamental bid price (the intrinsic value) plus the value of a call option on future land price increases,

$$S(\pi, \alpha, r, t) \equiv V(\pi, \alpha, r, t) + F(V(\pi, \alpha, r, t)) \qquad (1)$$

where π is the annual cash flow to the owner of the land, α is the growth rate of the annual cash flow, r is the annual discount rate, and t is the index for time. In (1) we interpret S( ) as the seller's ask price, V( ) as the buyer's bid price (as determined by the conventional PVM), and F(V( )) as the value of the real call option to the seller. In order to determine the economic value to the seller, we assume that the bid price and the value of the real option are additive as in (1). Thus, the asking price could deviate from the bid price if the value of the real option to the seller is not zero.

Conceptually, the real option value (to the seller) can be illustrated as in Figure 1 [11]. First, we identify the value of returns at which the NPV of the land is zero (point a). At this point there is value to the seller to postpone the sale, since the real option value is positive. The rationale for a nonzero real option value is that there is value in waiting for the underlying cash flows of the project to increase. Assuming that cash flow is uncertain, it can be shown that there is a positive probability that the cash flows and the net present value of the investment will increase. Of course, there is also a possibility that the cash flows will fall below point (a), where the net present value is negative. However, in this latter case the real option still has a positive value, since there is a value to postponing the sale of the land. At point (c) the cash flow of the asset has reached the "trigger" level where the real option value curve is tangent to the NPV of the asset. At this point there is no longer an incentive for the seller to postpone the sale. Thus, this option value diagram illustrates how the timing of the land transaction is a value-maximizing problem where the right to make an irreversible investment is delayed from the present to the future and there is no obligation associated with this right.

---

[11] In the real option case maturity occurs at the end of the seller's planning horizon, but the real option approach makes no specific assumption about the date of maturity. In contrast the date of maturity on a financial option is specified in the option contract.

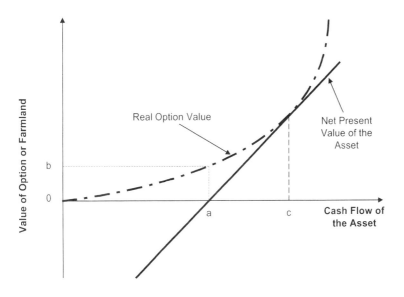

Figure 1. Real Option Value of Land

Uncertainty may cause behavior in the land market that is inconsistent with traditional net present value rules of investment. That is, land prices could adjust while land rents are not adjusting, or vice versa. In the real option framework, uncertainty causes a wedge to form between the ask price of the seller and bid price of the buyer[12]. In order for a transaction to take place today the buyer must compensate the seller for the value of the real option, before it is optimal to do so. In such a game both the seller and the buyer are better off, since the expected net present value will be positive to the buyer, while some portion of expected future capital gain is paid to the seller.

3. A MODEL

As suggested by Dixit and Pindyck, we use a dynamic programming approach to price the real option value of farmland. First, we determine a formula that describes the value of land as a function of the stochastic variables. In this case, the value of land is denoted by $V(\pi)$ where π is random. Second, we determine the real option value, $F(V(\pi))$.

The annual cash flow or return, π, is assumed to evolve over time according to geometric Brownian motion with drift,

---

[12] We note that transaction costs could also produce a wedge between the seller's ask price and the buyer's bid price. When land prices are rising the marginal transaction costs may be significant and lead to a postponement of a land transaction until prices are expected to rise sufficiently to cover the transaction costs (Chavas and Thomas).

$$d\pi = \alpha\pi dt + \sigma\pi dz \qquad (2)$$

where α is the drift parameter and σ is the standard deviation of the return. The term dz is the increment in a Wiener process. [13]

Following the present value model framework, the value of farmland equals the sum of the discounted returns to land. If the returns grow at a constant rate, α, and the interest rate is fixed at r, the result is,

$$V(\pi) = \frac{\pi}{r - \alpha} . \qquad (3)$$

Then, by substitution into (2), the price of farmland evolves according to the following process,

$$dV(\pi) = \alpha V(\pi)dt + \sigma V(\pi)dz \qquad (4)$$

$$dV(\pi) = \frac{\alpha\pi}{r - \alpha}dt + \frac{\sigma\pi}{r - \alpha}dz \qquad (5)$$

In (4) the current cash flow, π, is a known value. However, the future cash flows are uncertain [14]. Thus, we define the expected value (E) and variance (Var) as,

$$E(d(V(\pi))) = \alpha V(\pi)dt = \frac{\alpha\pi}{r - \alpha}dt \qquad (6)$$

$$Var(d(V(\pi))) = \sigma^2 V(\pi)^2 dt = \frac{\sigma^2\pi^2}{(r - \alpha)^2}dt . \qquad (7)$$

3.1.   The Dynamic Programming solution

In this framework, the question is: for what value of $V(\pi)$ is it optimal for a seller to complete the transaction (i.e., sell the land)? The solution to the problem is based on the notion that the real option value fluctuates with time and the changing volatility of returns.

Let $F(V(\pi))$ be the value of the option to sell. The objective of a seller is to maximize this value subject to (4). In a deterministic framework, the seller could choose the time

---

[13] $dz = \varepsilon_t \sqrt{dt}$ and $\varepsilon_t \sim N(0,1)$. The distribution of ε_t implies that $E(dz) = 0$ and $Var(dz) = E(dz^2) = dt$.

[14] From simple Brownian motion in (2) we know that percentage changes in cash flow are normally distributed and absolute changes in cash flow are log-normally distributed with a variance that increases linearly with the time horizon (Dixit and Pindyck). This appears to be a reasonable assumption for the cash returns to land.

period at which it is optimal to complete the transaction. However, in a stochastic frame-work this is not as straightforward. Rather, the seller must choose the value of $V(\pi)$ for which the objective function is maximized. Formally, this problem can be specified as,

$$F(V(\pi)) = \max_{\pi} E\left[(V(\pi) - P\right] \tag{8}$$

subject to (4). In (8), E denotes the expectation operator, P denotes the current market price of land (in option terminology, the strike price). If the value of land is greater than the optimal level, $V(\pi^*)$, the seller completes the transaction. Otherwise, the decision is to postpone the sale.

At an optimal solution, conditions (9) and (10) must hold. Condition (9) is the value-matching

$$F(V(\pi^*) = V(\pi^*) - P \tag{9}$$

$$F'(V(\pi^*)) = 1 \tag{10}$$

condition. It says that at an optimal level of cash flow (the trigger cash flow, π^*), the value of the real option equals the net present value of the investment. Condition (10) is the smooth-pasting condition. It says that the optimal cash flow is reached when the incre-mental gain in option value exactly equals the incremental gain in net present value. Thus, at some point the marginal change in the real option value equals the marginal change in the trigger investment value of the land, $V(\pi^*)$. Thus, the smooth-pasting condition identi-fies the trigger cash flow ($\pi*$) where the option value curve is tangent to the option payoff line (point c in Figure 1). It is at that point that the value of the real option is equal to the intrinsic value of the land.

In order to extend the maximization problem to many periods, we rewrite (6) as the Bellman equation,

$$F(V(\pi)) = \max_{u}\left\{\pi(x,u) + \frac{1}{1+r}E\left[F(V(\pi'))|x,u\right]\right\}, \tag{11}$$

where $F(V(\pi))$ is the value function, π' is the cash flow next period, $\pi(x,u)$ is the cash flow in the current period, u is a control variable (which takes the value 0 for waiting and 1 for deciding to sell) and x represents the information that is available at the time. Since $\pi(x,u) = 0$, [15] we obtain

$$rF(V(\pi))dt = E(dF(V(\pi))). \tag{12}$$

By using Ito's Lemma to expand $d(F(V(\pi)))$ and using (6) and (7), we obtain

---

[15] The cash flows associated with the real option (to sell) are zero in all periods except the period in which the optimal conditions are met. In that period the cash flow is the net payoff from the sale.

$$dF(V(\pi)) = F'(V(\pi))dV(\pi) + \frac{1}{2}F''(V(\pi))[d(V(\pi))]^2$$

$$E(dF(V(\pi))) = F'(V(\pi))V(\pi)dt + \frac{1}{2}F''(V(\pi))V(\pi)^2 dt \qquad (13)$$

Dividing (13) through by dt and combining it with (12), we get

$$\alpha V(\pi)F' + \frac{1}{2}\sigma^2 V(\pi)^2 F'' - rF = 0 \qquad (14)$$

The general solution to the ordinary differential equation in (14) is

$$F(V(\pi)) = A_1 V^{\beta_1} + A_2 V^{\beta_2} \qquad (15)$$

where A_1 and A_2 need to be determined. In this problem, we impose a restriction that

$$F(V(0)) = 0 \qquad (16)$$

which implies $A_2 = 0$. This in turn means that

$$F(V(\pi)) = AV(\pi)^{\beta} \qquad (17)$$

$$F(V(\pi)) = A\left(\frac{\pi}{r-\alpha}\right)^{\beta}. \qquad (18)$$

In summary, we have 4 equations [(9), (10), (14), and (17)] to solve for 4 unknowns $[V(\pi^*), F(V(\pi^*)), A, \text{ and } \beta]$. To solve, we use (14) and (17) to obtain

$$\beta\alpha + \frac{1}{2}\beta(\beta-1)\sigma^2 - r = 0 \qquad (19)$$

Solving (19) for a positive root,

$$\beta = \frac{1}{2} - \frac{\alpha}{\sigma^2} + \left[\left(\frac{\alpha}{\sigma^2} - \frac{1}{2}\right)^2 + \frac{2r}{\sigma^2}\right]^{\frac{1}{2}} \qquad (20)$$

From (10) and (17), we get

$$A = \frac{1}{\beta}V(\pi^*)^{1-\beta} \qquad (21)$$

We can solve for the bid price of farmland at the trigger cash flow, π^*, using (9) and (21).

$$V(\pi^*) = \frac{\beta}{\beta-1}P \tag{22}$$

We use (21) and (22) to solve for A,

$$A = \frac{(\beta-1)^{\beta-1}}{\beta^\beta}P^{1-\beta} \tag{23}$$

Using (17), (22) and (23), we solve for the real option value of farmland,

$$F(V(\pi^*)) = \frac{1}{\beta}V(\pi^*)^{1-\beta}V(\pi^*)^\beta$$
$$F(V(\pi^*)) = \frac{1}{\beta-1}P \tag{24}$$

Finally, we note that from (3) and (22), the trigger cash flow is

$$\pi^* = \frac{\beta(r-\alpha)}{\beta-1}P \tag{25}$$

4. DATA

The real option model of farmland pricing is illustrated using equations (22) - (25). Several scenarios are developed assuming different values for the key sources of land price uncertainty. Those variables include the volatility of cash flows (σ), the rate of growth of the cash flows (α), and the discount rate (r). The effects of these sources of uncertainty are evaluated at different assumed levels of expected cash flow per acre (π).

We use net cash rent per acre as a measure of the after-tax cash flow (π) to the land-lord. For this purpose we use the historical series of cash rents in Iowa during 1970-1998 (USDA). Net rent per acre is defined as the gross rent per acre less the real estate tax per acre.

4.1. Parameters

The parameters of the model could be developed either by attempting to model the stochastic process in (2) econometrically, or by the more direct statistical calculation of the means and standard deviations of the parameters from the time series. We use the latter approach and set the expected net cash rent (π) equal to the value in 1998, in order to simulate the option value to the seller in a recent year. The expected net cash rent in

1998 is about $97.50 per acre. The standard deviation of net cash rent during 1970-98 is about $23 per acre. Thus, we use $51.50 - $189.50 per acre in order to illustrate the model.

Similarly, the volatility (σ) of net cash rent is assumed to vary between 0.10 and 0.30. The volatility of the net cash rent series can be calculated in either of two ways, as an annual measure or as a total measure. Annual volatility can be calculated as the standard deviation of the one-year changes in net cash rent (e.g., $(R(t) - R(t-1)) / R(t-1)$). The annual volatility of net cash rent was about 0.10 during 1970-98. The total volatility of net cash rent can be calculated as the standard deviation of net cash rent divided by the mean net cash rent during 1970-98. The estimated total volatility of net cash rent is about 0.30. Thus, we set the base value of volatility at 0.20 and the range is 0.10 to 0.30. Our selection of 1970-98 as the period for parameter estimation is arbitrary, but that era does capture the significant volatility of cash rents and land values in the U.S. in recent years.

The growth rate of net cash rent (α) is defined as the compound rate of increase in net cash rent between 1970 and 1998. The rate of growth during those years was about 5 percent. Longer time intervals (such as 1900-1998) indicate that the average rate of growth was about 3 percent. The rate of growth of net cash rent during 1980-98 was relatively low at about 1 percent. We set the base value growth rate at 3 percent and the range at 0 percent - 5 percent.

The hurdle rate (r) is the rate at which cash flows are discounted to a present value. The discount rate theoretically reflects the risk-adjusted, marginal cost of funds (debt and equity) for the investor. It is the minimum rate of return that the investor would require in order to invest in a project. The hurdle rate is a subjective rate of return that an investor would require for projects of comparable risk. The hurdle rate may reflect several factors: the desired level of long-term economic return, the level of interest rates, and the investor's average cost of debt and equity capital. Thus, the hurdle rate may exceed the risk-adjusted discount rate.

The hurdle rate is calculated from the total rate of return on Iowa farmland. That is, the annual total rate of return is equal to the annual net cash rent plus the annual increase in land price (e.g., $(\pi(t) + P(t) - P(t-1))/ P(t-1)$). The mean total rate of return was about 12.5 percent during 1970-98 and about 7.5 percent during 1980-98. We set the base hurdle rate at 12.5 percent and the range at 7.5 percent - 17.5 percent.

5.   RESULTS

We illustrate our numerical results with tables that indicate the effects of uncertainty on real option value and, thus, land values. We report four key variables: the trigger cash flow, the land value at the trigger cash flow, the bid price, and the real option value. These sensitivity results illustrate several important relationships in the real option approach to land valuation. While they indicate the conditions that would need to exist for a sale to be optimal for the seller, they say nothing about how long a sale of land would be postponed by the seller.

First, we illustrate the smooth-pasting condition by varying the values of cash flow per acre (net cash rent) and the level of assumed volatility in Table 1.

Table 1.   Real Option Values for Different Levels of Cash Flow and Volatility

Cash Flow per acre (π)	51.50	74.50	97.50	120.50	143.50	166.50	189.50
Bid Price, V(π)	542.11	784.21	1026.32	1268.43	1510.53	1752.63	1994.74
Current Bid Price, P	1026.32	1026.32	1026.32	1026.32	1026.32	1026.32	1026.32
Net Present Value (NPV)	-484.21	-242.11	0.00	242.11	484.21	726.31	968.42
Assume $\sigma = 0.10$:							
Trigger Cash Flow (π^*)	144.15	144.15	144.15	144.15	144.15	144.15	144.15
Land Value at Trigger Cash Flow, V(π^*)	1517.34	1517.34	1517.34	1517.34	1517.34	1517.34	1517.34
Option Value, F(V(π))	20.41	63.87	144.69	282.25	484.24	766.60	1143.47
Assume $\sigma = 0.20$:							
Trigger Cash Flow (π^*)	174.73	174.73	174.73	174.73	174.73	174.73	174.73
Land Value at Trigger Cash Flow, V(π^*)	1839.26	1839.26	1839.26	1839.26	1839.26	1839.26	1839.26
Option Value, F(V(π))	51.25	118.16	217.19	350.71	520.70	728.88	976.78

Assuming $\alpha = .03$ and $r = .125$. Land prices and cash flow amounts are in $ U.S.

We observe that uncertainty creates value. The option value increases as the level of volatility increases, while holding constant the level of cash flow per acre and other model parameters. As the level of cash flow increases we observe that the NPV and the option value F(V(π)) begin to converge. At a net cash rent of \$143.5/acre, the NPV and the real option value are nearly identical when the volatility is 0.10. At higher levels of volatility (e.g., 0.20) the net cash rent amount must increase for the smooth pasting condition to hold.

5.1.   Sensitivity to initial net cash flow

As the level of the initial cash flow increases, the option value increases as well (Table 2). We note that the trigger cash flow increases in proportion to the initial cash flow. The reason for this relationship is that the trigger cash flow (π^*) is a function of beta, alpha, the hurdle rate and the bid price of land. In turn, the bid price of land is a function of the initial cash flow level, as in (3).

5.2.   Sensitivity to hurdle rate

There is an inverse and nonlinear relationship between the option value and the hurdle rate as shown in (Table 3). As the hurdle rate rises (holding other parameters constant) the bid price of land falls as in (3) and β increases as in (20). By using V(π) in place of P in (24), we see that the option value of land at the trigger cash flow falls faster than does the bid price. This illustrates the intuitive result that an upward shock in the opportunity cost of funds, due to increases in interest rates, has a potentially significant effect on both the real option value and land value.

Table 2.   Sensitivity of Real Option Value to Initial Cash Flow a/

Cash Flow, π	51.5	74.5	97.5	120.5	143.5	166.5	189.5
Trigger Cash Flow, π^* b/	92.3	133.5	174.7	215.9	257.2	298.4	339.6
Land Value at Trigger Cash Flow, $V(\pi^*)$	971.5	1405.4	1839.3	2273.1	2707.0	3140.9	3574.8
Bid Price, $V(\pi)$	542.1	784.2	1026.3	1268.4	1510.5	1752.6	1994.7
Option Value, $F(V(\pi^*))$	429.4	621.2	812.9	1004.7	1196.5	1388.3	1580.0

a/   Assuming $\sigma = .20$, $\alpha = .03$ and $r = .125$. Land prices and cash flow amounts are in $ U.S.
b/   The trigger cash flow is calculated with $V(\pi)$ in place of P in (25).

Table 3.   Sensitivity of Real Option Value to Hurdle Rate a/

Hurdle Rate, r	0.075	0.100	0.125	0.150	0.175	0.200
Trigger Cash Flow, π^*	236.3	195.0	174.7	162.5	154.2	148.2
Land Value at Trigger Cash Flow, $V(\pi^*)$	5250.6	2785.7	1839.3	1354.2	1063.7	871.9
Bid Price, $V(\pi)$	2166.7	1392.9	1026.3	812.5	672.4	573.5
Option Value, $F(V(\pi^*))$	3083.9	1392.9	812.9	541.7	391.3	298.4

a/   Assuming $\sigma = .20$, $\alpha = .03$ and $\pi = \$97.5$/acre. Land prices and cash flow amounts are in $ U.S.

5.3.   Sensitivity to growth rate

In Table 4 we see that even when we assume no growth in the cash flow stream, uncertainty has value due to volatility of the cash flows.

The zero growth rate assumption and higher growth rates illustrate an important feature of the stochastic specification of the cash flows in (2) and the underlying logic of real option value. Zero growth of cash flow implies zero drift, yet the cash flow series still exhibits volatility, which has value in this framework. Moreover, the relationship between real option value and the growth rate is nonlinear and increasing. An increase in the growth rate affects the bid price positively as in (3). Also, as the growth rate increases, β becomes smaller. As a result, when $V(\pi)$ is used in place of P in (24), the effect of a change in α is greater on the option value at the trigger cash flow than it is on the bid price.

In Table 4 we see that a significant upward shock in the cash flow growth rate expectations of investors can lead to a sharp increase in real option values and land prices in excess of the implied increases in bid prices that are predicted by the traditional PVM.

5.4.   Sensitivity to volatility of net cash rent

In Table 5 we see the effect of increasing volatility on real option value and, thus, land prices. These increases in volatility of cash flows numerically illustrate the impact of the stochastic term in (2) on cash flows and the real option value of land. As volatility increases, the real option value of land for the seller increases nonlinearly. This nonlinear relationship is reflected by the role of sigma in determining beta as in (20). As volatility increases, the value of land at the trigger cash flow, $V(\pi^*)$, and the bid price of land, $V(\pi)$, diverge as the value of the real option increases.

Table 4.   Sensitivity of Real Option Value to Growth Rate a/

Growth Rate, α	0.00	0.01	0.02	0.03	0.04	0.05	0.06
Trigger Cash Flow (π^*)	145.1	151.7	160.5	172.7	190.7	220.2	276.8
Land Value at Trigger Cash Flow, $V(\pi^*)$	1160.6	1319.2	1528.6	1817.9	2244.0	2935.8	4258.4
Bid Price, $V(\pi)$	780.0	847.8	928.6	1026.3	1147.1	1300.0	1500.0
Option Value, $F(V(\pi^*))$	380.6	471.4	600.0	791.5	1097.0	1635.8	2758.4

a/   Assuming $\sigma = .20$, r = .125 and π = $97.5/acre. Land prices and cash flow amounts are in $ U.S.

Table 5.   Sensitivity of Real Option Value to Volatility a/

Volatility, σ	0.07	0.10	0.15	0.20	0.25	0.30
Trigger Cash Flow (π^*)	279.1	161.50	161.0	172.7	187.9	205.5
Land Value at Trigger Cash Flow, $V(\pi^*)$	2938.0	1700.0	1694.9	1817.8	1978.4	2163.5
Bid Price, $V(\pi)$	1026.3	1026.3	1026.3	1026.3	1026.3	1026.3
Option Value, $F(V(\pi^*))$	1911.7	673.7	668.6	791.5	952.1	1137.2

a/   Assuming $\alpha = .03$, r = .125 and π = $97.5/acre. Land prices and cash flow amounts are in $ U.S.

We note that the above tables illustrate the singular effects of these variables. However, it is also possible to illustrate the combined effects of shocks to factors such as the rate of growth and volatility of cash flows per acre. These alternative scenarios raise the prospect that the real option framework could provide a useful interpretation of the phenomenon observed in U. S. land prices during the late 1970s and early 1980s.

6.   CONCLUSIONS

Real option theory may provide a way to better understand the role of uncertainty in land pricing decisions and the resulting behavior of land prices. The real option value framework provides a more useful framework than the standard PVM. Increasing land prices are a fundamental characteristic of markets where the real option has value to the seller. We can observe also that land prices rise even when there is no growth in returns to land, due to real option value. Shocks to investor expectations about the growth rate or volatility of cash flows can have significant impact on land values. Thus, the real option approach provides an alternative explanation for the observed divergence of land rents and land prices.

Additional research on land valuation seems to be warranted given the relatively new directions in which land pricing studies have been pursued in recent years. The real option approach requires further development and testing. One such direction for research might be to specify the stochastic process of cash flows and the character of the volatility of cash flows. Research efforts could also usefully explore the role of real option theory in the presence of transaction costs, as another plausible approach to solving the present value puzzle. In our view, transaction cost theory encompasses the real option value framework and these two approaches could be usefully integrated to improve our understanding land price behavior.

7. REFERENCES

Aiyagari S.R. (1993), Explaining Financial Market Facts: The Importance of Incomplete Markets and Transaction Costs, *Quarterly Review, Minneapolis Federal Reserve Bank*, **17**: 17-31.

Alston J.M. (1986), An Analysis of Growth of U. S. Farmland Prices, 1963-82, *American Journal of Agricultural Economics*, **68**: 1-9.

Bjornson B. – Innes R. (1992), Another Look at Returns to Agricultural and Nonagricultural Assets, *American Journal of Agricultural Economics*, **74**: 109-119.

Burt O.R. (1986), Econometric Modeling of the Capitalization Formula for Farmland Prices, *American Journal of Agricultural Economics*, **68**: 10-26.

Campbell J.Y. – Shiller R.J. (1987), Cointegration and Tests of Present Value Models, *Journal of Political Economy*, **95**: 1062-1088.

Chavas J.P. – Thomas A. (1999), A Dynamic Analysis of Land Prices, *American Journal of Agricultural Economics*, **81**: 772-784.

Clark J.S. – Fulton M. – Scott J.T.jr. (1993), The Inconsistency of Land Values, Land Rents, and Capitalization Formulas, *American Journal of Agricultural Economics*, **75**: 147-155.

Copeland T. – Weston J. (1988), *Financial Theory and Corporate Policy*, Addison-Wesley, Reading, Mass.

Dixit A.K. – Pindyck R.S. (1996), *Investment Under Uncertainty*, Princeton University Press, Princeton, NJ.

Dixit A.K. (1992), Investment and Hysteresis, *Journal of Economic Perspectives*, **6**: 107-132.

Falk B. (1991), Formally Testing the Present Value Models of Farmland Prices, *American Journal of Agricultural Economics,* **73**: 1-10.

Falk B. – Lee B.S. (1998), Fads versus Fundamentals in Farmland Prices, *American Journal of Agricultural Economics*, **80**: 696-707.

Featherstone A.M. – Baker T.G. (1987), An Examination of Farm Sector Real Asset Dynamics, 1910-85, *American Journal of Agricultural Economics*, **69**: 532-546.

Hanson S.D. – Myers R.J. (1995), Testing for a Time-Varying Risk Premium in the Returns to Farmland, *Journal of Empirical Finance*, **2**: 265-276.

Just R.E. – Miranowski J.A. (1993), Understanding Farmland Price Changes, *American Journal of Agricultural Economics*, **75**: 156-168.

Lence S.H. – Miller D.J. (1999), Transaction Costs and the Present Value of Farmland: Iowa, 1900-1994, *American Journal of Agricultural Economics*, **81**: 257-272.

Luttmer E.G.J. (1996), Asset Pricing in Economies with Frictions, *Econometrica*, **64**: 1439-1467.

Pope R.D. – Kramer R.A. – Green R.D. – Gardner B.D. (1979), An Evaluation of Econometric Models of U. S. Farmland Prices, *Western Journal of Agricultural Economics*, **4**: 107-119.

Trigeorgis L. (1997), *Real Options: Managerial Flexibility and Strategy in Resource Allocation*, The MIT Press, Cambridge, Massachusetts.

Turvey C. (1999), *Real Options and Agricultural Land Values*, Paper presented at the NC-221 Regional Research Group meeting, Ontario, Canada, October.

Weersink A. – Turvey C. – Clark S. – Sarkar R. (1999), The Effects of Agricultural Policy on Farmland Values, *Land Economics*, **75**: 425-439.

AGRICULTURE & RURAL DEVELOPMENT

ASSESSING STRATEGIC PROGRAMS FOR SUSTAINABLE LOCAL DEVELOPMENT: A CASE STUDY

Stefano Stanghellini and Tecla Mambelli [*]

SUMMARY

In recent years in Italy, public and private sectors have been stimulated by the National Government to cooperate in urban renewal, thanks to public funds committed through competitive procedure. In 1998, the Ministry of Public Works promoted new Programs for Urban Renewal and Sustainable Regional Development (Italian acronyme: Prusst). This kind of program puts together a package of public and private projects to promote economic and sustainable development. The main axes are networked systems, productive centres, urban rehabilitation and environmental preservation. New attention is given to cooperation and negotiation among local authorities and private operators.

The research deals with a critical analysis for the evaluation methods used to select the projects to include in the programs, with the aim to propose the Community Impact Evaluation (CIE) as a more appropriate method for their evaluation. CIE takes account of the total costs and benefits for the community involved, with particular attention to the impact of such costs and benefits on the various community sectors.

In addition, the authors try to integrate the CIE with other sectoral assessment methods necessary to obtain public funding, such as Social Financial Analysis and Social Impact Assessment. The methodology proposed is finally tested in the program carried out by the towns of Forlì and Forlimpopoli, in the Emilia Romagna region.

The purpose is to define an "accounting" matrix capable of emphasizing the advantages for the public and private sector participating in the program and of selecting a set of projects that, in minimizing negative impacts, respond to efficiency, sustainability and distributive equity criteria.

---

[*] Stefano Stanghellini, professor at the Planning Department of the Venice Universitary Institute of Architecture, responsible for sections 1 to 3. Tecla Mambelli, contract researcher at the Planning Department of the Venice University Institute of Architecture, author of sections 4 to 5.

Economic Studies on Food, Agriculture, and the Environment
Edited by Canavari *et al.*, Kluwer Academic/Plenum Publishers, 2002

171

1. INTRODUCTION

The increasing competitiveness and interdependency between local productive and organizational systems, in addition to the globalization, require that the decision-making processes used in siting, financing and building new regional infrastructures establish greater and more effective opportunities for co-development and participation than in the past.

The growing difficulty in finding financial resources is leading local public institutions to seek opportunities for public or private intervention and investment emanating from regional, national and, with ever-increasing weight, community policies. These opportunities are often decisive in attracting other public and/or private local resources. To take full advantage of this financial support and ensure that its despersement is related to local and regional effectiveness, it is necessary that all actors be involved in a clear and well-structured decision-making process from the beginning.

They strive to introduce a strategic-negotiation approach into the Italian planning system that addresses the economic-financial feasibility and assessment of the sustainability of renewal efforts. The public authorities need to identify resources and public or private entities that can be mobilized. The Programs for Urban Renewal and Sustainable Regional Development (otherwise known as Prusst) were created with the aim of packaging together public and private projects. They favor regional economic development in accordance with regional planning and environment compatibility.

Prusst represents an opportunity for defining priorities among development and infrastructure choices, and for obtaining a sufficient level of social consensus regarding these choices. Selecting which projects are to be financed and involving private investors are pre-conditioned by the interests that emerge in determining the advantages and disadvantages (in terms of social, economic and environmental impact) to the entities called upon to participate in the decision-making process and the social interest groups affected by these projects.

This chapter focuses on the role of assessment in determining the feasibility and sustainability of strategic programs for regional development and proposes an assessment methodology for selecting the strategic alternatives for up-grading the abandoned areas of the city of Forlì. Its aim is twofold: to examine the importance of assessment by considering the major methods available for evaluating strategic programs for sustainable regional development and to demonstrate the use of this methodology as applied to the Prusst "Inter-modal Corridor of Forlì-Forlimpopoli".

2. THE ASSESSMENT OF SUSTAINABILITY AND FEASIBILITY IN REGIONAL RENEWAL PROJECTS

By its very nature the planning process has always been full of assessments. This is especially true when planning has to reconcile renewal efforts (i.e. infrastructure) with the goals of defense. In this case assessment helps decision-makers establish the compatibility of proposed transformations with defense objectives and search for acceptable solutions.

In the strategic programs for regional development (Prusst), the choices that must be made imply formulating value judgements on all the resources at stake, given that the decisions have to be made based upon merit of use. Faced with the need to choose among alternative options in relation to a determinate set of goals, it is necessary to make assessments that estimate monetary values in terms of costs and benefits to the parties and community involved.

To assess whether a variation in the current resource allocation is efficient (or if it produces an increase in social well-being), economic analysis suggests measuring the costs and benefits and calculating the net social benefits resulting from the investment or legislative decision. The planned modification can thus be compared to the status quo.

Financial assessments define the physical-morphological, functional and economic contents of a proposed change. They not only identify the economic parameters (prices and costs) that delimit the field within which satisfactory solutions are sought; they also influence the choices regarding project quality, including the market desirability of the goods and services produced, cost control and scheduling, financial investment monitoring, and the management of goods and services over time.

From a market perspective and in the context of decentralized decision-making, economic feasibility assessment requires that the flow of costs and revenues be measured for those operators or social groups using or supporting them. The formulation of this accounting framework serves to verify the advantages each social group - or single operator - will obtain through the realization of the plan, and thereby allows the degree of participation or opposition to be foreseen. On the basis of this information, the plan can then be modified and individual advantages created, thus maximizing those effects that determine social advantage.

In order to provide satisfactory results, rather restrictive assumptions have to be respected in regard to monetary commensurability. This necessitates a substantial reduction in the scope of investigation and requires that the results of the economic analysis be integrated with an impact matrix which explains the effects of the project that are difficult to put into monetary terms.

Cost-benefit and cost-revenue analyses have been demonstrated to have significant limitations in the case of regional and urban plans, in as much as they do not permit the consideration of distributional aspects. As a result, some recent experiments have been aimed at modifying their structure, so that they can also take in account the distribution of the effects of the provisions and regulations among the different groups making up the involved community. These experiments are set up in strict harmony with the Community Impact Evaluation, proposed by Lichfield in the mid-1960s. This method constitutes an evolution of the Cost Benefit Analyses particularly well-suited to evaluating a plan or project's economic, social and environmental effects by estimating the costs and benefits associated with them and their distribution among social groups.

3.   THE ROLE OF EVALUATION IN PROGRAMS FOR URBAN RENEWAL AND SUSTAINABLE REGIONAL DEVELOPMENT

With the Ministerial Decree of 8 October 1998, the Ministry of Public Works promoted Prusst innovative programs. The goal was to endow these integrated programs with tools for pursuing economic competition among cities, by satisfying the demand for social sustainability and increasing the value given to the environment. Municipal to subregional in scope, these programs have sought to promote opportunities for sustainable development from an economic, environmental and social point of view, by creating individual or networked facilities, or through an integrated system of activities aimed at economic and productive renewal.

The programs are promoted by either the municipalities or the province and the region through prior agreement with the interested municipalities. Proposals to these entities can be formulated by other entities (proponents), such as regional public entities, other administrations, public agencies, or private entities.

Prusst enriches Italian urban planning by shifting the emphasis from physical structures to entities, resources and actions. Municipalities with a recent land use plan, or one whose basic premises, even if not yet in force, have already been defined, have interpreted Prusst programs as tools for accelerating the plan's implementation. This can occur to the degree in which they involve and/or stimulate the entrepreneurial, financial and real estate resources, etc. of a given region.

The stimulus toward sustainable development is an important result of the experiments underway. It is, perhaps, for the very first time that many municipal administrations have tried to solicit the collaboration of entrepreneurs, industrialists, banks, contractors and private operators in general to build projects of common interest.

The Ministry also introduced another innovation with the financial procedure that integrates public and private resources. The allocation of funds is governed by certain constraints regarding a minimum percentage of investment for private projects (of the overall investment) and by the presence of project financing (for which the share is established by the promoting entity).

The programs are evaluated for financing by the National Commission on the basis of indicators identified by Ministry of the Public Works up to 80 points (see Table 1) and by each Region up to 20 points.

The following technical documents are required for proposal evaluation: the identification and description of the specific program area, pre-feasibility studies, a description of the facilities and the system of the activities foreseen, a description of the Prusst's compatibility with the regional planning and programming tools, analyses of the consequences on the environmental and employment with the necessary verifications of environmental sustainability, a verification of costs, and the financial coverage of the public and private interventions. Finally, a written description detailing all the interventions foreseen is aimed at highlighting relationships of synergy and integration between various projects.

As underwritten in the agreed protocol of art. 8, para.3 in the call for proposals, the programs' promoting entities can request financing to cover the costs related to technical assistance. These activities are finalized as follows:

-   in the program definition phase, to the predisposition of pre-feasibility and economic-financial feasibility studies and the assessment of the action's effectiveness;
-   in the phases required to orchestrate the various entities and for project approvals, verifications of program compatibility and legal-administrative feasibility;
-   in the phase of program implementation, monitoring of the status of the planned activities.

In all phases technical support is provided for the entities involved with the aim of guaranteeing the program's adequate conclusion according to the established schedule.

Table 1.   Maximum Scores Attributable for Ministerial Assessment and Selection

	Max.score	Unit Score
1. Sectional integration policies	40	
1.1 Infrastructure rehabilitation deficit	10	0.1 for every billion ITL of investment on individual and networked infrastructures
1.2 Rehabilitation, conservation and en-hancement of the natural environ-mental patrimony	15	0.2 for every billion ITL designated to recovering abandoned areas, reduction of pollution in currently operating activities, hydro-geological works, increasing the value of the patrimony through *non-profit* revenue
1.3 Social goals	10	0.2 for every billion ITL designated to productive activities with stable effects on employment
1.4 Partnership, co-development and lo-cal orchestration	5	0.05 for every billion ITL designated to projects already launched through negoti-ated planning tools (regional pacts, area contracts)
2. Implementation in relation to financial coverage	20	
2.1 Already available funding	10	0.1 for every % point of total funding available
2.2 Private contribution over 1/3 of the total investment	3	0.1 for every % point over 1/3 of total investment
2.3 Public interventions done with pri-vate resources	5	0.1 for every % point of private interven-tion
3. Design quality [a]	20	- ecological-environmental quality - landscape quality - urban quality (accessibility and safety) - morphological quality and that of the urban fabric (continuity, typological mix, conservation)

Source: Dicoter
   [a]   Project quality is also to be considered in relation to the attention given to and quality of the docu-mentation presented (art.7 of the call for proposals).

During the realization of the projects, the "Modality of the use of funds" emphasizes the importance of assessments for sustainable development, which embrace economic-financial, environmental and social aspects. The assessments of economic-financial feasi-bility in urban environs include complex techniques for demonstrating the repercussions of costs, revenue and benefits on the community involved, by pre-figuring different sce-narios of intervention. The assessments for sustainability can cover a wide range of uses, ranging from identifying the significant indicators that launch phenomena of economic and regional development, to selecting design criteria suitable to achieving environmental quality and increasing the value of the architectural patrimony, and, finally, to monitoring capable of verifying the status of implementation during its realization.

4. CASE STUDY: THE SUSTAINABLE TRANSFORMATION OF COMPLEX AREA AROUND FORLÌ'S RAILWAY HUB

4.1. Goals and methodology

In the ambit of the "Complex areas – railway system" renewal, the Prusst's "Intermodal Corridor Forlì-Forlimpopoli" includes a group of abandoned and unused areas of public and private property, along the functional and morphological spine made up by the railway lines, between the historic center and the future axis of the city's ring road. This area is intended to be used in creating a new system of central urban and regional functions The renewal program encompass projects for reorganizing the infrastructure, urban facilities, and service sector activities not as yet adequately present in Forn or concentrated in the city center therein contributing to its congestion. These areas can be classified in three major zones: the ex Bartoletti area south of the railway station, the ex Eridania area northwest of the rail line, and that made up of the areas of the ex Foro Boario, Via Pandolfa, and the Ultragas complex north of the railway station.

In the phase of the program's realization, Forlì's Municipal Administration posed the problem of prioritizing the areas selected for public financing to facilitate the urban renewal process. This selection procedure, however, had to be based on an economic-financial feasibility analysis of the proposed projects for all the entities involved and, at the same time, on an analysis of the environmental and social consequences of the various projects on the different sectors of the community.

In compliance with the criteria of economic efficiency, the enhancement of environmental value, and social equity advocated by the ministerial call for proposals, an assessment methodology was devised based on the Community Impact Evaluation (Lichfield, 1996), so as to merge a purely financial analysis with the assessment of the impacts on the different sectors of the community.

The method is made up of various phases that range from the description of project alternatives to the identification of those sectors of the community involved in defining the impacts on the social sectors. It allows decision-makers to express preferences between the options on the basis of impacts and sectional preferences.

4.2. Description of the renewal alternatives for the railway area

4.2.1. Ambit A

The first intervention alternative involves renovating the ex Foro Boario area, upgrading the areas of the Consorzio Agrario and Ultragas, and constructing the new street, foreseen by the land use plan, and the railway station underpass.

The project provides for renovating Foro Boario and the area facing it (now occupied by an abandoned gas station and parking lot, which are very degraded) owned by the municipality, into a public park open to neighborhood activities. There will be a new layout of pedestrian spaces and links to the new neighboring urban areas. In addition to itinerant markets of different sorts (antiques and others), the park could also host cultural and recreational activities managed by the inhabitants themselves (art exhibitions, outdoor performances, etc.). The central area provides space for a new road that will continue that of the Station, ending in a building complex designated primarily for subsidized housing and service sector use (public and private).

In the private area of the Ultragas establishment [1], the project provides for the construction of a new mixed-used development (of prevalently residential use) organized into two, connected half-courts: an east-west pedestrian connection linked the with central section and the reorganization of the local road and parking network.

4.2.2. Ambit B

The urban renewal of the ex Eridania area is also subdivided into two sections: one for private and one for public intervention.

The western section will be transformed by a private project aimed at the realization of a healthcare facility for a total surface area of about 14,900 square meters. The Center will furnish hotel services, as well as basic day and night healthcare including general and specialized medical and nursing assistance, rehabilitation and pharmaceutical services.

The eastern section renewal, the property of which is to be conceded to the Municipal Administration (60,400 sq.m of the total area), includes the restoration of the existing buildings, to be re-used for activities linked to music, performing arts and leisure. Further east, this section also contains mixed-use development, including commercial, administrative, residential, and artisan service spaces for a total of 21,800 sq.m of gross usable surface area. Part of the residential space inside this section (2,000 sq.m) will be subsidized. Public parking and parks are also foreseen in the area.

4.2.3. Ambit C

The third region is located in a strategic position contiguous to the historic center of Forlì and the railway station. This area creates a corner between Via Leonardo da Vinci, parallel to the railway line, and Piazzale Martiri d'Ungheria, in front of the entrance to the Railway station.

The project includes the reorganization of a vast area made up of both public and private property. The former, which falls under the public land use plan and is known as the Integrated Transportation Center, is designated for public parking, a bus station and public buildings; the latter is part of the up-grading of the present seat of the industrial complex, Bartoletti S.p.a., now being moved to a new local area. This area has been designated for residential and administrative use.

The private area has an overall surface area of 46,648 sq.m. The project provides 20,000 sq.m of total gross usable surface area for residential use, while commercial and administrative uses have been estimated at 5,000 sq.m. Fifty percent of this area is devoted to commercial use and the remaining 50% for service sector and administrative use.

---

[1] For the area, the Municipal Administration has already defined the procedures required to relocate the facilities in the productive district of Villa Selva, given the need to distance this dangerous use from the residential areas.

Table 2.   Community sectors

Producers/operators		Consumers/users	
Description	*Options*	*Description*	*Options*
A1. Local Government	A, B, C	B1. New occupiers:	
		b.1.1 - building sector	A, B, C
A2. Public Transport Agency	C	b.1.2 - other activities	A, B, C
		B2. New inhabitants:	
A3.Development Agency	A	b.2.1 - public housing	A, B
		b.2.2 - private housing	A, B, C
A4. Landowners	A, B, C	B3. Users of green areas:	
		b.3.1 - inside the ambit	A, B
A5. Financial promoters	A, B, C	b.3.2 - outside the ambit	A
		B4. Users:	
A6. Builders	A, B, C	b.4.1 - health services	B
		b.4.2 - cultural facilities	A, B
A7. Industrialists	A, C	b.4.3 - public transport	C
		b.4.4 - offices.	A, B, C
A8. Business men	A, B, C	b.4.5 - trade centers	A, B, C
		B5. Pedestrians	A, B
		B6. Motorist	A, B

4.3.   Identification of the community groups involved and their sectional goals

The various sectors affected by the project (producers, social operators, final users) are taken into consideration through the CIE, which provides an impact analysis for the entire community. The method makes the repercussions of the plan or project explicit for every sector or group of the collectivity affected, by compiling a complete social accounting.

The sectors of the community affected by the three renewal projects are divided into producers/operators, individuals or groups who would play a part in creating and operating the services to be realized from the proposals, and consumers, those who would use these services (Table 2).

The check list was integrated with a more thorough analysis of the stakeholders (Mac Arthur, 1997) that highlighted the importance of considering not only the producers/operators and on-site users, but also those coming from nearby communities. This evaluation places greater emphasis on underprivileged populations, including those individuals who will take advantage of the subsidized elderly health-care residence, parks and public facilities, and subsidized housing.

The sectional preferences were identified through social survey techniques (Rosenberg, 1968; Marradi, 1980), using direct interviews with representatives of determinate economic categories, associations and interest groups. A questionnaire was also filled out by a sample of the population residing in the area of Foro Boario di Forlì, next to three zones under consideration. The sample used in the study was based on the census (Istat, 1991) of residents living in the three areas, equal to 2,696 persons, stratified with respect to different age groups. Within these, the percentages of each age sub-group were calculated and projected on the sample by sex (M= 47% and F= 53%). The second phase of the sampling calculated the overall number of subjects to whom the questionnaire was to be directed. Two-hundred and eighty-four subjects were interviewed, equal to 10.5% of the resident population, respecting the proportions of sex and age. [2]

4.4.   The assessment of the the economic, environmental, and social effects

Two distinct phases were identified to assess the effects of the three alternatives: design/planning/construction and management. The goal of this subdivision was to be able to analyze project impact over a medium- and long-term time span. Every project was further subdivided into characteristics, for which the important effects were identified from an economic, environmental and social perspective.

The alternatives were assessed first through a Social Financial Analysis that keeps track of the cash flows relative to the realization of the projects and the management of the finished good for a period of eight years, [3] for the main public and private actors involved. The annual growth of all cost and revenue items was determined over the indicated time span for every intervention alternative. The economic-financial information was obtained by merging the information obtained from indirect sources for the city of Forlì (real estate consultants, Gabetti real estate agents, etc.) and those obtained directly through investigations with local building contractors and real estate experts.

Total annual costs and revenue were later brought up-to-date by multiplying them by the construction coefficients, on the basis of an interest rate of 8%. The difference between current revenue and costs furnished the Net Present Value (NPV) of the operation, which represents one of the three indicators of the economic effects of the alternatives.

Moreover, with the threshold rate for evaluating the advantage of the operation with respect to alternative investments is fixed at 15%. The Internal Rate of Return (IRR) was calculated and used as a further indicator by comparing it for across the three alternatives for the private investors. Finally, the costs of maintaining and managing public spaces (green and open spaces, etc.) were also taken into consideration.

The environmental effects were expressed through the variation of traffic (indirect indicators for acoustic and atmospheric pollution) based on the three projects, the presence of barriers for acoustic and visual protection and, finally, the "greater adoption of standard" per inhabitant in each area (the percentage of green and public space in relation to the uses planned).

The data on the presumed traffic for the three project areas under examination was deduced from a study conducted while formulating the Plan for Urban Traffic (PUT) the city's new land use plan, and the Prusst. The demand for mobility was quantified on the basis of the traffic generated in the urban planning previsions for an average working day, and further elaborated through the application of suitable parameters for defining with

---

[2] The results of the questionnaire, made up of 23 questions, all multiple choice with the exception of two open-ended questions, are reported in M. Merelli, *Il quartiere Foro Boario di Forlì, Problemi, valutazioni, aspettative dei suoi abitanti* (The Neighbourhood of Foro Boario in Forlì, Problems, Assessments and Resident's Expectations), Forlì, 1998.

[3] This period, which includes the time required to obtain the concession and approval of the projects (bureaucratically), was estimated as the time necessary to complete the intervention and sell the real estate built.

greater precision the magnitude of traffic for each means of transportation and the typical rush hour. [4]

Finally, the social effects were examined through the occupational increment (held to be one of the primary objectives of the ministerial request for bids for Prusst financing) and the provision of subsidized housing (a municipal priority to meet the needs of the population by obtaining regional financing).

The employment impact of the new facilities was estimated for each area, distinguishing between temporary employment, based on the construction work, and more long-term employment, bound to the management of public facilities (recreation centers and the assisted elderly health-care residence) and private development (service sector and commercial activities). This was done through the indexes from the National Association of Builders, which express the number of jobs generated by the investment in building (either through direct employment or in related fields), and indexes taken from official publications of the National Institute of Statistics (Istat) and other sources. [5]

Since the new facilities will, in part, house firms relocated from less functional sites, it was reasonable to estimate the jobs created ex novo equal to 65% of those necessary to run the new facilities.

The information generated from the sectional assessments are summarized in Table 3, which reports economic, environmental and social effects for each area.

4.5.   The evaluation of sectional preferences

Forecasting impacts is a very delicate operation in that the impacts regarding the project variables are weighed with respect to the sectional objectives determined by social surveys. The assignment of weights has to reflect the relative importance of the various types of impacts considered for the three major social groups, re-aggregated in local government, producers and consumers.

A rating method, commonly used in planning (Lichfield et al., 1975; Miller, 1980), was used to estimate the weights for this project. [6]

With the method of standardization $E_{ij}=(e_{ij} - \min e_j)/(\max e_j - \min e_j)$ [7], alternative B turns out to be preferable for the local government, which attributes great importance to the criterion of social equity, protecting the disadvantaged social classes looking for housing (at subsidized prices), and temporary and long-term employment. At the same time, residents of the middle class opt for better environmental conditions, in terms of increased green space, reduced pollution and acoustic barriers. For this group and, more generally, for consumers (broadly understood as the users of the area amenities, as resi-

---

[4] Between the hours of 7:30 – 8:30 a.m. and 17:30 – 18:30 p.m. The choice of means of user transport was taken from the data of the Istat 1991 census, re-calibrated to keep in mind the trend of motorization and the variations in the use of public transport from 1991 to 1996.

[5] Istat reports, for the various categories of surface area dimensions, provide data regarding the number of local units and workers. From there, the average number of workers by surface area category can be deduced yielding an average sq.m/worker. This data discounts a certain approximation that increases as the size of the local unit increases, but which seemed acceptable in light of the scope of this study.

[6] This kind of method requires the interlocutor to assign an established number of points (e.g. 100) among the identified criteria so that points assigned to each criteria reflect its relative importance (Voogd, 1983).

[7] Where E_{ij} is the data corresponding to criterion j and to the standardized alternative i, e_{ij} the data before standardization, and $\max e_j$ and $\min e_j$ the maximum and minimum values observed for the criteria j among all the other alternatives i (i=1,2.....I).

Table 3. Synthesis of the economic, environmental and social effects

Effects	Unit of Measure	Ambit A	Ambit B	Ambit C
Economic				
E1 - Public management costs for 8 years	ITL x 1,000/sq.m	84,785	291,965	177,968
E2 – IRR for private entities	%	18.7	19.2	18.6
E3 - NPV for public operator	ITL x 1,000/sq.m	- 4,144,000	- 12,832,000	- 8,900,000
Environmental				
AU1 - Vehicular traffic based on future users of the areas	no. autos in both directions during rush hours	431	1,010	535
AU2 - Acoustic and atmospheric barriers and protection of the major axis	ml of protected green space	4,400	2,800	0
AU3 - Green space and open areas with respect to the population to be accommodated (residents and workers)	sq.m green spaces / resident and working pop.	157	67	27
Social				
S1 - Housing rights for less-privileged categories	sq.m of subsidized housing	6,855	2,000	0
S2 - Long term employment	New employment created by the activities foreseen	86	287	83
S3 - Temporary employment	New employment in the building industry	330	752	501

dents, insiders, users of the facilities or green space) the preferred alternative is A. Alternative C, which maximizes the criterion of economic efficiency, is preferable for producers/operators (Table 4).

From this analysis, it is not easy to formulate a univocal conclusion regarding the preferred alternative because each one satisfies some community groups more than others regarding economic efficiency, environmental protection and social solidarity. It is interesting however to note that alternative B represents the second choice option for both producers, with a minimum difference with respect to option C, and consumers.

Using a different standardization method, $E_{ij}=e_{ij}/\max e_j$, all three sections of the community prefer the same alternative, B. This order of preferences was verified through a sensitivity analysis, by assigning different weights to the criteria (economic efficiency, social equity and environmental protection) to understand the range of variation that makes hypothesis B preferable to the others.

5. CONCLUSIONS

The conceptual framework reported here represents a new approach to Community Impact Evaluation, through a relatively simple analysis of the relevant effects on the dif-

Table 4.   Synthesis of the sectoral preferences

Effects	Local Government				Producers				Consumers		
	A	B	C		A	B	C		A	B	C
E1	10	0.00	10.00	4.50	0	0.00	0.00	0.00	5	0.00	5.00	2.25
E2	10	0.00	7.49	10.00	65	0.00	48.66	65.00	0	0.00	0.00	0.00
E3	5	5.00	0.00	2.26	0	0.00	0.00	0.00	5	5.00	0.00	2.26
AU1	15	15.00	0.00	9.91	5	5.00	0.00	3.30	20	20.00	0.00	13.22
AU2	5	5.00	3.18	0.00	0	0.00	0.00	0.00	10	10.00	6.36	0.00
AU3	10	10.00	3.08	0.00	5	5.00	1.54	0.00	15	15.00	4.62	0.00
S1	15	15.00	4.38	0.00	5	5.00	1.46	0.00	15	15.00	4.38	0.00
S2	20	0.29	20.00	0.00	5	0.07	5.00	0.00	25	0.37	25.00	0.00
S3	10	0.00	10.00	4.05	15	0.00	15.00	6.08	5	0.00	5.00	2.03
		50.29	58.12	30.73		15.07	71.65	74.38		65.37	50.36	19.76

ferent sections of the community. The great utility of the method is that it links the direct effects of building and operating a facility to the probable social and economic responses in the areas examined. It does this by combining the established methods of Community Impact Evaluation with new conceptual links to Social Impact Assessment and Social Financial Analysis. Thus an integrated set of economic, environmental and social effects can be described, linking sectional groups in the local study areas to project-induced effects.

The methodology employed in this approach uses many theories and techniques that have been developed in recent years and introduces social surveys to identify the community sectors and the sectional preferences for project options. Nonetheless, the approach has to be ameliorated in its conceptualization of the sectional groups as the basic component for defining social change. This would allow the research team to create an accurate description of the impact area's social structure and to divide the community in groups non *a priori* defined.

6. REFERENCES

Brosio G. (1992), *Economia e finanza pubblica*, La Nuova Italia Scientifica, Roma.
Finsterbusch K. – Llewellyn L.G. – Wolf C.P. (eds., 1983), *Social Impact Assessment Methods*, Sage Publications, London.
Fusco Girard L. – P. Nijkamp (1997), *Le valutazioni per lo sviluppo sostenibile delle città e del territorio*, Franco Angeli, Milano.
Lichfield N. (1969), *Cost Benefit Analysis in Urban Expansion*, Pergamon Press, London.
Lichfield N. (1996), *Community Impact Evaluation*, London.
MacArthur J. (1997), Stakeholder analysis in project planning: origins, applications and refinements of the method, *Project Appraisal*, **12**(4): 251-265
Marradi A. (1980), *Concetti e metodi per la ricerca sociale*, La Giuntina, Firenze.
Merelli M. (1998), *Il quartiere Foro Boario di Forlì, Problemi, valutazioni, aspettative dei suoi abitanti*, Forlì.
Prizzon F. (1995), *Gli investimenti immobiliari*, Celid, Torino.
Realfonzo A. (1994), *Teoria e metodo dell'estimo urbano*, Nuova Italia Scientifica, Roma.
Rosenberg M. (1968), *The logic of Survey analysis*, Basic Books, New York.
Rostirolla P. (1992), *Ottimo economico: processi di decisione e di valutazione*, Liguori editore, Napoli.
Simonotti M. (1997), *La stima immobiliare*, Utet, Torino.
Soderstrom E.J. (1981), *Social Impact Assessment*, Praeger publishers, New York.

Stanghellini S. (ed., 1996), *Valutazione e processo di piano*, Alinea, Firenze.

Taylor N.C. – Goodrich C. – Bryan H. (1995), Issues-oriented approach to social assessment and project appraisal, *Project Appraisal*, **10**(3): 142-154 .

REINTERPRETING STRUCTURAL CHANGE IN U.S. AGRICULTURE

Philip M. Raup [*]

SUMMARY

The structure of U.S. agriculture has been profoundly changed by the growth in specialized production units in animal agriculture, by monoculture and duoculture in field crops, and by increased dependence on export markets. Land use choices are increasingly internationalized, with emphasis shifting from domestic to global markets.

The importance of equity capital in land ownership declines, as emphasis shifts to operating capital, with growing risk exposure by the commercial banking sector. The typical farm now involves both owned and rented land, reflecting heavy migration out of agriculture and parcelization of land ownership.

Traditional family farms, once diversified, participate in specialization by converting to two-earner families, with diversified income sources from both off farm work and non-farm work on the farm. This may prove to be a surprisingly durable transition.

1. INTRODUCTION

The structure of U.S. Agriculture was profoundly changed in the second half of the 20th century. In the sense used here, structure refers to the components of agricultural production and rural life, including but not limited to farm size, organization, ownership and control, management and operating systems, geographic distribution, and interrelations with farm families, households, rural communities, and national and international economies.

In the limited scope of this paper it will not be possible to explore all of these dimensions. Selected elements of structure will be emphasized, with the choice determined by a judgement regarding their importance in terms of underlying trends, and possible consequences.

The focus will be on five major characteristics of contemporary U.S. agriculture:

[*] Professor Emeritus, Department of Applied Economics, University of Minnesota.

Economic Studies on Food, Agriculture, and the Environment
Edited by Canavari *et al.*, Kluwer Academic/Plenum Publishers, 2002

185

1. Specialization and concentration in production.
2. Separation of animal agriculture from crop farming.
3. Expanded dependence on export markets for crops.
4. Demographic transition in farm family and household structure.
5. Transformation in the sources and magnitude of risk in farming.

2. A PRELUDE TO SPECIALIZATION IN AGRICULTURAL PRODUCTION

The history of agriculture in the United States covers a transition over three centuries in the driving force of markets. In its early, colonial era, agriculture was heavily dependent on exports. The colonies of the Atlantic seaboard exhibited a sharply bi-polar rural structure of large estates and plantation, existing symbiotically with struggling subsistence farms. This era was short-lived in the north, with the end in sight well before the Declaration of Independence from English rule in 1776.

It persisted in the south throughout the 19th century, despite the major disruption of the American Civil War of 1861-65. The decisive force shaping the structure of U.S. agriculture came with the availability of land for settlement, dramatized by the Louisiana purchase in 1803, and the subsequent expansion of the domestic market. The focus shifted from exports to internal supply.

The landmark policies generating and reflecting this shift were a «Preemption Act» in 1841, giving any head of family over 21 years of age the right to settle on 160 acres of land in the public domain, and to buy it without competitive bidding at the minimum price (initially $1.25 an acre). This was followed by a «Homestead Act» of 1862 which gave the land free to any settler who would live on and develop it for 5 years. The land boom was launched.

The growth of the internal market was phenomenal. In approximate terms, the U.S. population doubled from 1825 to 1850, doubled again from 1850 to 1875, and again to the outbreak of the World War of 1914-18. The rate of increase slowed in the interwar years, and especially in the 1930's, but resumed after 1945. The current U.S. estimated population of 276 million is roughly double its level at the end of the Second World War in 1945.

3. THE RISE OF SPECIALIZATION IN ANIMAL PRODUCTION

The growth in demand for the products of agriculture was both quantitative and qualitative. Per capita consumption of calorie-rich food staples steadily declined (grains, potatoes, other root crops), parallel with increases in protein-rich additions to the diet, especially animal products.

This gave impetus in farm management and research to the perfection of animal breeding, nutrition, and disease control. Much attention in recent years has been given to «miracles» in crop farming (hybrid corn, semi-dwarf wheats, high-yielding rice)–the «Green Revolution.» Less public acclaim has followed the no less revolutionary progress that has been made in the breeding, care, feeding, and health of animals. Red meats, poultry, and more recently, some types of fish, have been transformed from luxury items into foods within the economic reach of virtually all consumers.

This has not been achieved without cost.

Throughout history, a major goal of agriculture has been to escape the constraints of climate. In the animal sector in the United States this is now being achieved. A sharply declining fraction of the labor force in agriculture finds it necessary today to check the weather upon rising in the morning, before deciding what work to do.

Apart from herding and ranching, virtually all of the livestock sector is increasingly insulated from the weather: Poultry, dairying, beef-feeding, hogs. In these weather-independent sectors, it is possible to organize work by using the industrial production line as a model. Work in these sectors of agriculture is becoming a full-time job throughout the year.

One consequence of this change has been the rise of animal production and marketing under contracts. In production contracts, the producers often do not own the animals, but typically own the buildings in which they are housed. This generates new forms of risk.

One risk involves the changed role of the price system in guiding production decisions. In traditional, small scale, family-farm types of enterprises in animal production, diversified farmers have typically had more than one source of income. They may have been primarily dependent on animals, but (with the exception of dairying) if price-cost relationships were unfavorable they could usually decide to reduce animal output without a necessarily direct reduction in labor income. They had other things they could do. Adjustment was painful but possible.

In factory-type animal production, under contracts, the nominal producer becomes in fact a custodian. Decisions to adjust output to demand are transferred from the producers to those in charge of the marketing decisions in the supply chain, typically the contractors. For the producer, output decisions are transformed into those of a «go or no-go» type, i.e., full speed ahead or stop.

This results from the fact that the fixed costs of constructing and maintaining buildings, and servicing the debt in single-purpose animal enterprises are inflexible, and usually borne by the producer. Labor costs are also transformed into a type of fixed cost, in that the producer must remain in production, or is unemployed.

Adjustment of supply to demand becomes «sticky», in that contracts are usually written for several production cycles, often for two or three years, or longer. Demand-side changes in product prices are only slowly reflected back to producers. When they are, they are apt to be traumatic, in that contracts are not renewed, or are offered at prices that involve a substantial change in producer income.

The seeming income certainty of producing animals under contract may not reduce economic risk to the producer, but simply repackage it into different time-units. Short-run volatility in income may be reduced, at the cost of an increase in longer-run amplitude. We have too little experience with production of animals under contract to judge whether or not ultimate risk to the producer has increased or decreased. When producers typically are the nominal owners of the housing units, as in poultry and hogs, long-run risk has almost certainly increased.

Where production under contract predominates, conventional or «spot» markets lose their power to inform. The concept of «market price» can no longer be used as a reliable reference point, even though many marketing contracts may contain pricing formulas that are based on market price. This is one of the most serious concerns expressed by those who question the growing prevalence of farming under contract.

There is another and related concern. Not only does price reporting grow more «opaque» in sectors dominated by contracts, but at higher levels of concentration the reporting of price and production data tends to cease altogether, except at state or national levels. This is now occurring.

A distinctive characteristic of the collection and reporting of statistical data on U.S. agriculture has been the reliance on voluntary reporting by producers. They were so numerous, so widely distributed over the landscape, and so relatively small in scale that the risk of falsified reporting in order to achieve market advantage has been low.

Production under contract, and concentration in a small number of large, vertically integrated units, change the basic assumptions underlying the data collection and reporting system. From the earliest days of agricultural statistics reporting, there has been one strictly observed rule: There will be no disclosure of data on individual farms or firms. This has been enforced by observance of the «rule of three» in data reporting. For data to be reported in a disaggretagtion by regions or classes there must be at least three observations before they can be averaged and reported in any one cell in the data distribution. For most types of data, no one firm should account for more than 60 percent of the total being reported. The reasoning is that, if averaging is based on two or fewer observations, the promise of privacy is violated. The data are then omitted in the disaggregated reporting, but included in the next level of aggregation if they meet the «rule of three.»

This has long been a problem in obtaining public access to data for small regions (townships, counties), or for specialty crops or animal production (canning crops, rabbit production, goat milk). It is now becoming a problem at successively higher and higher levels of aggregation or product class. For example, statistics on meat production become proprietary data, even at the level of states or major regions, when meat packing is dominated by one or a few firms.

The resulting «empty cells» in cross tabulations of food and agricultural data are increasingly common in current data sets. When classified by farm size, for example, data on land use and production tend to disappear at any level below the state, for farms in the largest size classes. The data are included in totals, but not published in disaggregated form.

Not only are data on production, farm size, or land use becoming less «transparent» in agriculture but product or activity data are disappearing for the largest farms and firms at any level below the state. Our agricultural data are deteriorating, in published form that can be used for small-area analysis, e.g., by counties.

These changes in animal agriculture have occurred so rapidly and so recently that statistical reporting has lagged behind. The U.S. Department of Agriculture, in its *Farm Costs and Returns Survey* in 1993, published data on the percent of output from farms involved in either production or marketing contracts. The highest frequencies were in poultry (89 percent), peanuts (65 percent), dairying (48 percent), vegetable, fruit, and nursery crops (47 percent), cotton (33 percent), rice (20 percent), and cattle, hogs, and sheep (19 percent). For corn and soybeans, only 12 percent of production came from farms using marketing or production contracts, and only 7 percent of wheat.

Since 1993 events have outpaced the official statistics. Especially rapid expansion has come in hogs. Using their own surveys, *Pork* (April 2000), a leading trade journal in the hog industry, estimated that «in January 2000 74 percent of all hogs were sold under some sort of contract agreement,» compared to 64 percent in 1998, and 57 percent in 1997.

The rise of «tied producers» of animal products, in large-scale single-purpose units, has also disclosed another cost that is more difficult to quantify. This concerns the increased threat of water and air pollution from high-density animal populations. This dimension of factory-type animal production is most apparent to the general public, and has generated intense opposition to its expansion.

It is ironic to note that, in containing urban sprawl of dense human populations, one of the most effective tools of urban planners has been control of sewer line extensions and regulation of the location of septic tanks and drainage fields. A similar driving force is emerging in concentrated animal agriculture, with waste disposal as the primary focus of public concern.

This has long been a problem in factory-type poultry production and some progress has been made in reducing it to a manageable scale. The simple fact of volume of waste creates a more difficult problem with pork and beef production. The longer run consequences seem likely to be a regrouping of large-animal production in areas more remote from urban populations.

This is already well under way in beef feeding. It is estimated that over one-half of all the fed beef produced in the U.S. comes from feedlots in the high plains of Texas, Oklahoma, Kansas, Colorado, and Nebraska. These areas have some of the lowest densities of human population in the U.S.

4. SPECIALIZATION AND CONCENTRATION IN FIELD CROPS [1]

A different pattern has emerged in field crops of food and feed grains and oilseeds. This can be illustrated with data from Minnesota, historically one of the most diversified agricultural states in the U.S.

In 1950, corn (maize), wheat, and soybeans were grown on 40 percent of the total cultivated land in Minnesota. This increased to 75 percent in 1990-94, and in 1998, these crops accounted for over 85 percent of all acres of harvested cropland. Oats in 1950 occupied almost five times the acreage planted to soybeans, and the acreage in flax exceeded the acreage in soybeans. Today, oats and flax have become minor crops on Minnesota farms. The state's agricultural economy is heavily dependent on just three crops, each of which is highly sensitive to export markets (Table 1).

It has been estimated (reliable data are almost impossible to assemble) that in the 1950's some 80 percent of the maize and a higher percentage of the oats (then the two major crops in terms of land use) were fed to livestock on farms in the areas in which they were produced. Apart from wheat (Minnesota grows no cotton), there was a minor exposure to world markets. Today, and given its geographic location as the major grain producing state most distant from ocean ports, Minnesota has become one of the agricultural

---

[1] This section and the following one have been adapted from Philip M. Raup, «Dynamics and Interdependence Between Rural Real Estate and Financial Markets in the U.S.A.», a paper prepared for XXIX Incontro di Studio, Centro Studi di Estimo e di Economia Territoriale (Ce.S.E.T.), Università degli Studi di Padova, 8 October, 1999.

Table 1.   MINNESOTA: Shifts in Acres Harvested of Principal Crops, 5-Year Averages, 1950-54 through, 1990-94

Crop(s)	ITEM	1950-54	1960-64	1970-74	1980-84	1990-94
		000 Acres or Percent				
Corn	ACRES	7,721	9,235	11,156	13,761	13,669
Soybeans						
Wheat	% of Total	40.17	51.73	60.22	68.22	74.78
Barley	ACRES	6,443	4,211	3,564	2,432	1,290
Oats						
Rye	% of Total	33.52	23.59	19.24	12.06	7.06
All Hay	ACRES	3,807	3,637	3,130	2,862	2,310
	% of Total	19.81	20.37	16.90	14.19	12.64
Flax						
Sunflowers	ACRES	1,249	771	675	938	699
Sugar Beets						
Potatoes	% of Total	6.50	4.32	3.64	4.65	3.82
Dry Beans	ACRES	(a)	(a)	(a)	178	312
Sweet Corn						
Peas	% of Total				0.88	1.71
TOTAL, All	ACRES	19,200	17,854	18,525	20,171	18,280
Listed Crops	% of Total	100.0	100.0	100.0	100.0	100.0

(a) Not separately published.
Source: Minnesota Department of Agriculture, and U.S.D.A., *Minnesota Agricultural Statistics 1999*, and earlier years.

states in the U.S. most exposed to variability in world demand for grains. Agricultural land use decisions in Minnesota have been internationalized.

What have been the driving forces behind these changes? There are no simple answers, but one cause has clearly been the doubling in world trade in food and feed grains in the past half century. Coupled with domestic U.S. farm price support programs (which favored grain crops), mechanization which released farm labor, and many other interacting causes, the result has been a shift within farms in the attractiveness of grain (including soybean) production vs. livestock. This has shifted the market orientation of U.S. farms from a primary focus on domestic markets to a focus that today includes a large element of risk exposure to international markets.

In Minnesota, for example, the sale of livestock and products primarily into the domestic market accounted for 71 percent of gross farm receipts in 1950 and barely 50 percent in 1998. With variations, this history characterizes other major grain states in the Midwest. The element of volatility introduced by this expanded international market exposure has dominated the evolution of the U.S. agricultural structure for the past three decades.

Following the collapse of the USSR, the mix of major grain importing countries changed dramatically in the 1990s. As recently as 1992/93 the former Soviet Union (FSU) imported 24.3 million metric tons of wheat and 11.2 million tons of coarse grains, for a total of 35.5 million tons.

For 1999/00 the forecasts for FSU imports are 8.3 million tons of wheat and 2.0 million tons of coarse grains, for a total of 10.3 million tons. This reflects a decline in seven

Table 2.   Comparison of Trends in Wheat Imports by China, the Former Soviet Union, North Africa and the Middle East

Country or Region	1979/80[a]	1989/90[b]	1994/95[c]	1999/00[d]
	Million Metric Tons			
China	8.9	12.8	10.3	0.8
FSU	12.1	20.4	8.3	8.3
Combined	21.0	33.2	18.6	9.1
North Africa	10.1	14.2	15.9	14.9
Middle East	7.1	15.9	9.7	15.7
Combined	17.2	30.1	25.6	30.6

[a]   USDA, FAS, Foreign Agricultural Circular, Grains, FG, 28-83, September 15, 1983.
[b]   USDA, FAS, *Grain: World Markets and Trade*, FG 4-94, April 1994.
[c]   USDA, FAS, *Grain: World Markets and Trade*, FG 1-99, January 1999.
[d] Forecast in: USDA, FAS, *Grain: World Markets and Trade*, FG 05-00, May 2000.

years in FSU imports of wheat of 16.0 million tons, and a drop in coarse grain imports of 9.2 million tons, for a total decline in wheat and coarse grain imports of 25.2 million tons.

To give these figures a sense of scale, the seven-year decline in FSU imports from 1992/93 to 1999/00 is equivalent to 15.3 percent of total world wheat trade and 9.4 percent of total world trade in coarse grains, as forecast for 1999/00 (USDA, FAS, *Grain: World Markets and Trade*, FG 05-00, May 2000).

In quantity terms, over this seven-year period, the decline in annual FSU wheat imports of 16.0 million tons was slightly below average annual Canadian wheat exports, and the decline of 9.2 million tons in coarse grain imports equaled average annual Argentine exports of coarse grains. No drop in grain trade of these magnitudes has been recorded since the Second World War.

The historic admission in 1971 that the Soviet Union could not produce the meat it wanted from its own resources was ideologically devastating in the long run, but helped give it two more decades of life. It also gave the Common Agricultural Policy of the European Community an important outlet for its agricultural surpluses, especially feed quality wheat, and it supported, not solely but very prominently, the continuation for another two decades of agriculture policy in Canada and the United States, in a relatively unchanged form. By deciding to import grains in 1971/72, the Soviet Union, in a desperate attempt to avoid internal unrest, served to underwrite contemporary agricultural policy in a substantial portion of the rest of the trading world. This has now come to an end.

One consequence has been a far-reaching readjustment in world grain trade. Consider the North African states that include Morocco, Algeria, Tunisia, Libya and Egypt. Add the states of the Middle East, including especially Iraq, and Iran. The states of this contiguous area are peculiarly vulnerable to climatic shifts, are almost uniformly grain deficit, and exhibit one of the highest rates of population growth in the world. Their principal food staple is wheat, not rice. The result is that North Africa and the Middle East combined have consistently had a higher wheat import tonnage in the 1990s than the states of the former Soviet Union plus China, and they are far higher than the FSU and China in the forecasts for 1999/00 wheat imports. The data are shown in Table 2.

To summarize: The drastic reduction from 1992/93 to 1999/00 in grain imports by the states of the former Soviet Union removed 25.2 million tons from the demand side of world grain trade, an annual volume approximately equal to the total wheat exports of Canada plus the total coarse (feed) grain exports of Argentina, if averaged over those years. World trade in wheat and coarse grains declined from 206.1 million tons in 1992/93 to 187.2 million tons in 1995/96, and slowly recovered to 202.1 million tons in 1999/2000. It has taken almost a decade for the expansion of demand by other grain importing countries to replace the lost demand from the FSU. This is the basic explanation of much of the economic distress now being felt by farmers in grain surplus producing countries..

But it is not the only explanation. The more than doubling of world trade in grain in two decades, from 101 million tons in 1971/72 to peaks of 208 million tons in 1989/90 and 207 million tons in 1991/92, fostered a belief in seemingly unlimited market expansion potentials.

This was fueled by forecasts of world population growth, and by extrapolations of demand for food as developing countries raised their levels of consumption. The result has been the emergence of monoculture or duoculture in the American grain belts. In the Corn Belt, this involves primarily corn and soybeans, in the Hard Winter Wheat Belt, wheat and sorghum, in the Hard Spring Wheat regions, a virtual monoculture in spring wheat. As noted above, there has been a drastic decline in livestock keeping on grain farms. Much of the animal population on grain-producing farms today could properly be classified as companion or «hobby-farm» animals.

The opportunity for full employment in farming has fallen with the decline in animal keeping. With no animals, there has been a sharply reduced opportunity for year-round or day-to-day self-employment on the farm. This changes the status of many grain producers to that of part-time farmers, measured by the number of months in the year in which they are fully employed in farming. Specialization in field crops has increased underemployment in U.S. agriculture.

5. SPECIALIZATION, CONCENTRATION, AND LAND TENURE

One dimension of agriculture in the United States deserves emphasis. In contrast to the explosive growth in farm population from about 1820 to 1935, the countryside since the Second World War has undergone the most severe pattern of off-farm migration in its history. Millions of people have moved out of agriculture in what ranks as one of the biggest internal migrations in any country. One consequence is that the effective ownership of assets in farming is increasingly held by non-farmers. Much of the ownership of farm land moved off the farm with the migrants. This is changing the nature of the farm problem in America, especially as it affects the distribution of agricultural income and wealth.

There are several dimensions of migration out of agriculture. One has involved multiple job-holding, with one job in farming and one in some non-farming activity. The census terminology used to describe this occupational migration is «off-farm work.» Of the 1,911,859 farms enumerated by the U.S. Census of Agriculture in 1997, a total of 1,042,158 reported some work off farm in the previous year, and 709,279 or just under 40 percent reported 200 days or more. Viewed in this light, many farms have become a part of the non-farm labor support base.

This trend is not entirely due to farmers seeking additional income to augment their low family farm income. A large but difficult-to-measure fraction of those reporting off-farm work have made a choice of life styles, to permit rural instead of urban living. But they can still be legitimately included in the farming population, and classified as «part-time farmers.»

This melding of country and city is of course most characteristic of rural areas surrounding large industrial and commercial job markets. It is not surprising that, for most of the period since 1950, the U.S. state with the largest proportion of its agricultural land included in part-time farms is Michigan, the traditional home of the automobile industry.

What is surprising is that the era of electronic commerce has opened up non-farm employment opportunities for farm residents in remote areas, far outside usual commuting belts around cities. The most striking examples are in the grain-crop regions of the upper Mid-West, and Great Plains. Mechanization has expanded the area that can be managed by one operator, and the move out of livestock keeping has reduced demands on the operator's time.

Equipped with the needed computer skills, the grain farmer can assume full time responsibilities for a non-farm job that can be fulfilled without leaving home. This possibility is being achieved in a small but increasing number of remote farm communities. The essential base is a high quality communications system, by telephone or otherwise. Where that exists, the initiative for this type of on-farm performance of non-farm work has in many cases been taken by the prospective employers, for an unusual reason: They can save on industrial and commercial building costs.

If supervisory and monitoring tasks are manageable, i.e., through independent contracting, the employer of on-farm workers linked electronically with the home office can eliminate the cost of an expensive office building. The «office in the farm home» becomes a realistic possibility. Alternatively, the «office in a small town» has attraction similar to those of off-shore financial havens in the Caribbean or Channel Islands, and for the same reasons: Tax reduction or avoidance.

The unexpected consequence is that some of the more remote rural communities in one-crop regions see new potentials for diversification, through non-farm work on the farm.

A phenomenon also associated with unprecedented internal migration out of agriculture has been the increase in what have been classified in the U.S. Censuses of Agriculture as part-owner farms. A customary classification of the tenure of U.S. farms has been tripartite, as operated by full owners, part-owners, or full tenants. For the U.S. as a whole, the class of fully tenant-operated farms (the operator rents or leases all of the land farmed) has been relatively stable for 3 decades, at 11 to 13 percent of the total area of land in farms. The area operated by full owners (those who own all of the land they farm) has also been stable for three decades. It was 32.7 percent in 1978 and 33.9 percent in 1997.

The most significant change has involved farm operators who are part-owners, but are also part tenants. This today is the dominant tenure class in American agriculture. It is also the source of considerable confusion in interpreting trends in land tenure. From one perspective tenancy plays an important role in U.S. agriculture. For the U.S. as a whole, the proportion of agricultural land operated by tenants stood at 39.6 percent of all land in farms in 1978 and 40.6 percent in 1997. Full tenants in 1997 accounted for 11.6 percent, and rented land operated by part-owners for 29.0 percent. The distribution of owned and rented land in farms by tenure class is shown in Table 3, for 1978, 1987, and 1997.

Table 3. Distribution of Owned and Rented Land in Farms. United States, 1978, 1987, 1997[a]

Tenure Classification (Acres)	Acres of Land in Farms and Percent		
Owned Land in Farms	**1978**	**1987**	**1997**
Full Owners	**331,920,878**	**317,787,149**	**316,044,548**
%	32.7	33.0	33.9
Part Owners	281,452,255	244,363,607	237,660,722
%	27.7	25.3	25.5
Subtotal Owned	613,373,133	562,150,756	553,705,270
%	60.4	58.3	59.4
Rented Land in Farms			
Part Owners	279,686,464	275,450,916	270,012,622
%	27.6	28.6	29.0
Full Tenants	121,717,637	126,868,953	108,077,363
%	12.0	13.1	11.6
Subtotal Rented	401,404,101	402,319,869	378,089,985
%	39.6	41.7	40.6
Total Land in Farms	1,014,777,234	964,470,625	931,795,255

[a]   U.S. Department of Commerce, Bureau of the Census, *1982 Census of Agriculture*, Vol. 1, Part 51, Table 5, p. 173, *1987 Census of Agriculture*, Vol. 1, Part 51, Table 48, p.49; and *1997 Census of Agriculture*, Vol. 1, Part 51, Table 46, p 57. Data cover all farms with annual agricultural product sales of over $1,000.

From another perspective, the percentage of all land in farms owned by the operator has been relatively stable for the past half century. Full owners in 1997 accounted for 33.9 percent of land in farms, and owned land included in part-owner operation for 25.5 percent, making a total of 59.4 percent of all land in farms owned by the operators. For the country as a whole in 1997, for every two acres of rented land there were three acres of owned land in farms.

More relevant is the fact that land operated by full tenants, at 11.6 percent, is near its all-time low (of 11.5 percent in 1982). Read in another way, 88.4 percent of all land in farms in 1997 was in the hands of operators who own some or all of the land they farm, i.e., have the mind-set of owner-operators. This is one of the highest percentages on record.

The emergence of farm operations combining some owned and some rented land (the «part owner» class) is a direct consequence of the heavy migration out of agriculture noted above. A representative family farm today consists of a family member who has taken over the operation of the parental farm and who shares inherited ownership with siblings. Some of the land is owned outright, some is rented from siblings. With the passage of time, the farm operator may want to buy their shares, but in the meantime remains a renter. This demographic process accounts for much of the increase in the part-owner tenure class described above, and has facilitated the accompanying increase in size of farm operating units.

It also has played an often-decisive role in the market for farm land. A large part of the supply of land for possible purchase is dependent on the resolution of the process of land transfer through inheritance. We are now in a stage in this process of demographic-induced transition in which the part-owner farm plays a pivotal role in the land market.

A farm operator who already owns some farm land is in a strategically favorable position in bidding for any near-by land that is for sale. The part-owner can use owned land to increase collateral in arranging mortgage credit, and to spread fixed costs of management and household maintenance over more acres, thus lowering total unit costs of production. Part-owner operators are likely to be better informed about local soil, climate, and drainage conditions than a more distant buyer. They have become vigorous participants in the market for farm land.

Tracing this influence through an analysis of individual sales is difficult, and expensive. Proxy data are available from land market studies in Minnesota, showing that three-fourths of all sales of farm land in recent years have been to buyers living within 10 miles (15 kilometers) of the land purchased. As the quality of land increases, this distance decreases. Over 80 percent of all acres sold in the best farming regions in the 1990's were to buyers living less than 10 miles away. The market is very local, and neighbors are the principal buyers.

These changes in land tenure and the mix of farm enterprises have altered fundamentally the relationship between the roles of equity and debt capital in American agriculture. In the past, the major debt exposure of farmers had been to credit secured by land. Credit for operating (seasonal or short-term) capital had always been important, but it was for most farms a lesser fraction of total capital requirements.

Concentration and specialization in livestock production are converting animal agriculture into quasi-industrial types of organization. These operations typically require little land, and the capacity of the operator to contribute equity through land ownership is sharply reduced.

In contrast, the demand for operating capital is increased. With over half of total cost of production in agriculture now accounted for by purchased inputs and services, there is a diminishing opportunity for farm operators to capture value-added in the form of economic rent for their land and rewards for labor and management.

This is transforming the managerial and financial structure of agriculture. From its earliest days as a nation, land-based credit has been the financial backbone of American agriculture. Recognition of this fact was the earliest guide for public policy supporting agriculture, leading to federal government support for the establishment of a system of Federal Land Banks, beginning in 1917. Originally drawing on government-supplied capital, these have now been converted into privately financed institutions, owned by their member-borrowers, with policy coordinated and supervised by the federal Farm Credit System.

Beginning in the depression years of the 1930's and accelerating after the end of war in 1945, what is now the Farm Credit System added farm operating credit (production credit) to its range of farm credit activities. At its peak in 1981-84, the system held over one-third of all farm business debt and over 43 percent of farm real estate debt. Commercial banks held only 7 percent of farm debt secured by real estate mortgages. (USDA, ERS, *Agricultural Income and Finance*, AIS-71, February 1999, and earlier reports).

Following the collapse in 1981-82 of a devastating inflation in farm land prices (land prices in Corn Belt states had increased 5-fold from 1971 to 1981), and driven by the demand for farm operating capital, the supply sources for farm credit have reversed. Commercial banks in 1998 held 41 percent of all farm business debt, and the Farm Credit System only 26 percent. The most dramatic change has been in the proportion of real estate debt held by commercial banks, rising from 7 percent of the total in 1982 to over 30 percent in 1998.

This underscores the extent to which risk in farm real estate lending has shifted to the private commercial sector. Any weakening in the fortunes of farming will now have a much larger impact on the commercial credit structure than at any time in the past three decades. The exposure of the commercial banking system to farm real estate credit risk has sharply increased.

The expansion of urban-type land uses into rural areas has created a new problem in rural credit. The collapse of U.S. farm land prices after 1981 had many causes, but one prominent one was weak standards for appraisal and valuation that fed the land price boom in the 1970's. Following the collapse of farm land prices in 1981-82, renewed emphasis was placed on appraisal and valuation practices for farm land, using improved data on soil productivity, erosion susceptibility, water regime, and location with respect to agricultural markets. The focus was still on agricultural use of the land.

For a growing fraction of the area of agricultural land, and a much larger fraction of its value, the «highest and best use» of farm land, as measured in land values, is not in farming. This can be illustrated by the problem of valuing the farm home.

The vast migration out of agriculture and the decline in number of farms has left many rural communities with an over-supply of farm dwellings, and associated land and buildings. In many farming areas farm houses or farm buildings add nothing to the value of the land. They may detract, since it will be costly to demolish them and return the land to cultivation.

In contrast, in an ever-widening belt around cities, and even smaller rural towns, the dwelling and the dwelling site are the principal component of the farm land's value. This reversal can extend great distances from core cities, reaching 75 to 100 miles. In these areas of the «urban shadow», the value of land in agricultural use is often a minor factor in an appraisal for farm land credit. The sectors of housing finance and farm finance are blending.

Even in strictly agricultural uses, the problem of building valuation has acquired a new urgency. Poultry production, hog production, beef cattle feed lots, dairy production, and nursery and horticultural crops all now require specialized buildings and handling equipment. They are single-purpose structures, and have virtually no resale value except in their intended use. Paralleling the rise of monoculture in field crops has been this rise in single-purpose capital in animal production. Much of the value of rural real estate is now dependent on the market for a single product, or for single-purpose buildings. Risk in rural lending has increased.

Cost of land is still a major component of total costs of production in some types of farming. This is not true in poultry, beef-feeding, pigs, and some types of dairying. The decline in family farms has been especially concentrated in these types of farms.

Where land is still a major capital cost, the family farm retains an advantage. Families will apparently hold land at lower rates of return on capital than will corporate investors or non-family and non-farming owners. This tends to be true in farms producing grain crops (wheat, corn, soybeans, sorghum, barley), and to a lesser extent in cotton, sugar beets, tobacco, and hay and pasture lands.

Why is this so?

Capital invested in farm land is relatively immobile, and illiquid. The transaction costs in converting farm land capital into financial capital are high and the process is both cumbersome and time-consuming. The market for farm land is thin, and localized. In the major farming regions, the turnover in ownership of land in farms rarely exceeds one to two percent of the total area, in any one year. Investing in farm land is a capital commitment for the long run, with the prospect of a slow payout, or an expensive exit.

It remains to be seen whether commercial non-farm capital will find this attractive over the long run. The evidence to date is that, in grain crop farming, many non-farm corporate investors do not want to own farmland. It commits too much capital, the rate of return is low and highly variable, and it is difficult to withdraw.

The expansion of non-farm capital in agriculture in the U.S. especially since the mid-1980's, has occurred during a period of general prosperity. Capital costs have been relatively low, and farm land prices have been recovering, from the low points reached in most regions in 1986-87. The commercial banking system is more heavily committed to the fortunes of farming and to rural real estate than at any time since 1970. And the exposure of American agriculture to the vagaries of foreign markets has never been higher. The commitment of non-farm capital to agriculture has not been tested by a severe slowdown in the domestic non-farm economy. It is premature to predict the demise of the family farm, but it is clearly changing.

6. WHAT IS CHANGING THE FAMILY FARM

A distinctive feature of contemporary U.S. agriculture is that it is almost completely monetized. In the principal agricultural regions, there is very limited production for subsistence. In major farming regions, it is rare to find farm families producing food for their own use. Examples of home production that have virtually disappeared include: Home butchering of red meats, poultry and egg production, butter making, milk and cream for home use, and any processing of grains. Some subsistence production of fruit, vegetable, and «garden crops» remains, but on a sharply reduced scale. Home canning has almost become a «hobby farm» activity.

The monetization or commercialization of U.S. farming was under way throughout the 20th century, but at an uneven pace. It was retarded by the depression years of the 1930's, and by the Second World War. One indicator is that the peak in number of farms in the U.S. occurred in 1935 at 6.8 million, and the subsequent decline was modest until the 1950s. It accelerated after about 1960, falling to an estimated 1.9 million in 1997. Why?

A major reason is the changing structure of families and households, throughout the economy and especially in farming. One key to this change is the role of women. Historically, and well past the mid-20th century, women were a crucial element in diversified farming. They provided most of the care for small animal production, and virtually all of the labor in home food processing. The nostalgic image of the self-sufficient farm would have been impossible to maintain without the labor input of farm women.

Another distinctive characteristic of farming in the U.S. has always been the dispersed location of farmsteads. With random exceptions, there are few examples of village agriculture in major farming regions, in which farm homes and service buildings are grouped together rather than scattered over the landscape. This promoted a culture of independence and self-sufficiency. It also meant that there were few alternative jobs available to women and youth, other than on the farm or in the household.

This changed dramatically, with the widespread availability of the automobile and better roads. Throughout the first half of the 20th century in most agricultural regions there were massive investments in better rural roads. «Getting the farmer out of the mud,» and «farm-to-market roads» were prominent slogans. The dispersed location of farmsteads meant that it was relatively easy to mobilize political support for rural road improvement, often financed with local (country or township) revenues from the property tax. This changed with widespread use of the automobile, and the rise in revenue from taxes on motor fuels.

Rural political support was influential at state levels for the adoption (often by amending state constitutions) of the principle that taxes on motor fuels should be dedicated to the construction and maintenance of roads, streets, and highways. Motor fuel taxation in the United States occurs prominently at the state level of government. Currently in the U.S. as a whole, the average state tax on gasoline is over 19 cents, plus a federal tax of 18.4 cents.

Initial rural support for dedicated motor fuel taxes was driven by a desire to lower the cost of transporting crops and livestock to markets. As a consequence, transport has become a much more critical production and marketing component in U.S. agriculture in the past three decades. This trend has been especially apparent since completion of the Interstate Highway System in the 1970's. The contemporary structure of U.S. agriculture can be regarded as «road-dependent,» to a degree never before experienced.

A second major consequence, little discussed at the time these policies were adopted, was to expand the labor market available to rural residents. Good roads made it possible to introduce realistic estimates of the opportunity costs of rural labor, on the farm and in the household.

It is difficult to over-estimate the significance of this shift, in explaining structural change in American agriculture in the second half of the 20th century. This is especially true of the role played by rural women and rural youth. It is the major explanation for the decline in food processing at the farm level, and for the reordering of production labor responsibilities on the farm.

With rare exceptions, it was virtually unknown to pay family labor on farms prior to the Second World War. Farm youth and women were essentially outside of the remunerated labor force. This was altered dramatically with the advent of war-time labor shortages, and with the parallel impact of related changes in rural economic and social life.

These changes are too numerous to inventory in this paper, focused on an attempt to understand structural change in agriculture, but three stand out. As noted above, one was the explosive expansion in the network of all-weather rural roads. This was closely related to the consolidation and centralization of rural schools. The third was the extension of electric power grids to serve almost all farms.

The principal impact of these changes has occurred in recent decades. In 1940, only 30 percent of all farms in the U.S. had central station electric service. This doubled to 61 percent in 1947, reached 95 percent in 1957, and 98 percent in 1966 (U.S. Dept of Agriculture, *Agricultural Statistics*, 1947, 1957 and 1967).

Rural school consolidation peaked in the decades following the baby-boom of 1946-1964. Good roads made it feasible to transport children longer distances to schools, lengthening the hours that they would be unavailable for work on the farm or in the home. Larger schools led to an expansion in after-school activities. School bus transport accelerated the expectation that all youth would attend high school (grades 9 through 12) in addition to primary school (grades 1 through 8).

Nation-wide, the percentage of the total population age 25 to 29 who had completed high school almost doubled from 38.1 percent in 1940 to 75.4 in 1970, and continued to rise to 87.4 in 1997 (U.S. Dept. of Education, *Digest of Education Statistics, 1998*, Table 8). A study in 1991 of farm and non-farm residents, age 25 and over, showed that a higher proportion of farm residents had completed high school than had non-farm residents, and proportions were higher for females than for males (U.S. Bureau of the Census and U.S. Dept. of Agriculture, *Current Population Reports*, p. 20-472, August 1993).

This expansion in educational opportunities is relatively recent, and has had a greater impact on fam than on non-farm women and youth. One of the most important explanations of the changing role of the family farm in U.S. agriculture is that farm residents now know the opportunity cost of family labor on farms. The educational attainments of rural youth opened up alternative job opportunities. The work ethic fostered by participation in a family business makes workers with a farm background highly competitive in the labor market. In asking «What is changing the family farm,» it now seems more relevant to ask, «What is changing the farm family?»

One change is common to farm and non-farm businesses. While industry and farming in the United States are linked in many ways, one of the least noticed is the growing similarity between the two sectors in their struggles with problems of succession in ownership and control of assets. This seems at first glance to be an unlikely pairing.

The dominant form of organization in commerce and industry is the corporation, in which shareholders are often widely dispersed, and have limited liability for debts of the corporation. In farming, the dominant organizational structure has been individual proprietorship, with land ownership as a driving goal, and personalized responsibility for debt.

Although very different in organization and structure, many corporations and farms share a common problem: They are now moving into their third or fourth generation of inherited ownership, and many current owners want to cash in and get out. Apparently the attraction of a family business structure begins to weaken after about the third generation.

Family-owned businesses of all types, farm and non-farm, feel this pressure. The crucial problem is how to arrange an equitable buy-out of the heirs who want out by those who want to continue the family business. The key controversy typically involves the valuation of the assets of closely-held family business. Since there are usually no shares publicly traded, there is no market-determined process to derive value.

In non-farm family businesses, a frequent solution is to sell the business, or to «go public,» and offer some of the shares to non-family members. This involves painful decisions, and sometimes ends in family quarrels. There have been recent and well-publicized examples of the pain and strain involved in this devolution, involving family-held newspaper publishing and communication firms

The problem is similar in family farms. Many of the best farms have been in the same family for more than a century. Current heirs are often three or more generations removed from the original founder, are widely dispersed, and the tie to the family farm is weak. Many want to sell out. But how to agree on a price?

This is the root of one of the most intractable problems with family farms in the U.S. The interim solution in many families has been to divide ownership of the land legally, and then rent the land to the one or several family members who what to continue in farming. This accounts for much (perhaps most) of the increase in rented farm land in areas once characterized by individually owned and operated family farms. Owners are the victims of generational partitioning, and much of the land is in the hands of individuals who made little sacrifice to acquire it. This situation is inherently unstable.

But it is not hopeless. Off-farm work, and non-farm work on the farm, are currently the support base for many family farms. Two-earner families are becoming the farm family norm.

Until well after 1950, the principal prescription to reduce risk in farming was to diversify. This was prominently defined to mean mixing animal product output with field crops, and avoiding monoculture.

As emphasized above, the trend today is sharply in the opposite direction. The diversified farm is disappearing, as it was once defined. It is reemerging, with diversification redefined to mean a mixing of farm and non-farm income sources. In this rejuvenated form, the diversified family farm may prove to be a remarkable durable institution.

RISK MANAGEMENT:
FROM THE RESEARCHER TO THE FARMER

Vernon Eidman and Kent Olson [*]

SUMMARY

The agricultural industry is in the midst of a major period of structural change creating a great deal of uncertainty for producers concerning the appropriate portfolio of production and marketing activities for the future. Our challenge as academics and advisors of producers is to produce research and educational programs that will assist farmers manage risk and remain competitive in this increasingly dynamic industry environment. We briefly review the current research on risk management, the sources of risk for farmers, and the alternative strategies for farmers to manage and control their exposure to risk. We then describe the current approach to providing educational programs on risk management and the available evidence that farmers are adopting recommended strategies. Our analysis suggests more farmers are adopting tools to shift some of the short-run risk, but that they have been less receptive to adjustments in their strategy to deal with longer-run changes. We also suggest some implications that seem to follow from this overview: (1) extension education programs to assist farmers choose among the crop insurance alternatives seem to be working well, (2) educational efforts on the use of the futures and options markets are being widely accepted and the use of these markets by farmers is increasing, (3) the strategic planning approach has the most potential to help producers evaluate and manage longer-run or strategic risk but educational programs have not been well attended. We suggest three types of assistance to enhance farmer participation in strategic planning: (a) public assessment of the forces driving change and the implications of those changes for farms, (b) provision of efficiency standards or benchmarks to help farmers appraise whether they can compete with other producers, and (c) provision of multiple-day workshops on strategic planning spread over several months to allow time to reflect and prepare for each step in the strategic planning process.

---

[*] Department of Applied Economics – University of Minnesota.

Economic Studies on Food, Agriculture, and the Environment
Edited by Canavari *et al.*, Kluwer Academic/Plenum Publishers, 2002

1. INTRODUCTION

The U.S. agricultural industry is in the midst of a major period of structural change, creating a great deal of uncertainty for producers of agricultural products. Globalization of the food and fiber markets is the major force driving this change. Retailers search world wide to provide the greatest value for their customers, creating uncertainty for their traditional suppliers of agricultural products. The food industry is supplying a wider range of products to satisfy consumer demands for convenience, safety, and specific character-istics that cater to specific market segments. Information technologies are being applied to coordinate product flows and to cut costs out of the supply chain. Biotechnology is begin-ning to open new markets for crops and livestock with specific output traits, requiring supply chains that can preserve the identity of products from producer to consumer. Con-sumer demand for food products manufactured from nongenetically modified ingredients is further enhancing the use of identity preservation in marketing chains. As these tech-nological, social and economic factors interact, governments continue to shape markets through changes in commodity policy, trade policy, environmental regulations, food safety, and intellectual property rights, adding to the uncertainty producers face in mar-keting their products.

This period of structural change is resulting in wide price swings for several major commodities, and considerable uncertainty among farmers about the appropriate portfolio of production and marketing activities for the future. While farm audiences are likely to applaud anyone suggesting that the wide fluctuations in commodity prices over the past four years are due to the 1996 FAIR Act, many farmers understand that the rapid changes in market conditions are being driven by a broader set of forces. The current political climate suggests the U.S. will not return to an agricultural policy providing high guaran-teed commodity prices that reduces the risk producers face. Our challenge is to produce research and educational programs that will assist farmers manage risk and remain com-petitive in this increasingly dynamic industry environment.

In this paper, we first review the current research on risk management, the sources of risk for farmers, and the alternative strategies for farmers to manage and control their exposure to risk. In the rest of the paper, we describe the current approach being taken to provide educational programs on risk management, and to review the available evidence that farmers are adopting recommended strategies. Our analysis suggests more farmers are adopting tools to shift some of the short-run risk, but that they have been less receptive to adjustments in their strategy to deal with longer-run changes. We also suggest some im-plications that seem to follow from this overview.

2. RESEARCH ON RISK MANAGEMENT

A large body of conceptual and applied research on quantifying and managing risk in agriculture has been developed. Eidman provides a summary of the research literature and also notes some of the early educational efforts to move risk management concepts into use by farmers. A more recent review of research on decision making under uncertainty and risk aversion for agricultural producers is provided by Moschini and Hennessy. This body of literature documents the sources of risk farmers face, evaluates alternative meth-ods of quantitatively describing risky outcomes, and provides alternative decision criteria for use by farmers. It includes a rich set of empirical studies evaluating alternative strategies farmers might use to reduce and/or shift risk to another entity. This literature also provides an excellent source of hypotheses to test with respect to the decisions farmers make under risk.

The literature identifies five sources of risk for farmers. They are production, marketing, financial, political/legal, and human resources. We subdivide human resources into two parts to emphasize the importance of moral risk.

Production risk is not knowing for certain what the quantity and quality of crop yield, animal productivity, or other production will be. The major sources of production risk are weather, pests, diseases, imperfect knowledge about the technology being used, genetics, machinery efficiency and reliability, and quality of inputs.

Price or marketing risk results from unanticipated changes in the price of outputs or of inputs that may occur after the farmer has made a commitment to production. Many factors affect the supply and demand for feed and other inputs as well as products being produced. These changes result in shifts in prices, unless the producer has taken action to lock in a price for the production period.

Financial risk is the added variability of net returns to owner's equity that results from the financial obligation associated with debt financing. The use of debt capital to leverage the owner's equity multiplies the potential gain and loss to equity capital from good and poor operating performance. The major components of financial risk are the chosen level of debt financing, the uncertain interest rates and the uncertainty of loan availability.

Political/legal risk results from changes in policies and regulations that affect the farm operator and the business. The risk of changing policies and the regulations required to implement the laws may affect the availability of an input, such as a pesticide, restricts the methods used to store and apply manure, or restricts the use of land. The legal issues most commonly associated with agriculture fall into four main areas: business structure and tax and estate planning, contractual arrangements, tort liability, and statutory compliance, including environmental issues. Each of these contributes to the potential risk exposure of the farm business.

Human resource risk is common to all types of business. Disruptive changes may occur due to injury, divorce, or death of one of the principal members of the business. If people are not managed well, risk will increase due to improper operation and application of production and marketing procedures, for example. However, the proper hiring, training and supervision, and appropriate use of marketing resources are examples of using human resources to reduce risk.

Moral risk is part of human risk, but we list it separately because moral risk is due to the behavior of people outside as well as inside the firm. Moral risk results not just from the potential for corrupt and criminal behavior; it also results from the potential for devious and less than truthful behavior of individuals and companies.

Each of these sources of risk has potential uncertainty for the longer-run as well as the current period. However, empirical applications prepared estimates of risk primarily for the current production/marketing period. Furthermore, they have focused on price, production and, to some extent, financial risk. The emphasis on price, production, and interest rate risk for the current period is not surprising given the relative ease of describing these risks using available quantitative methods and the available data. Providing reliable data to quantify the other sources of uncertainty for the current period, and for all five sources for future periods is more difficult.

Many in the agricultural community argue that the longer-run sources of uncertainty have become more important over the past 15 years. Changes in government policy, along with the increased openness to world markets, growing international demand, and more global competition to supply these markets increased the importance of longer-run risk in planning the farm business. Research and educational work during the 1990s has responded by placing increasing emphasis on applying concepts from the strategic management literature to farm businesses (Miller, Boehlje, and Dobbins, 1998). Strategic management focuses attention on the longer-run forces driving change in an industry, the impact such forces are likely to have on the firm's competitive position, and changes the firm can make to sustain/enhance its competitive position. The focus of strategic management is on the uncertainties arising from the macro environment and the competitive conditions in the industry. We refer to these longer-run uncertainties as strategic risk. The sources of strategic risk are unanticipated changes in (1) the political and legal environment, (2) the social environment, (3) demographics, (4) technology, (5) the macroeconomic environment, and (6) the impact of these and other factors on the competitive environment in the industry. Some examples of strategic risks are given in Table 1.

It is easier to conceptualize the magnitude of price, production and interest rate risk for the current production/marketing period, and more tools are available to shift these risks, such as insurance, forward contracts, and options. The impact of most strategic risks on the individual business is difficult to analyze, and the strategies to manage these risks are often less familiar. What will the changing structure of the swine business mean for my ability to be competitive over time? If I sign a production contract with a price guarantee so that I can remain in the swine business, I will replace price risk with contractual risk. But what if the contractor goes bankrupt? One of the methods of analysis commonly used is what-if analysis. That is, what if the processor goes broke? Some strategies to analyze draw on flexibility, adaptability and diversification. These and other strategies to use in dealing with strategic risk are described below.

3. ALTERNATIVE STRATEGIES FOR FARMERS TO DEAL WITH RISK

Farmers have many alternative strategies they can use to manage and control their exposure to risk and make decisions in a risky world. These alternative strategies include both institutional instruments, management alternatives, and tools for strategy development. Institutional instruments are those management tools made available by the government and institutions such as insurance companies and banks for farmers to manage their risk exposure. Management alternatives are methods that farmers themselves can use to reduce and control their risk exposure. Tools for strategy development include external and internal analysis, scenario development, and crafting strategy.

Table 1.   Sources of Strategic Risk in Agriculture

Source	Examples
Government Policy/Regulation	- Enactment of an agricultural policy or emergency legislation.
	- Changes in efforts to liberalize trade.
	- Instability in foreign financial markets, or changes in trade barriers.
	- Changes in environmental regulations concerning the use of fertilizer or manure.
Social	- Changes in the acceptability of genetically modified plant and animal products in human foods.
	- Changes in what is considered humane production conditions for livestock.
	- Shifts in tastes and preferences for foods perceived to be more healthy.
Demographic	- Changes in the age and cultural mix of the population shifts the demand for various food products, and the demand for away from home consumption.
Technology	- Continued deterioration in the productivity of existing pesticides, varieties, and hybrids.
	- Availability of new varieties, pesticides, and other inputs subject to new testing requirements.
	- Development of new technologies that provide a competitive advantage for production in another region of the country, or world.
Macro-Economic	Changes in interest rates, exchange rates and policies limiting inflation.
Competitive Conditions	- Changing structure of the industry that changes your ability to be competitive.
	- Changes in availability of inputs and services needed to be competitive.
	- Changes in availability of markets for products.
	- Development of substitute products or the development of production in a new area of the country/world that can supply the market more efficiently.

Source: Based on Miller, Boehlje, and Dobbins, 1998, p. 16.

3.1.   Institutional instruments

Protecting prices is a major part of risk management and garners a major share of both farmers' time and attention. Many **marketing tools** are available from exchanges such as the Chicago Board of Trade, local elevators, and other marketing organizations. **Hedging** is trading cash price risk for basis risk. Besides knowing the market conditions, a farmer also needs to be aware of the transaction costs and potential margin calls before making the final decision to hedge or not. **Forward contracts** usually have both a fixed quantity and price in the contract. A farmer usually does not contract his/her whole crop due to the potential of low production that would require the farmer to buy on the spot market at a price probably higher than the contract price to fill the contract quantity. **Options** allow a farmer to buy a price minimum without potential margin calls; they allow a farmer to protect his/her ability to costs.

Insurance is paying someone else to take your risk. Several types of insurance are used to transfer different types of risk: crop, health, fire, life, and liability. According to some interpretations, farmers can also be said to have price insurance in terms of government loan rates in the U.S. and the resulting loan deficiency payments. The cost of purchasing options on the futures market to set a minimum price could also be viewed as a form of insurance.

The decision to buy insurance involves estimating potential losses, the probability of losses, the cost of the insurance, and both the ability to bear the loss and the impact on the business. The ability to bear the loss refers to whether it will cause business failure. While the impact on the business considers cash flow deviations and other changes that could cause undesirable deviations from the original plans, but not complete failure.

Currently, the Federal Crop Insurance Corporation (FCIC) provides several **crop insurance** programs through local crop insurance agents: Multi-Peril Crop Insurance (MPCI), Catastrophic Risk Protection (CAT), Crop Revenue Coverage (CRC), Revenue Assurance (RA), Income Protection (IP), and Group Risk Protection (GRP). These programs are described briefly after some general information.

Yield levels for all FCIC crop insurance coverage is based upon actual production history (APH), a percentage of an established county yield, or a combination of both. APH requires a minimum of four years, to a maximum of ten years, of production records. Farmers who have less than four years of production records are required to use a county or area yield for those years without actual production records.

Each of the FCIC crop insurance coverage programs has a premium and an administrative fee. These premiums and fees vary by program, county, crop, and level of coverage. Premium costs are partially subsidized by the federal government. In 1999, there was a minimum 30 percent discount in premiums as part of a USDA disaster assistance package; the discount could be more depending on the type of program chosen by the farmer.

Unit structure is an insurance coverage selection that enables the farmer to combine crops for coverage purposes. Thus a farmer could evaluate the riskiness of different crops and production areas and select a type of unit that balances the costs of coverage and risks of losses. Four types of unit structures are available: Basic, Optional, Enterprise, and Whole farm. In Basic Unit Structure, all 100% share land is grouped together in one unit. Land shared with each different landowner is another basic unit of its own. In Optional Unit Structure, acreage of a given insurable crop can be divided into separate units such as by section, rented acres grouped by landowner, owned land divided by section, fields with different planting patterns as long as the pattern does not cross section line, etc. Optional units must have separate production records. In Enterprise Unit Structure, all insurable acres of the same insurable crop are lumped together into one unit regardless of site location and rental arrangements. In Whole Farm Unit Structure, all insurable acres of all insured crops are lumped together into one unit, regardless of land location and rental arrangements. Premiums vary and obviously decrease as the unit increases in area and crops covered.

Catastrophic Risk Protection (CAT) is the minimum level of crop insurance coverage. It covers losses to crop yield only. The APH and/or County Yield are used in the coverage calculation. There is no premium for CAT coverage, but there is an administrative fee of $60 per crop per county. CAT insurance will cover yield reductions that are 50 percent or more below established yields. The payment rate will be 55% of Market Price guarantees.

Multi-Peril Crop Insurance (MPCI) provides comprehensive protection against weather-related causes of loss and certain other unavoidable perils. It protects against losses to crop yield only. Coverage is available on over 60 crops in primary production areas throughout the U.S. at 50% to 75% of the APH for the farm. An indemnity price election from 60 to 100% of the FCIC expected market price is selected at the time of purchase. Premium amounts per acre vary by crop and county and are partially subsidized by the federal government. MPCI includes the choice of either Basic or Optional unit structure.

Crop Revenue Coverage (CRC) protects against losses from both yield and price fluctuation by converting the bushel guarantee per acre to a dollar guarantee per acre. A loss results when the calculated revenue is less than the final guarantee. The difference between these two figures times the insured's share, results in a payable indemnity. Losses are based on the minimum or harvest guarantee (whichever is higher) and the calculated revenue. CRC coverage includes the choice of Basic, Optional, or Enterprise Unit structures. CRC yield coverage choices are the same as for standard MPCI, from 50% to 75% (85% in some counties). APH and/or County T Yields are used in the coverage calculation for CRC.

All price levels used in CRC calculations are average Chicago Board of Trade (CBOT) futures prices on specific contract months depending on the crop. Local elevator prices have no impact on CRC coverage guarantees or payments, nor do the farmers' actual sales. Sale of the crop is not even required.

The minimum revenue guarantee with CRC coverage is established by the March 15th closing date. That guarantee cannot be reduced, but it can be increased if harvest CBOT futures prices are higher than the base CBOT price. If the harvest CBOT price is higher than the initial base price, the higher level becomes the new revenue guarantee with CRC coverage. If the harvest CBOT price is lower than the initial base price, the bushel threshold increases because losses are determined by a revenue guarantee. The calculated production revenue is the actual production times the final calculated harvest price based on CBOT futures. It the calculated production revenue is lower than the final guaranteed revenue, an indemnity payment is made on the CRC policy.

CRC premiums are partially subsidized by the federal government. They are based upon the Base Price and remain the same even if the Harvest Price is higher than the Base Price. Administrative fees for CRC are the same as for MPCI coverage.

Revenue Assurance (RA) insurance coverage is new in Minnesota for 1999. RA is similar to CRC and provides coverage to protect against loss of revenue caused by either low yields, low prices, or a combination of both. Currently, RA coverage applies to corn and soybeans only. It is available in any percentage from 65% to 75% for Basic, Optional, and Enterprise Unit structures, and from 65% to 80% for the Whole Farm Unit structure. The calculations for APH and the use of County T yields with factors are the same for both RA and MPCI.

The basis for RA insurance coverage guarantee is the APH yield times the coverage level (%) times the projected harvest price, which is based on CBOT options contracts. RA, unlike CRC, has no maximum upward price movement for insurance coverage guarantee. Thus, the RA insurance coverage guarantee may increase during the insurance coverage period.

If the fall harvest price is greater than the projected harvest price, an indemnity payment is made. Therefore, yield variability is based on APH rules and price variability is based on CBOT options contracts. Local elevator prices, basis levels, and individual farmer's grain sale prices have no impact on guaranteed price levels for RA coverage.

RA premiums are partially subsidized by the federal government. Administrative fees for RA coverage are the same as for MPCI options.

Income Protection (IP) insurance coverage protects against reductions in gross income when either a crop's price (projected spring commodity price) falls or yield declines from early-season expectations. IP is similar to CRC and RA.

Group Risk Protection (GRP) insurance coverage protects against yield and price loss. Coverage is based on an index of expected county yield for a given crop. When the county yield for the insured crop falls below the yield level chosen by the farmer, the farmer receives a loss payment. GRP coverage is similar to MPCI except that GRP targets farmers without production records, and thus, APH.

Participating in government programs has been a major source of avoiding the risk of extremely low incomes for U.S. farmers in recent years. The Freedom to Farm Bill passed in 1996 was intended to gradually decrease government payments to farmers for income support. To do this, the former program of target price supports was replaced with fixed, declining payments until 2002 when the payments were scheduled to end. However, the bill still contains the loan program that provides farmers with price protection through loan deficiency payments when the market prices falls below the government loan rate for certain grains. This loan deficiency program does not have a maximum production limit nor a maximum payment limit.

Also, even though the 1996 bill was supposed to decrease government payments, the extremely low prices and thus low incomes starting in 1998, created enough political pressure for Congress to spend several billion dollars for farm income support. Even before the 1996 bill, participation in insurance, conservation, and other government programs became required as part of the eligibility requirements for receiving price support payments in earlier bills.

Contracts have been used in farming for a long time and in many ways: land rental agreements, marketing agreements, forward contracts, futures contracts, security agreements, loans, contracts for deed, labor contracts, contracts to produce vegetables and seed crops, contracts to delivery sugar beets as part of a processing cooperative, etc. Farmers have entered into contracts for income stability, improved efficiency, market security, and access to capital. Processors have entered into contracts to control input supplies, improve responses to consumer demand, reduce price and financial risk, and expand and diversify operations.

In the U.S. in recent years, production contracts have become increasingly common and important for livestock (especially hogs), grains to meet specific characteristics, new venture cooperatives for grain processing, new products, etc. Production contracts specify the quality and quantity of the commodity to be produced, the quality and quantity of production inputs to be supplied by the contracting firm (e.g., processor, fed mill, or other farm operation), the quality and quantity of services to be provided by the grower (i.e., contractee), and the type, magnitude, and schedule of compensation that the grower will receive. The manner in which contracts are constructed affect the legal relationship between the producer and the contractor as well as the sharing of risk and income.

Sales contracts or market-specific production contracts are very similar to forward contracts. With sales contracts, the farmer produces the crop or livestock and agrees to sell the product at harvest to the contractor. The farmer retains ownership until harvest and subsequent sale. According to legal consul, these contracts are usually subject (in the U.S.) to the provisions of the Uniform Commercial Code (UCC) relating to sales contracts. A buyer in a sales contract benefits mainly because delivery schedules are specified and thus more stable allowing the buyer to stabilize further processing or transportation and take advantage of efficiencies not available when deliveries are not stable. The seller benefits from risk reduction of finding a market and usually receives premiums above a spot-market price. The market-specific or sales contract usually transfers minimal control across stages (that is, between the producer and processor).

A *production-management contract* increases the control of the buyer (say a vegetable processor or seed company) over the production process. Production-management contracts are often used when the seller (i.e., producer), through production decisions, can affect the value of the product to the buyer, or when the seller, through marketing decisions, can affect the value of the product to the buyer. In this form of contract, buyers gain control over decisions once made in open production, such as the timing of planting, control methods, harvest timing, etc. By taking more control than assumed in a market-specific contract, the contractor or buyer usually takes on more of the producer's price and production risk and benefits from the control of the quality and timing of the product. For instance, a vegetable processor can gain control over the timing of the planting of specific seed varieties and the timing of harvest and, thus, presumably, benefits from a more stable flow of a product with a more consistent quality.

A production-management contract may be considered a *bailment* in the U.S. a bailment is the legal relationship that exists when someone else is entrusted with the possession of property, but has no ownership interest in it. A common bailment is a grain storage contract in which the elevator stores but does not own the farmer's grain. Crop production contracts structured as bailments provide the contractor with additional protection against the unauthorized distribution of seeds, crops, livestock, and genetics that remain the property of the contractor — a major concern with some of the new genetic material.

Production contracts may also be structured as *personal service contracts* or *resource-providing contracts*. These contracts specify that the producer is to provide services, not commodities, to the contractor. For example, the producer will provide the services of a hog finishing building and management knowledge to the owner of the hogs. The hogs will reside in the producer's building but the contractor will retain ownership of the hogs. According to legal consul, the UCC provisions in the U.S. relating to sales of commodities will not be applicable to a personal service contract. Some production contracts may be *leases* of facilities, especially if the contracts relate to production of livestock.

3.2. Management alternatives

Vertical integration through sole or shared ownership of businesses can be used to control the supply of inputs and access to markets. Vertical integration reduces the risk of losing input supplies and product markets, helps control input and product prices, and provides a greater degree of control over the farmers' products and income through ownership of a larger portion of the supply chain. Many view the main benefit of vertical integration as controlling and capturing a greater share of the potential value of the raw product; this view is essentially the same as reducing income risk by protecting the income of the producer or farmer.

A farmer's **choice of activities** can have a large impact on the amount of risk taken on by the farmer. **Diversification** decreases risk by choosing activities that will have good and poor years in different years. The success of diversification depends upon the correlations between the choices and the variations of the potential returns. Diversification can be done by choice of commodity, variety or breed, geographical location, time (e.g., planting date), nonfarm activities and investments, or any combination of these choices. **Crop and variety selection** can provide drought and disease resistance and other tolerances.

By **renting versus owning** land, machinery and livestock, a farmer decrease the commitment of term debt and capital and decreases risk by increasing flexibility. This is especially true when a farmer rents land and buildings compared to the magnitude of the debt required for buying. **Sharecropping instead cash renting** transfers some production and price risk from the tenant to landowner.

Retaining experts to monitor business increases management information and decreases surprises. This includes: veterinarians and crop scouts on scheduled visits, retaining a lawyer for legal concerns, and management consultants to view the business from another viewpoint. A farmer can also reduce the risk of poor decisions by organizing a Board of Advisors consisting of people with financial, legal, production, and other knowledge areas that are needed for the farm. This Board of Advisors can evaluate management's plans for the future in much the same way a large company can benefit from the wisdom of its Board of Directors.

Obtaining more and better information by subscribing to market news services, attending meetings, calling those who (should) know, and trying other methods of obtaining information (rather than relying on current sources) can reduce the risk of surprises and future unknowns.

Shortening lead times in production allows a farmer to recapture costs sooner. Examples of shorter lead times are: annual crops versus permanent crops, hogs versus cattle, and buying heavier feeders versus lighter feeders.

Learning new skills and knowledge can help a farmer achieve earlier awareness of important internal and external changes and conditions. For example, gaining a better understanding of trade issues and the countries involved can help a farmer understand the importance, potential impact, and likely direction of trade negotiations.

Environmental control can decrease risk by controlling both the microclimate for production and the macro environment of the farm. Examples of controlling the microclimate include irrigation, frost control, hot caps, plastic mulch, and greenhouses. Strip farming and contour plowing are examples of environmental control techniques for reducing the risk of losing soil productivity.

Redundancy of resources can reduce the risk of losing productive time and resources. Examples of redundancy include having two smaller tractors instead of one big tractor or a backup generator for livestock confinement buildings.

Financial control is the process of determining and implementing the necessary actions to make certain that financial plans are transferred into desired results (or as close as possible to the desired results). Financial control takes place in three parts: establishing standards, measuring performance, and taking corrective actions, if needed. Examples of financial standards include: expected interest rates, expected rates of return, timing of borrowing needs, etc. Measuring financial performance takes place in two ways: monitoring financial conditions and recording actual performance. Corrective actions can be in one of three forms: 1) rules to change the plan if conditions change (e.g., repair instead of replace a combine if interest rates increase); 2) rules to change the implementation of the financial plan (e.g., change the amount of equity capital if new sources are found); 3) change the goals and standards (e.g., change the expected interest rate as the Federal Reserve changes its policies).

The three general types of control (preliminary, concurrent, and feedback) are used in financial control also. For example, a projected cash flow statement can show the expected size and timing of needed operating loans, so preliminary control can take place by negotiating for interest rates before the need is immediate. Monitoring actual cash flow through the year and comparing it to the planned cash flow is one example of concurrent control that allows early detection of deviations from expectations and the need for implementing corrective actions. Year-end financial analysis of actual performance provides feedback information for planning for the future.

Financial reserves have been and are used to reduce the financial impact of poor years by saving in good years. This use of financial reserves is accomplished with after-tax money. The use of "cash on hand" had a much higher use among farmers than forward contracting, hedging, and diversification according to a recent USDA study (Harwood et al.). To provide additional incentives for maintaining financial reserves, the U.S. Congress is debating the creation of Farm and Ranch Risk Management (FARRM) accounts which would allow farmers to place income into these accounts before calculating tax requirements, thus reducing their taxable income by that amount in "good" years. Then, during a "bad" year, the money could be withdrawn and would be subject to taxation in that year — possibly at a lower tax rate. Currently, the proposal includes the restriction of not being able to keep the money in the FARRM for more than five years.

Flexibility in plans means a farmer is willing and able to change as new conditions and opportunities arise. Examples of flexibility include: buildings designed for flexibility not specialty, leasing/renting versus owning, and growing annual crops versus permanent crops. Flexibility also means knowing where your production and marketing decision points are and being willing to change the original plan. For example, a farmer may ask, "Do I feed my animals to a heavier weight as planned or do current market conditions indicate I should sell them now?" Flexibility can also be seen in the reversibility of decisions and commitments, such as the ability to buy back a contract if the economic environment changes after the initial decisions were made.

In actuality, **combinations of the above methods** are used. For example, one farmer may specialize in one product and use detailed forecasting and marketing tools to reduce the risk of unforeseen price changes. Another farmer may choose to be more diversified or to contract for price protection and not spend so much time and resources on forecasting the future.

3.3. Strategy development

Industry analysis is a critical part of strategic management and positioning the farm within the industry. Understanding the chosen industry and what is expected of firms within that industry is, as mentioned earlier, a critical piece of risk management in the future. Industry analysis can help a farmer understand the magnitude and trends in the external sources of strategic risk mentioned in Table 1.

Benchmarking a farm's internal production methods and costs can help a farmer determine whether he is an efficient producer compared to other producers in the industry. Understanding and keeping abreast of changes in technology and trends in costs of production will help a farmer choose the appropriate timing of investments and changes needed to maintain a competitive position. A farm that does not benchmark increases the risk of becoming obsolete and high priced compared to industry expectations.

Developing scenarios and crafting strategy. As mentioned earlier, failing to understand and adjust a business to long-term strategic risks is potentially a large source of risk. A viable option to using complicated models to make decisions is to evaluate alternative plans for the future in the light of scenarios or descriptions of the future. Rather than just choosing a set of variables in a seemingly random or professionally random process, good scenario development will help a manager develop a set of more rational, internally logical, and consistent view of the future. Developing scenarios and crafting strategy on the basis of those scenarios can help a manager (1) identify which factors and forces are and will be important, (2) focus on the forces and trends in the marketplace and other environments, (3) see the future even with imperfect information, and (4) not blindly accept one view of future.

We know how analysis takes place with our models for individual plans and scenarios, but how should a manager compare scenarios, their results, and their impacts on choices. The first step in strategy formulation is trying to decide which scenario is most likely to happen or is actually happening. Some common methods for choosing a strategy include these listed below.

Once a good set of scenarios are developed, they can be used to evaluate alternative plans for the farm business, modify these plans, develop new plans, and, finally, to choose which plan, i.e., strategy, best fits the farm, the people and the future. There are several ways in which this decision process and final strategy formulation can take place.

1) Bet on most probable scenario. Decide which scenario is the most likely to happen and craft a strategy to position the business in the best way for the events thought likely to happen. However, choosing a strategy based on only one scenario may be expensive in the sense of opportunities passed up by not following other strategies, and it adds the risk of choosing a scenario that is far from what actually occurs in the future.

2) Preserve flexibility. If two (or more) scenarios are considered to have equal or similar chances of occurring, craft a strategy that allows flexibility and ability to make adjustments as the future unfolds. Also, a manager can craft a strategy for what appears to be the most likely future scenario but still retain the ability to make strategic changes by continuing to evaluate what is happening and be ready to change if needed.

3) Influence the outcome (i.e., what happens). Use advertising and public relations to influence and change consumers' opinions and choices. If policy changes are needed, work with legislators and Congress to encourage those changes be made. This influence is, of course, assumed to be legal; some companies and persons have gotten into trouble by trying to influence outcomes in illegal ways such as price fixing, collusion, and other unlawful activities.

Combinations. Bet on the most probable scenario and try to influence the outcome, but keep an eye on the future as it unfolds and be ready to move and change if necessary.

A robust strategy is viable despite what scenario happens. However, a robust strategy may be very expensive due to the costs of remaining viable for many options, so financial goals may not be met. Also, developing a strategy of being ready to move as needed in all or several scenarios may seem robust, but the business may lose by ending up stuck in the middle.

4. EXTENSION EDUCATIONAL PROGRAM CONTENT

Our extension education programs emphasize the application of tools to deal with production and price risk for the short run, but they provide more limited coverage of strategic planning and strategic risk. Some programs in risk management education for farmers introduce strategic planning and place some emphasis on legal and human resource risks. The use of contracting to market grain and livestock has increased in recent years, and the increasing use of hired labor on the larger farm units has heightened interest in legal and human resource risk. However, it has been difficult to engage farmers in assessing the changing competitive position of their business and to develop a strategic plan that will enable them to manage strategic risk.

The extension service provides short programs, typically about two hours in length, to discuss the market outlook and considerations in pricing the major crop and livestock products during the current marketing year. These programs frequently discuss the use of futures and options, but the opportunity to offer in-depth training is limited by the time available. A second type of program, involves six days of in-depth training on the use of marketing tools, and the development of a marketing plan. This longer program also encourages farmers to form marketing clubs, groups of 15 to 25 farmers who meet regularly to learn marketing techniques and study market dynamics.

In addition to the marketing oriented programs, the extension service has begun promoting risk management education. The content of a typical risk management program is shown in Table 2. The topics listed are adapted from the website of the Farm Financial Management Center in the Department of Applied Economics, University of Minnesota. The program is intended for delivery within one day to extension educators and others working with producers. More time would be required to teach these topics to farmers.

The program begins with an overview of the sources of risk and strategies that farmers can use to manage the various sources. The second presentation describes the strategic planning process and the importance of this topic for a farm business to be successful over time. The next four segments describe price risk, the steps in developing a marketing plan, and the use of various tools to shift price risk for major commodities using contracts, options and futures markets. Then the program focus moves to the use of crop insurance to shift production risk. The final two parts of the program describe some of the major types of legal and human resource risk, and strategies farmers can use to manage these risks.

The Extension Service in several North Central States have offered intensive strategic planning workshops for farmers, but they have found few farmers willing to invest the time and effort required to complete the strategic planning task. One might argue that managers have little incentive to be concerned about their competitive position and risk when profits are strong, as they were during 1996 and 1997. If this reasoning is correct, the lower incomes of the past two years should increase producers' interest in developing

Table 2.   A Generic Risk Management Educational Program

• Managing for the New Millennium	• Provides of the major sources of risk and strategies that may be useful to manage them.
• Strategically Positioning the Farm Business	• Emphasizes the importance of both operational efficiency and strategic positioning of the business, and outlines the strategic planning procedure.
• Price Risk	• Explains the opportunities to achieve better prices than many farmers receive by following a disciplined and marketing plan.
• How to Develop a Marketing Plan	• Describes a procedure to follow in developing a market plan.
• Marketing Tools	• Describes the alternative ways to price commodities using cash sales, forward contracts, hedge to arrive contracts, options and basis contracts.
• Fundamental Analysis	• Describes how to access data on the major components of supply an demand and analyze how changes observed are expected to impact market price.
• Crop Insurance	• Describes how to choose the appropriate type of insurance to shift production risk.
• AgRisk Software	• Illustrates how to use a Windows program designed to assist corn, soybean, and wheat farmers to manage harvest-time revenue risk.
• Legal Risk	• Describes how to prepare for several of the more common areas of legal risk. These include marketing contracts, use of the most common types of insurance (property, medical, and liability), and safeguards to prevent environmental claims.
• Human Resource Risk Management	• Describes the types of human resource risks and actions to help prevent their occurrence.

Source: Adopted from Integrated Risk Management Education "Train the Trainer" Sessions, www.agrisk.umn.edu. Farm Financial Management Center, Department of Applied Economics, University of Minnesota.

strategies that will increase the competitive position of their business. However, extension educators in the North Central Region report little response to educational offerings on strategic planning throughout the past year.

5.   FARMERS REPORTED USE OF RISK MANAGEMENT STRATEGIES

An appropriate method to evaluate the progress being made in risk management education would be to survey the use of risk strategies by farmers on a periodic basis and track the changes over time. However, the available surveys track the use of only a limited number of strategies, primarily the use of crop insurance and methods of pricing products. USDA's Risk Management Agency compiles data on the use of crop insurance by state and by crop. These data indicate that crop insurance has been widely used at about the same level in both the U.S. and Minnesota each year from1995 through 1999. There has been some substitution of multi-peril crop insurance for catastrophic risk pro-

Table 3.   Farmers' Reported Use of Risk Management Strategies (% using by sales class)

Strategy	<$50,000	$50,000 - $249,999	$250,000 - $499,999	$500,000+	All U.S.
Forward Contracting	22%	51%	58%	60%	30%
Diversification	30%	46%	56%	55%	35%
Cash on Hand	65%	83%	90%	91%	70%
Hedge in Futures	17%	30%	42%	42%	27%

Source: Harwood et al.

Table 4.   Use of Forward Pricing by Minnesota Farmers, 1997

Gross Farm Income	Percent Using Forward Contracts	Percent Using Futures Contracts	Percent Using Options Contracts
$0 to $ 100,000	62	9	6
$100,001 to $250,000	74	20	15
> $250,001	84	37	32
All	74	22	18
Average Percent of Total Grain Production Contracted	35	22	19

Source: Hanson and Pederson

tection over time, indicating more producers are choosing higher levels of insurance coverage over time.

Two studies cited here indicate the use of pricing strategies by farmers. Results of the 1996 Agricultural Resource Management Study (ARMS) conducted shortly after the passage of the 1996 Farm Act provides a benchmark at the national level (Harwood et al.). The results shown in Table 3 indicate operators with sales above $250,000 are the most likely to use each of the four strategies covered in the survey. Approximately 60 % of producers with gross sales over $250,000 used forward contracting, and 42% indicated they hedge in the futures market. However, keeping cash on hand for emergencies and to take advantage of good buys was the number one strategy for all size categories.

A random sample survey of 800 farmers in southern Minnesota was conducted in March of 1997, approximately one year after the ARMS study (Hanson and Pederson). Eighty percent of the respondents indicated that at least half of their total farm sales were from grain (principally corn and soybean). About 74 % of the respondents used forward contracts, 22% used futures, and 18% used options to control price risk (Table 4). As in the federal survey, a larger proportion of the farmers in the higher sales categories used these tools. About 84 % of respondents with gross sales of $250,000 or more used forward contracts. Of that group, 37 % used futures contracts and 32 % used options. The study found that the proportion of respondents using forward pricing increased with educational level and decreased with age.

Data from the Southeastern and Southwestern Minnesota Farm Business Management Associations show an increase in the use of crop insurance between 1993 and 1995 (Figure 1). Anecdotal discussions and observations suggest this increase may be more of a response to a requirement to buy crop insurance in order to be eligible for government program benefits and less of a strategy to reduce exposure to yield risk. After 1995, the

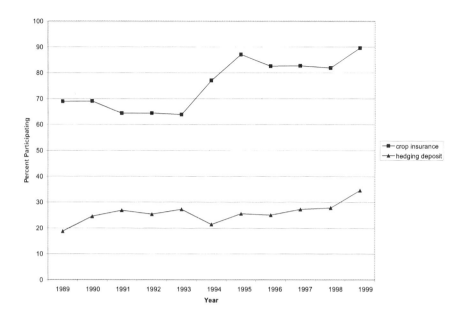

Figure 1. Farmer Participation in Crop Insurance and Hedging Activity

level of participation in crop insurance has remained at the higher level. Farmers' use of hedging accounts has remained at a relatively stable level within these two associations. The apparent increase in 1999 is, as yet, unexplained and may not indicate a higher level of use; we will await the data for 2000 to see whether this continues or not.

The authors are unable to cite survey data on the use of strategies farmers are using to manage strategic risk, but anecdotal evidence suggests farmers use strategies to manage longer-run sources of risk less consistently than they use insurance and pricing strategies to shift production and price risk.

6. IMPLICATIONS

In discussions with farmers this past fall, Olson (1999) found that farmers themselves have a broad view of risk management. As one farmer said, risk management is "protecting income, protecting resources." This farmer and most others feel a need to protect their income in both the current year and in the future. They want and use more than the traditional tools of hedging, options, and insurance. The farmers were concerned about contract evaluation and negotiation techniques as contracts become more prevalent. Personnel selection and management was important not just for production management but also for risk management. Process control was also mentioned as another set of tools needed for the future to better ensure that the actual production process will provide the product desired (physically and economically), especially under contract production. Strategic management itself was seen as a risk management tool.

There are several implications that seem to follow from this overview of our research, extension efforts to help farmers manage risk, and the observations noted by Olson.

1. Our extension education programs that were developed to assist farmers choose among the crop insurance alternatives seem to be working well.

2. The educational efforts on the use of the futures and options markets to price commodities are being widely accepted and the use of these markets by farmers is increasing. The multiple-day workshop format currently being used seems to provide training in sufficient depth for a farmer completing the training to begin using these tools in managing price risk. If the farmer also participates in a marketing club, this group can provide answers to some questions, and if the extension service provides appropriate educational backup for the marketing club, producers with this training will be able to maintain and enhance their expertise in the use of these tools to reduce price risk over time.

3. The strategic planning approach has the most potential to help producers evaluate and manage longer-run or strategic risk. However, the research and extension community will need to provide some assistance for operators of most farms to find enough time to do a careful job of strategic planning. We visualize that three types of assistance will be important to enhance farmer participation in strategic planning. (i) One of the time-consuming aspects of strategic planning is developing an understanding of the forces driving change in an industry, and assessing the implications of those changes for the opportunities and threats facing firms in the industry. This is a daunting task, particularly for farmers who are busy managing their ongoing business. And a farm producing three commodities, e.g., corn, soybean and hogs, would need an industry analysis for each of the three. This type of analysis could be provided by the public sector. (ii) Farmers need efficiency standards to help them appraise whether they can compete with other producers supplying the same market. Providing such standards requires a database that will allow producers to benchmark their operation against other farms in the region, other regions of the country, and, for some products, other producing areas of the world. Providing benchmarking for farmers that permit these comparisons is another component that will make strategic planning more attractive to farmers. (iii) We need to develop multiple-day workshops on strategic planning, with the days of training and planning spread over several months, allowing participating farmers time to reflect on the materials before they are asked to move ahead with the next step of the process. The obvious problem with this approach is the amount of resources that will be required for these workshops, and whether the workshops can be delivered in a cost-effective manner.

7. REFERENCES

Eidman V.R. (1989), *Quantifying and Managing Risk in Agriculture*, Staff Paper P89-41, Dept. of Applied Economics, University of Minnesota, St. Paul, MN.

Hanson D.K.- Pederson G. (1998), *Price Risk Management by Minnesota Farmers*, Minnesota Agricultural Economist, No. 691, Department of Applied Economics, University of Minnesota, St. Paul, MN.

Harwood J. – Heifner R. – Coble K. – Perry J. – Somwaru A. (1999), Farmers' Reported Use of Risk Management Strategies, in: *Managing Risk in Farming*, Agricultural Economic Report Number 774, ERS, USDA, Washington, D.C., pp. 59-64.

Hill C. – Jones G. (1998), *Strategic Management Theory*, Houghton Miflin Co., Boston.

Miller A. – Boehlje M. -Dobbins C. (1998), *Positioning the Farm Business,* Staff Paper 98-9, Dept. of Agricultural Economics, Purdue University.

Moschini G. – Hennessy D. (1999), *Uncertainty, Risk Aversion and Risk Management for Agricultural Producers*, Staff Paper 319, Dept. of Economics, Iowa State University.

Olson K. (1999), The Farm Manager of the Future, in: *Proceedings of the 1999 Crop Pest Management Shortcourse*, Minnesota Extension Service, University of Minnesota, St. Paul, MN, pp. 69-74.

NEW PARADIGMS IN RURAL DEVELOPMENT: SOME LESSONS FROM THE ITALIAN EXPERIENCE

Donato Romano [*]

SUMMARY

The purpose of this paper is to analyze the Italian rural development experience, emphasizing its strengths and weaknesses on one side, and its peculiarities and possibility of generalization on the other.

Such peculiarities call for a critical appraisal of the main theoretical approaches for the analysis of rural development. For example, a mere sectoral approach - i.e. based on the "rural/urban" dichotomy - does not explain satisfactorily the recent Italian experience, while the "local/regional" approach seems to explain much more. Therefore, an attempt is made to interpret Italian rural development patterns adopting the analytical categories proposed by the Italian school of industrial economics. The two main conclusions are that the development process: (*i*) is the outcome of the interplay between socio-economic variables, territorial characteristics (that is, history and geography), and institutions, and (*ii*) doesn't show a unique sequence, but it can be characterized by a plurality of organizational forms and many different development paths.

Building on this, the paper tries to answer a set of relevant questions. First is whether or not agricultural, agro-industrial, or rural local development systems do exist. If so, how can these constructs help to analyze different development patterns, i.e. how many different "agricultures" do exist in Italy? Having acknowledged this plurality of rural development patterns, and typologies and roles of agriculture within them, a fundamental question is how a given (rural) local development system does change. Different typologies of transformation are proposed (evolution, restructuring and metamorphosis) and critical points for agricultural change dynamics in the Italian context are analyzed. Finally, the consequences for the policy-makers are emphasized, focusing both on alternative strategies for local development policies at large, and for rural development policy in particular.

---

[*] University of Florence.

Economic Studies on Food, Agriculture, and the Environment
Edited by Canavari *et al.*, Kluwer Academic/Plenum Publishers, 2002

1.   THE EVOLUTION OF RURAL DEVELOPMENT POLICY IN THE EU

> Diversity is one of the main features of European agriculture. It is also becoming one of
> the keywords in the debates on Common Agricultural Policy. Any European perspective
> on rural development must be grounded on the recognition of such diversity and must
> necessarily build upon it in order to maintain the agriculture required by Europe's people
> (Long and van der Ploeg, 1994).

This quotation was reported at the very beginning of the paper that I presented at the Abano Conference (Romano, 1996). Four years later we can say that what used to be only a scholars' remark has been translated into legal acts. Agenda 2000 recognizes that the challenge of world trade globalization and the shift of focus from production to consumption affect not only agricultural markets, but also local economies in rural areas. It explicitly states that the future of the agricultural sector is closely linked to a balanced development of rural areas.

The EU rural development policy is now the "second pillar" of the Common Agricultural Policy (CAP), alongside the market measures and the requirements of a competitive European agriculture. It is aimed at meeting the various needs of the rural world, together with the expectations of overall society and sound environmental standards.

As a consequence, the new CAP is based on the following principles:

a)   the multi-functionality of agriculture, i.e. its varied role beyond the production of foodstuffs. This implies the recognition and encouragement of the services provided by farmers;

b)   a multi-sectoral and integrated approach to the rural economy in order to diversify activities, create new sources of income and employment, and protect the rural heritage;

c)   subsidiarity and decentralization, aimed at stimulating a key role by regional and local communities, and at transparency in drawing up and managing programs.

These reforms stemming from Agenda 2000 simply follow the development seen in recent years in the EU, and they seem to fit very well with most Italian rural area (see below).

2.   ECONOMIC DEVELOPMENT: THE ITALIAN EXPERIENCE

The process of economic development in Italy has been, as compared to earlier cases in other industrialized countries, slower, highly specific and more spatially differentiated. To understand this, let us summarize very briefly the most important phases in Italian development [1]:

a)   **From the creation of the Italian State (1860) to the end of the nineteen century.** In the second half of nineteenth century, Italy participated in the process of economic development initiated by the industrial revolution by exporting agricultural products to the northern European countries. This early European divi-

---

[1]   This section draws heavily on Saraceno (1992).

sion of labor lasted until the arrival of American grains. It led to the accumulation of resources in the hands of land owners but did not result in the take-off of Italian industry.

b) **From the beginning of the twentieth century to the Second World War.** The first half of the twentieth century was characterized by a low consumption equilibrium and direct state intervention in the process of industrialization (Bonelli, 1978). The government assigned to the agricultural sector the function of maintaining the growing population at subsistence levels, and thus of regulating the flow of labor into the very slow growing industrial sector. This "model" of development, which kept a high demographic pressure on land for a relatively long period of time, had a number of consequences for rural development: (1) a fragmented farm structure that still characterizes Italian agriculture; (2) the continued expansion of farming in marginal (mountain) and reclaimed land; (3) the permanence of the majority of the population in rural areas; and (4) the very low levels of consumption, in relatively poor, isolated local markets, satisfied by the supply of artisan preindustrial crafts and subsistence agriculture.

c) **The immediate postwar period.** In the 1950s and 1960s a liberalization of the economy only partly balanced previous state intervention and stimulated a rapid growth of manufacturing industries, led by exports (Graziani, 1972). This resulted in a sharp increase in the demand for labor concentrated not only in urban centers but in specific regions - the "industrial triangle", constituted by Lombardy, Liguria and Piedmont. The spatial impact of this new model of development was a significant and sudden migration towards this area of attraction, not only from rural areas but also from North Eastern, Central and Southern regions. It did not produce the social desertification of rural areas, partly because the excess population accumulated in the previous period was large and birth rates remained quite high. In addition many families had become farm owners and this reduced their willingness to migrate permanently; in fact, only higher mountain areas suffered from abandonment.

d) **From the end of the 1960s on.** Quite unexpected by policy makers and the scientific community, still another pattern was recognized, based on small and medium manufacturing enterprises, diffused among small towns and even in the countryside. This new spatial pattern could not be explained merely as the decentralization of congested urban industries in contiguous areas (i.e. filtering down) but rather as an endogenous and not necessarily dependent process, characterized by non traditional forms of cooperation between enterprises (industrial districts) and in the organization of production. Spatial systems of small and medium enterprises have achieved economies of scales comparable to those of large enterprises, operating both on the national and international markets, and have been responsible for the geographic expansion of industrial development from the industrial triangle to the Central and Northeastern regions (Fuà and Zacchia, 1983; Garofoli, 1981).

A few conclusion can be drawn from this very brief synthesis. First of all it shows that Italian development did not follow the classical sequence of stages highlighted by Rostow (1960). This is true in time series as well in cross section, e.g. diffused industrialization has not eliminated the competitiveness of urban industry; both systems are, rather, competing in the world market with the changing advantages and disadvantages that both types of locations have. If we accept the theoretical possibility that there are multiple paths to economic development, this cannot therefore be dismissed as an Italian, non-exportable, extravagance: on the contrary, the experience could contribute to development theory and its spatial implications for rural areas.

Second, rural areas have played different roles over time, e.g. first allowing for the subsistence of a growing population with slow top-down industrialization, then increasing agricultural productivity with fast urban industrialization, and finally providing capital, human resources (entrepreneurs and labor), and space for diffused industrialization. The social and economic organization inherited from that "model" appears now as a crucial factor for understanding the origins of diffused industrialization, i.e. rural areas didn't play only the classical function of foodstuff production, rather they have evolved as mixed economies.

Third, there have been important differences in the way in which rural areas have gone through the sequence of development phases just mentioned. Agrarian preindustrial relationships meant quite different things in the South, where latifundia and landless workers prevailed, and in Central Italy, where sharecropping dominated. In the South, the remittances from migration played an important role in sustaining family incomes and resulted in slower and more difficult access to peasant ownership. In Central Italy, the agricultural household became a tightly knit firm which maximized profit, even if this was quite modest, and provided one of the social bases for rural entrepreneurship. The period of intense urban industrialization of the 1950s and 1960s attracted laborers to the Northwest mostly from the rural and urban areas of the South, the Northeast, and mountains. In the last decades, diffused industrialization developed spontaneously in rural areas of the Central and Northeastern regions and only more recently in some rural areas of the South, however leaving most parts of the latter almost untouched.

It seems clear that there are two significant processes going on which are certainly affecting rural areas. First is the redistribution of employment, even more than population, in favor of medium/small towns; Second rural areas appear to be increasingly diversifying the economic activities that they offer to the population.

Such processes appear to be happening quite spontaneously, and this means that rural areas have very dynamic economies.

From the theoretical point of view, the rural/urban dichotomy (defined either by demographic size criteria, or by more refined indicators) does not seem to explain satisfactorily the recent trends. Problems of setting a threshold criteria between the two, of defining each concept, and of significant variability within each category, however defined, suggest strongly that we should reconsider why we want to use this dichotomy, since with increased diversification rural areas are becoming less agricultural. At least in the Italian case, the "local/regional" approach seems to explain much more than the "rural/urban" dichotomy.

3. ALTERNATIVE PARADIGMS FOR RURAL DEVELOPMENT ANALYSIS: SECTORAL VS. REGIONAL/LOCAL APPROACHES

The peculiarities of the Italian experience have put under stress the traditional development paradigms, which focus on spatial segregation of economic activities and on a hierarchy among economic sectors, such that we could predict a given sequence of stages of growth and define a development ranking among countries (Rostow, 1960; Geschenkron, 1962).

The classical coincidence of agriculture and rural areas on one side, and other economic activities and urban areas on the other, was not apparent in the Italian case. Or, more precisely, as stressed by Elena Saraceno,

> The rural/urban reading of spatial differentiation was meaningful when processes of urbanization and industrialization worked in the classical concentrated way that was typical of the first generation of developed countries (Saraceno, 1992: 467).

The spatial coincidence of both processes lasted only for a very short period in Italy (roughly the fifties and sixties). Then, the decline of such a coincidence (diffused industrialization, new leisure functions of rural areas, decentralization of public services) has progressively blurred the original homogeneity of these analytical categories, confusing the neat division of labor between rural and urban areas, thereby reducing its explanatory capacity.

Namely, rural areas are not a homogeneous category: they have played different roles in the economic development process, they are currently different from each other, and they are diversified, i.e. several different economic activities co-exist there.

Therefore, an alternative analytical approach has been proposed, the so-called "regional/local approach" (Saraceno, 1992), which aims at identifying compact territorial entities, functionally organized within themselves by some type of exchange (Becattini, 1979 and 1989).

Using, again, Elena Saraceno's words,

> In contrast to the rural/urban category, based on the homogeneity and non-contiguity of their spatial characteristics, the regional/local economy approach is based on the effects of heterogeneity and contiguity, being interested in describing the various forms of integration and exchange that develop among its spatial segments and sectors of activities (Saraceno, 1992: 468).

The thesis we would like to develop is that the regional/local approach is better suited for the analysis of rural development than the traditional one. Indeed, adopting the classical sectoral approach, there is no room for rural development by definition, since the latter implies economic differentiation and therefore, the areas earlier classified as rural, will become first semi-rural, then peri-urban or industrial. The problem is that rural is only defined in negative terms and it can remain rural only if it does not change or if it declines.

A second step is to analyze what the implications of adopting the regional/local approach are, both in theoretical and empirical terms, for the analysis of rural development (Romano, 2000). In other words, we'll try to verify whether concepts and models developed by the Italian School of Industrial Economics are useful in the interpretation of rural development.

4. LOCAL DEVELOPMENT SYSTEMS: THE CONTRIBUTION OF THE ITALIAN INDUSTRIAL ECONOMICS SCHOOL TO DEVELOPMENT THEORY

Italian industrial economists and policy-makers began to focus on local development patterns in the early seventies, when seminal studies on diffused industrialization, decentralized production, small and medium enterprises, etc. were published (Becattini, 1979 and 1987; Fuà and Zacchia, 1983; Garofoli, 1981; Brusco, 1989).

The studies on the "industrial districts" (Becattini, 1979; Brusco, 1989) and "peripheral development" (Fuà and Zacchia, 1983) broke away from a tradition of development models generally accepted by mainstream economics, but seldom verified in reality. Indeed, such studies challenged the principle of increasing return to scale, acknowledged the key role of the territory (i.e. of history and geography) in shaping development patterns, and suggested the existence of new determinants, not necessarily technical ones, which play a crucial role in decision making and local transformation dynamics (Becattini and Rullani, 1993; Garofoli and Mazzoni, 1994).

This marked a major shift from the traditional approach characterized by an "engineering" vision of development dynamics, toward the analysis of the relationships among development, territory and institutions. The offspring of such a shift was the acknowledgment of a plurality of organizational forms and of development paths, i.e. that it is not possible to identify "the most efficient" development path and sometimes it is difficult to even rank different patterns according to efficiency criteria.

In the last fifteen years, a hot debate on different development paths has developed in Italy, focusing on the contrast between "endogenous" and "exogenous" development models. Where the qualification of the two alternatives depends on the use of internal/external resources and knowledge transfers and on the internal/external control of accumulation and innovation processes. A general, quite robust conclusion, is that endogenous development patterns are better suited for igniting and sustaining (local) development processes.

Crucial in the definition of endogenous development patterns is the concept of a local development system (LDS), which can be defined as

> an organizational pattern of territorially based production, having strong relationships with the local socio-institutional system, and characterized by external economies. An LDS is derived from strong exchanges of commodities and information within the production system, as well as by continuous production and reproduction of specific knowledge, skills, local regulation mechanisms, and territorial specifics, which can not usually be exported to other contexts» (Garofoli e Mazzoni, 1994: 17).

Such a definition highlights the main characteristics which differentiate the local/regional approach from the traditional ones:

a) the emphasis on the main feature, intrinsically dynamic, of the development process, characterized by (social) production and reproduction of knowledge, skills and regulation mechanisms (Becattini and Rullani, 1993), and the non transferability of the process, which determine the peculiarity of each development path;

b) the emphasis on the links with the local environment, as a necessary condition for positive externalities (external to a given firm, but internal to the LDS), which determines a competitive advantage (Dei Ottati, 1987) with reference to different organization forms of production and/or territories;

c) the social nature of the development process, over and above its technical character.

Briefly, in a given LDS, any particular dimension – economic, social, territorial – cannot be isolated from the others, because of the close interplay of variables and their own mutual interdependence. Moreover, such a definition is broad enough to entail several "patterns" of local development: growth poles, company towns, industrial districts, etc.

In particular, the industrial district (ID), which has played a crucial role in Italian development, is a peculiar LDS which adds to the general characteristics of any LDS. Schematically, any ID is characterized by:

a) a given production orientation, very often characterized by a specific product;

b) spatial concentration of firms and residences;

c) presence of close and peculiar inter-industrial relations, which determine the existence of a "communitarian market" (Dei Ottati, 1987: 124);

d) presence of interpersonal networks, so that a peculiar "informational atmosphere" exists (Becattini, 1987: 47);

e) divisibility of economic processes;

f) firms which are mostly small and medium sized enterprises.

Characteristics a) through d) are common to all LDSs: namely, c) and d) determine a compact socio-economic fabric, which is the determinant of external economies (low transaction costs) and of knowledge dynamics which allow competitive advantages to emerge and to sustain the development process.

Characteristics e) through f) are typical of IDs which translate into competitive advantages stem from a specific organizational pattern: divisibility is a necessary technical condition for the "factory" organization of production processes (Georgescu-Roegen, 1982) to take place. Such an organization allow optimal use of funds, reducing idle time and lowering production costs. The presence of many small and medium enterprises means flexibility of the whole system, and organizational advantages with respect to large firms (e.g. lower costs for auxiliary processes, coordination economies, lower costs of production factors, etc., see Tani, 1987).

5.   ARE THERE "RURAL" LOCAL DEVELOPMENT SYSTEMS?

A relevant question, here, is to ascertain whether or not agricultural, agro-industrial, or rural LDSs do exist.

In the LDS literature it has been emphasized that to identify an LDS we must begin by identifying those characteristics that are determinants of competitive advantage for the LDS-candidate and that differentiate it from other territorial entities. Usually, there are three elements to look at:

a) production specialization of the local system, i.e. existence of a dominant eco-
    nomic activity and complementarity among the activities that take place within
    the LDS-candidate;
b) spatial proximity of economic activities (i.e. firms) and social units (i.e. families
    belonging to the same LDS-candidate);
c) presence of a compact social fabric, that allows a "communitarian market" and a
    peculiar "informational atmosphere" to appear.

In principle, the above mentioned criteria are verified in the rural case also, so that
the category of a local development system could be used for analyzing the rural development. The difficulty lies in the attempt to identify a "rural" local development system
(RLDS), since it is very difficult to define rural in terms of a production activities mix or,
even more difficult, in terms of production specialization of a given area.

More likely the rural qualification of a given LDS should be determined by other
means. We must recall that a given LDS is qualified by the presence of a peculiar production factor – a non-transferable collective good – which determines a long lasting
competitive advantage for the LDS. Such a good has been identified by Giacomo Becattini and Enzo Rullani as the so-called "contextual knowledge":

> Any local system integrates explicit knowledge (we call it "coded") and tacit knowledge
> (we call it "contextual")... The latter derives from the memories and the interpretation of
> personal experiences; it is essentially tacit and informal, and can be directly socialized
> only through long-lasting and costly processes of context and experience sharing (Becat-
> tini and Rullani, 1993: 29 e 36).

As stressed by Claudio Cecchi:

> A local system is defined as rural only if agriculture is the core of its contextual knowl-
> edge (Cecchi, 1998: 14).

However, such a definition does not entail all of the types of agriculture we could
find within any LDS. Indeed, it could be that a given "non-urban" territory is part of a
broader LDS, but this does not qualify such a system as rural, since agriculture is not the
key element of its contextual knowledge. In this case, agriculture is merely the
"periphery" of non-rural local systems (industrial, urban, etc.), though some agricultural
activities are carried out.

Therefore, we can accept the following definition of a rural LDS:

> A rural local development system (RLDS) is defined as an LDS whose social and eco-
> nomic environment is characterized by production and valorization activities based on
> the exploitation of renewable natural resources having a common territorial basis and
> that are the core of the system's contextual knowledge (Romano, 2000: 235).

It would be interesting to verify whether the analog of the industrial districts exist in
the rural context, i.e. what Romano (2000) called "agricultural" and "agro-industrial"
districts.

It is very hard to see how the former could exist. Indeed, recalling the indivisibility of
agricultural production processes, it is clear that an agricultural district (AD, i.e. based on
pure agricultural activities, that is without processing of agricultural products) cannot

exist. The AD can only be thought of as slightly more likely when "conventional" divisibility of the agricultural process takes place, as is the case when a market for some mechanical services develops and/or some animal husbandry activities are carried out.

However, since in reality agriculture is always associated with a certain degree of agricultural production processing, it is clear that, from a practical point of view, the AD category loses significance.

The case of agro-industrial districts (AID) is more interesting and exists whenever a RLDS is characterized by:

a)   the divisibility of the production processes,
b)   the presence of many farms and agro-industrial firms,
c)   agricultural activities representing the core of contextual knowledge,
d)   agricultural inputs to the agro-industry which are mainly local agricultural products.

The existence of an AID is more likely when agricultural and agro-industrial products have a high degree of specificity.

6.   HOW MANY AGRICULTURES ARE THERE?

The conceptualization of RLDS (and of its subsets, AD and AID) on the basis of contextual knowledge let us shed light on different phenomena in the Italian countryside. Let us provide some examples[2] (Figure 1).

The paradigmatic example of a RLDS is the Chianti region (Polidori and Romano, 1997). It contains all characteristics of an RLDS: presence of contextual knowledge based on many natural resource-based activities (agriculture, agri-tourism, wine production, etc.), many production units (farms and firms) producing typical products (Chianti wine, Tuscan "nice landscape", etc.) whose main inputs are strictly local, a self-contained labor market, acompact community fabric, and stability of relations over time. Of course, the Chianti region is also an example of AID, since the thousands of small and medium enterprises are linked to each other by an inextricable network of supplier-client relationships, thanks to the decomposability of production processes.

It should be stressed, however, that AIDs do not always show the complexity of economic activities seen in the Chianti region. This is the case in the Parmigiano-Reggiano cheese area, where although the contextual knowledge is based on agriculture, and cheese processing uses only locally produced milk (Bertolini, 1988; Giovannetti, 1988), some activities are not present. For example, while the strong role played by the Chianti landscape in attracting tourists is clear, we cannot say the same for the Parmigiano-Reggiano area.

---

[2] In Figure 1 ILDS stando for industrial local development system and SLDS stando for service local development system.

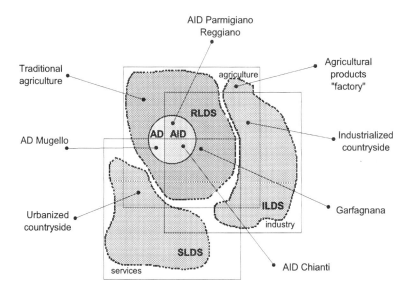

Figure 1. Economic loci of different types of agriculture, according to the local approach

The presence of agricultural activities (and services of technical assistance to agriculture) characterize the example we propose as an AD: the Mugello area (Cianferoni and Cecchi, 1987). Here, the production system is made up of many animal husbandry farms, specializing in milk/meat production. In such a case the processing phase takes place outside the Mugello area, i.e. at a Florence milk processing plant and several slaughters outside the Mugello area.

Of course, there are examples of RLDSs that are neither ADs nor AIDs. Again in Tuscany, this is the case in Garfagnana (Romano, 1989). Here an intelligent institutional capacity building activity, started by the Forest Service, has ignited a development process with valorization of local renewable natural resources (agriculture, forestry, agritourism, outdoor recreation, direct selling of locally processed agricultural products, etc.) without divisibility of the agricultural production processes (i.e. processing takes place only at farm level, using craftsmanship techniques).

On the other side, there could be local production systems that are in a regressive phase, rather than a propulsive one as is the case in standard RLDSs. This is true for most of the marginal Appennines areas, characterized by traditional organization of production. This production is based on small peasant farms that currently are not good examples of "development", and whose evolution could be characterized by either further marginalization or recovery (if they are able to exploit the opportunities offered by changes in consumption patterns).

Another relevant question is: do the agricultures that belong to an RLDS, and *a fortiori* to an AD or AID, cover the whole spectrum of current Italian agriculture? It does not seem so.

Again, it is the concept of contextual knowledge that helps in classifying other agricultural patterns which take place in "non-urban" spaces, as a consequence of joint action of both the structural transformation of capitalistic mode of production and of new consumption patterns. There are, indeed, local production systems in which some agricultural activities take place, but they are characterized by a non-agricultural contextual knowledge. Therefore, they cannot be classified as RLDSs; that is, agriculture is part of a non-rural LDS.

For example, there are areas where agriculture is "modern" agriculture, characterized by a dependency relationship (both technologically and economically) with national and international non-agricultural production sectors. Such areas are similar to an agricultural products "factory", characterized by farms integrated in the food chain but lacking organizational autonomy. A good example of such an agriculture is the sugar beet farms of the Pianura Padana, whose existence and organizational pattern depend on the contractual arrangement with the sugar processing firms.

The same can be said about the so-called "industrialized countryside" (Becattini, 1975). The classical example is the Prato textiles industrial district (Cecchi, 1988), where the contextual knowledge is an industrial one and it determines external diseconomies to the farms. In other words, the "dissonance" between the socio-economic environment and agricultural activities is such that agriculture does not fit into the complexity of activities and functions that characterize the industrial district. The relationships between agriculture and other activities are therefore conflicts rather than synergies.

Lastly, the so-called "urbanized countryside" does not conform to the rural requirements as defined above, since it shows mainly residential functions and it is characterized by an urban contextual knowledge which qualifies it as dependent development (the geographical and economic periphery of urban centers).

7. HOW DO RLDSs CHANGE?

Local systems are neither static nor self-sufficient; they react to external stimula and to internal evolutionary forces. Even more important,

> The local system *must* continuously change its inner structure, in order to react to changes in the competitive environment. Its products, its processes, its relation to external markets, its organizational patterns of production-distribution are "condemned" to be continuously modified (Becattini and Rullani, 1993: 31, emphasis added).

Let's therefore have a look at different typologies of RLDS dynamics.

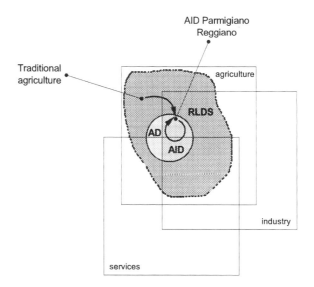

Figure 2. An example of "evolution": the case of Parmigiano-Reggiano AID

7.1.   Different typologies of RLDSs transformations

Different typologies of LDS transformations can be summarized as follows (Bellandi, 1994):

a) "evolution", typical of systems that are stable, viable and growing: the system's main production activity and its specialization sectors keep absorbing and orienting local production factors;

b) "restructuring", typical of systems where the main activity is substituted for some activities that used to be secondary: a new specialization sector grows and the specialization of the system changes;

c) "metamorphosis", typical of local systems that transform into another LDS form, for instance the evolution of an industrial district into a growth pole or a network-firm.

An example of evolution is the dynamic situation of the Parmigiano-Reggiano AID in the last decade (Figure 2), where production processes are

> locally ruled for decades through a strict production code, finalized at obtaining a high quality product which valorizes the milk produced in the area. The system is made up of some 15,000 farms, 1,000 dairy plants and many ageing/maturing (I would use the word ageing and add a footnote saying what the ageing process is designed to do such as age-ing cheese.) and marketing firms. Restructuring took place in most dairy plants, with heavy changes on the organization of production process, on plants localization and on the involvement of Consortium members. Even more important are changes on age-

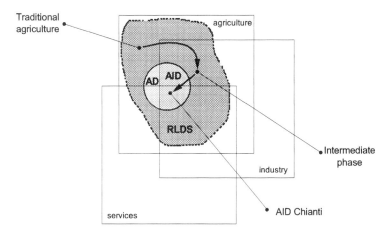

Figure 3. An example of "restructuring": the case of Chianti AID

ing/maturing and marketing firms, which have played a crucial role, because of very high investments as well as changes in their inner organization (Montresor, 2000: 200).

Restructuring is typical of the Chianti AID, that evolved from traditional agriculture to the current district form (Figure 3). Until immediate postwar years, the local production system was an agricultural one, centered on the typical Tuscany *fattoria*, which is based on sharecropping contractual arrangements. The structural transformation of the economy and the changes in consumption patterns brought about, first, a new organization of the main production system, with the development of extra-farm wine-making activities, and, more recently, the agro-tourism boom (Polidori and Romano, 1997).

An example of the metamorphosis typology is the case of poultry industry in Emilia-Romagna (Figure 4). Poultry industry in Emilia-Romagna is

characterized by a big firm and some medium enterprises (mostly cooperatives). Different segments of the poultry chain are not organized in a rigid vertical integration; many production activities take place on a WHAT IS THIS? basis, and the close intra-sectoral and inter-sectoral linkages determine the existence of a poultry district. Success determinants are the local job market, industrial relations, education system, local availability of advanced services, and collective action (Montresor, 2000: 203).

In the eighties, some major changes in the poultry sector (market power, cyclic over-production crises, changes in consumption patterns) reshaped the competitive conditions of the local system, and eventually determined the transformation of the leader firm into a global firm. This was accompanied by a metamorphosis of the local system which changed its form from district to industrial pole (Romano, 2000).

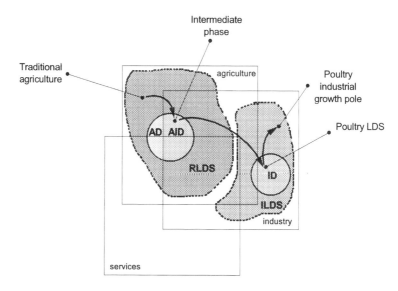

Figure 4. An example of "metamorphosis": the case of poultry LDS

7.2. Critical points of agricultural change dynamics

Critical points of agricultural change dynamics can be summarized in two different situations: traditional agriculture and what we defined as "agricultural products factory".

In the former case, the most likely path is the metamorphosis of traditional agriculture towards agricultural typologies within non-rural local systems, like the industrialized or urbanized countryside, or even the complete abandonment of agricultural activities (Figure 5). This is the case with some mountain farms, managed by aged people and characterized by scarce or null market involvement and socio-economic marginality.

In this case there is only one chance for a positive evolution, i.e. the exploitation of traditional agriculture potentials. Broadly speaking, traditional agriculture, untouched by agricultural modernization, is based on very high quality natural resources and in some cases - like in the case of Garfagnana – it could be possible to ignite endogenous rural development processes (Romano, 1996). Such restructuring could be an intermediate phase towards a more evolved form, like in the case of the Chianti AID.

The prospects of the "agricultural product factory" do not seem positive (Figure 6). Its metamorphosis toward an RLDS, though possible in principle, is not so likely due to a lack of the necessary conditions for the birth of an RLDS (high quality natural resources, compact social fabric, consistency between current contextual knowledge and that of an RLDS, etc.). It seems bound to compete on a mere production-cost basis, which forces it to adopt even more labor saving technologies. Likely outcomes could be either the survival of modern agriculture, where the gains in terms of cost savings and/or income inte-

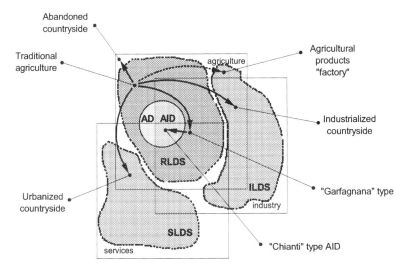

Figure 5. Likely paths of the critical point "traditional agriculture"

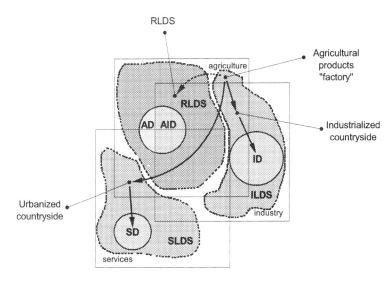

Figure 6. Likely paths of the critical point "factory of agricultural products"

gration allow it to survive, or the regression towards industrialized or urbanized country-side, not to mention the extreme case of complete abandonment of agriculture.

8. CONSEQUENCES OF ADOPTING THE REGIONAL/LOCAL APPROACH FOR THE ANALYSIS OF RURAL DEVELOPMENT

8.1. From static to dynamic competition: the concept of *milieu innovateur*

We already stressed that, in order to comply with an ever-changing competitive environment, the LDS is bound to change its inner structure. On the other hand, in order to let the LDS concept be operationally useful, it is clear that some "permanent characters" of the system should be identifiable. What is "permanent" in the LDS dynamics is its innovative capability and the complex interplay which characterizes the LDS "atmosphere". Both characteristics are subsumed in the concept of *"milieu innovateur"* proposed by regional economists (Aydalot, 1986; Camagni, 1994).

The concept of *milieu innovateur* can be viewed as

> the dynamic counterpart of similar constructs developed since the late seventies within the local, or bottom-up, approaches [3] .... What is different and novel in the [*milieu innovateur*] approach is its focus on *innovation processes, besides local efficiency factors*: imitation and "technological creation" processes, rapid reaction capabilities, resources re-allocation from declining sectors and products to new sectors, products that use the same basic know-how, regeneration and restructuring of local production fabric when hit by an external shock (Camagni, 1994: 28, emphasis added).

Schematically, any *milieu innovateur* can be identified by the following indicators (Camagni, 1994: 48):

a) an index of "local synergies", which shows local innovation potentials, through imitation processes, local actors interaction, private-public partnership on services and infrastructures projects, interaction between research centers and potential adaptors, etc.;

b) an index of "local innovation", which represents all innovative phenomena that can determine economic development, like Smithian division of labour, Arrowian learning by doing and learning by using, marshallian externalities, etc.

Cross tabulation of such indicators yields four different situations (Figure 7):

a) proper *milieu innovateur*, characterized by a high level of both local synergies and local innovation;

b) local innovation without local synergies, broadly referring to exogenous development patterns;

c) high local synergies without local innovation, which we can refer to as "potentially" innovative *milieu*; and

d) lack of both local synergies and local innovation, that is, no development.

---

[3] Among which we can recall, the concepts of "industrial district" (Becattini, 1979), "local context" (Johannison and Spilling, 1983), "local production system" (Scott and Angel, 1987) and many others based on the ideas of "bottom-up development" (Stöhr and Todling, 1977), "indigenous potential" (Ciciotti and Wettman,1981) and "flexible specialization" (Piore and Sabel, 1984).

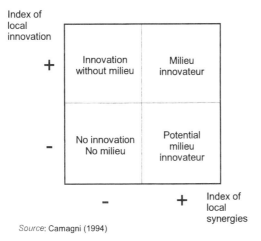

Figure 7. Identification of *milieux innovateurs*

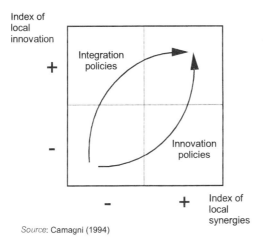

Figure 8. Alternative strategies for local development policies

This tabulation is useful from the point of view of policy design, since it allows us to identify two alternative strategies of local development (Figure 8): the first one is based on external innovation policies, that eventually integrate with the local social and production fabric (upper arrow), while the second one is founded mainly on the development of local synergies aimed at progressive upgrading of innovations and productivity (lower arrow).

The latter could be static as well as dynamic competitive factors, like efficiency, competence, flexibility, local innovation capability, local synergies, and linkages with the outside local system. These factors can be considered specific objectives of local development policies: infrastructures and cultural policies, human capital policies, compensatory policies for flexibility (for example, when unemployment increases), internal integration policies (e.g. participation projects through collective agents and facilitators) and external integration policies (e.g. inter-regional cooperation projects).

Pursuing such policies means following four strategic paths (Camagni, 1994):

a) integration of all policy instruments oriented to local environment (entrepreneurship, human capital, infrastructures, etc.);
b) geographical selection of intervention areas;
c) strengthening of already existing local know-how as well as of local production "vocations", i.e. turning specificities into assets;
d) signing cooperation and partnership agreements with firms and institutions of other regions, so that the local production system can capture the external technological and organizational know-how.

This means focusing primarily on such cases that permit the maximizing of net benefits of a given intervention. It involves focusing on potential *milieux innovateurs* and, where there is innovation without *milieu*, enriching the local environment through competence- and innovation-augmenting actions, or strengthening the linkages between external production units and the local system.

However, a crucial role should be played by institutions, since the new competitive factors can not spontaneously emerge, because: a) radical changes (those required by today's competition) are *per se* beyond the cultural horizons of the average small entrepreneur (i.e. its social culture and previous success stories could inhibit its capability to understand how crucial innovation is); b) there are economic barriers that, by and large, inhibit the access of small firms to large scale R&D and marketing; and c) new competitive generation factors (infrastructures, communication services, etc.) are *per se* beyond legal control of private actors (Bianchi, 1994).

Taking into account such a general picture, let us try to draw some conclusions for agricultural policies within RLDSs.

8.2. Consequences for RLDSs policy design

Using the local development approach has proven very useful in discriminating among different phenomena which take place in the Italian countryside. On operational grounds, this means that economic policies must either be modified, if sectoral ones, to take into account territorial specific, or be completely re-designed such that any sectoral dimension can be turned into a local one.

Taking into account such considerations, let us examine some important questions of local development policies, with special reference to agriculture:

a) the presence of many agricultures and, therefore, the need for diversified policies;
b) the need for an *ex-ante* identification of intervention areas, so that policies can be properly addressed;

c)   the eligibility of RLDSs for interventions.

8.2.1.   Many agricultures, many policies

The main conclusion of adopting the regional/local approach is that there are many agricultures that require different policies, according to their own characteristics. A first partition of the agricultural universe is the one between RLDSs and non-rural LDSs.

In the first case rurality, which can be highlighted through the existence of rural contextual knowledge, represents one of more significant phenomenon of current capitalistic restructuring in the countryside. Here, the most important feature is the intersectoral integration of activities within the local development system. This means that sectoral policies are not suited, and more general intervention policies are required, like the ones for local or rural development.

This is precisely what the EU has proposed, acknowledging the lack of the coincidence between rural areas and agriculture. What used to be "rural areas" are now subsumed under the "regions with structural problems" category, i.e. regions currently in a transition phase from the old rural development policy to socio-economic cohesion policies.

If agriculture in the RLDSs does not pose concerns, the more critical points are traditional agriculture and the factory of agricultural products, which correspond to a potentially innovative *milieu* and to an innovative environment without *milieu*, respectively.

The former needs policies aimed at upgrading traditional agriculture towards further production differentiation of the system, through the value of potentials deriving from the use of (high quality) natural resources. A possible strategy for the latter is the strengthening of forward linkages within the agro-food chain, aiming at raising the value of local agricultural products. Of course, this is not an easily transferable strategy; it is a valid option only for those situations where agricultural products are present whose quality can be effectively exploited.

On the contrary, where such a strategy cannot be pursued, it is very likely that the final stage is one of industrialized or urbanized countryside. In these situations, agriculture will be taken into account only within other policies if and only if it is consistent with the objectives of other policies (industrial, urban, etc.).

8.2.2.   The issue of identifying RLDSs

Any policy needs an *ex-ante* identification of the area upon which it should operate; that is, operational criteria aimed at identifying RLDSs are needed. Usually, Italian as well as EU institutions use the so-called "local labor markets" as proxies for RLDSs. This is probably not a sound practice because such a proxy is better suited for manufacture LDSs. They do not fit very well with the RLDSs, which are typically plurisectoral, characterized by integration and complementarity among different economic activities, i.e. the production specialization is not a good indicator for such systems.

Moreover, the "local labor markets" are based on static indicators, and they do not grasp the inner dynamic meaning of the *milieu innovateur*. Therefore, proxies for local innovation and local synergies should be identified.

Another important issue is the level at which such proxies should be evaluated, since those proxies are only relative to a given context. Those indexes should be evaluated at the regional level, since: a) there is a high variability at the national level and, therefore, assuming a national average as a benchmark will be useless; and b) the institutional architecture of Agenda 2000 assigns to regional administrations/bodies a key role in articulating and in maintaining consistency between the primary subject of programming (and its objectives) and the intervention area.

8.2.3. Are RLDSs eligible?

Last but not least is the issue of eligibility; that is, are RLDSs the right counterpart and target of economic policy? We think that the characteristics of RLDSs, or better the community living in a given RLDS, create an economic policy demand suitable to represent the needs of the area.

However, such conclusions cannot be maintained for other agricultures (i.e. the ones outside RLDSs), where interests of different social groups are highly diversified, if not conflicting. It is difficult to develop something that can represent competing interests. This calls for a strong political mediation, spatially articulated, between local demands and central supply.

Decentralization and governance are the two keywords for these situations, which calls for a strong economic programming, mainly as coordination among local demands rather than as planning acts.

The very last issue refers to the "location" of the RLDS. In other words, is a new institutional level needed? The Italian experience (L.n. 317/91) can be judged using the words of the father of industrial districts literature, Giacomo Becattini, who argues that

> an unprepared and ill-disposed policy-maker like the Italian one [lets] the industrial district be a target for industrial policy chaotically juxtaposed to the traditional ones (Becattini, 1993: 101).

This outcome must be avoided. There are objective difficulties that do not suggest the creation of a new institutional level for the RLDSs [4]. Again, it seems that the regional level, if flexible enough, can properly take into account the demands stemming from RLDSs.

9. REFERENCES

Aydalot Ph. (1986), *Milieux innovateurs en Europe*, GREMI, Paris.
Becattini G. (1975), Introduzione, in: *Lo sviluppo economico della Toscana con particolare riguardo all'industrializzazione leggera*, IRPET, Le Monnier, Firenze.
Becattini G. (1979), Dal settore industriale al distretto industriale. Alcune considerazioni sull'unità di indagine dell'economia industriale, *L'Industria. Rivista di economia e politica industriale*, **V**(1): 7-21.

---

[4] Examples of likely conflicts can be represented by the following questions: a) which relationship will the new institutional level have with the so–called "contractual programming" (area contracts, territorial pacts, program contracts, etc.)?; b) which relationship with Regions, Provinces, etc., after the so-called Bassanini act?; c) which relationships with other already existing zonings (EU structural funds, mountain and disadvantaged areas, etc.)?

Becattini G. (ed., 1987), *Mercato e forze locali: il distretto industriale*, Il Mulino, Bologna.

Becattini G. (ed., 1989), *Modelli locali di sviluppo*, Il Mulino, Bologna.

Becattini G. (1993), L'impresa distrettuale: alcune note a margine, *Economia e politica industriale*, **80**: 99-103.

Becattini G. – Rullani E. (1993), Sistema locale e mercato locale, *Economia e politica industriale*, **80**: 24-48.

Bellandi M. (1994), Struttura e cambiamento economico nei distretti industriali, in: *Sistemi produttivi locali: struttura e trasformazione*, Garofoli G. – Mazzoni R. (eds.), F. Angeli, Milano, pp. 217-238.

Bertolini P. (1988), Produzioni DOC e difesa delle economie locali: il caso del Parmigiano-Reggiano, *La Questione Agraria*, **30**: 97-122.

Bianchi G. (1994), Requiem per la Terza Italia? Sistemi territoriali di piccola impresa e transizione post-industriale, in: *Sistemi produttivi locali: struttura e trasformazione*, Garofoli G. – Mazzoni R. (eds.), F. Angeli, Milano, pp. 59-89.

Bonelli F. (1978), Il capitalismo italiano. Linee generali di interpretazione, in: *Storia d'Italia*, Annali, vol. I, Einaudi, Torino, pp. 1193-1255.

Brusco S. (1989), *Piccole imprese e distretti industriali,* Rosenberg & Sellier, Torino.

Brusco S. (1991), La genesi dell'idea di distretto industriale, in: *Distretti industriali e cooperazione fra imprese in Italia*, Pyke F. – Becattini G. – Sengenberger W. (eds.), Studi e Informazione, Quaderni, **34**: 25-34.

Camagni R.P. (1994), Il concetto di "milieu innovateur" e la sua rilevanza per le politiche pubbliche di sviluppo regionale in Europa, in: *Sistemi produttivi locali: struttura e trasformazione*, Garofoli G. – Mazzoni R. (eds.), F. Angeli, Milano, pp. 27-58.

Cecchi C. (1988), Distretto industriale: l'agricoltura dalla complementarità alla dissociazione, *La Questione Agraria,* **32**: 91-123.

Cecchi C. (1998), La ruralità nella periferia e nel sistema locale, Aestimum, **2**(Dicembre): 11-35.

Cianferoni R. – Cecchi C. (1987), Problemi e prospettive del settore lattiero in Toscana, in: Atti del Convegno su *L'uso pubblico dell'interesse privato: il caso della Centrale del latte di Firenze e Pistoia*, Firenze, 8.11.1985, Tip. Nuova Stampa, Firenze.

Ciciotti E. – Wettman E. (1981), *The Mobilization of the Indigenous Potential*, Commission of the European Communities, Internal Documentation on Regional Policy n. 10, Bruxelles.

Dei Ottati G. (1987), Il mercato comunitario, in: *Mercato e forze locali: il distretto industriale*, Becattini G. (ed.), Il Mulino, Bologna, pp. 117-141.

Fuà G. – Zacchia C. (eds., 1983), *L'industrializzazione senza fratture*, Il Mulino, Bologna.

Garofoli G. (1981), Lo sviluppo delle aree periferiche nell'economia italiana degli anni settanta, *L'industria. Rivista di economia e politica industriale*, **II**(3): 391-404.

Garofoli G. – Mazzoni R. (1994), I sistemi produttivi locali: un'introduzione, in: *Sistemi produttivi locali: struttura e trasformazione*, Garofoli G. – Mazzoni R. (eds.), F. Angeli, Milano, pp. 7-24.

Geogescu-Roegen N. (1982), *Energia e miti economici*, Boringhieri, Torino.

Geschenkron A. (1962), *Economic Backwardness in Economic Perspective*, Belknap, Cambridge.

Giovannetti E. (1988), Difesa dei sistemi regionali ed evoluzione delle forme concorrenziali o rendita di monopolio? I prodotti a denominazione di origine controllata (Doc), *La Questione Agraria*, **30**: 123-155.

Graziani A. (ed., 1972), *L'economia italiana: 1945-1970*, Il Mulino, Bologna.

Johannisson B. – Spilling O. (1983), *Strategies for Local and Regional Self-Development*, NordRefo, Oslo.

Long A. – van der Ploeg J.D. (1994), "Endogenous Development: Practices and Perspectives", in: *Born from Within. Practice and Perspectives of Endogenous Rural Development*, van der Ploeg J.D. – Long A. (eds.), Van Gorcum, Assen, pp. 1-6.

Montresor E. (2000), *I sistemi locali di produzione agro-alimentare*, in: *Secondo Rapporto sull'Agricoltura. L'agricoltura tra locale e globale. Distretti e filiere*, CNEL - Consiglio Nazionale dell'Economia e del Lavoro, Roma, pp. 179-218.

Piore M. – Sabel C. (1984), *The Second Industrial Divide: Possibilities and Prosperities*, Basic Books, New York.

Polidori R. – Romano D. (1997), Dinamica economica strutturale e sviluppo rurale endogeno: il caso del Chianti Classico, *Rivista di economia agraria*, **LII**(4): 395-427.

Romano D. (1989), *La valutazione economica dei servizi ambientali: il caso della ricreazione all'aperto*, Tesi di Dottorato di Ricerca (II ciclo) in "Economia e Pianificazione Forestale", Università degli Studi di Firenze, a.a. 1988/89.

Romano D. (1996), Endogenous Rural Development and Sustainability: An European (Non Orthodox) Perspective, *Proceedings of the Fifth Conference on "Agriculture, Food, and the Environment"*, Abano Terme (PD), 17th-18th June 1996, Working Paper WP96-4, Center for International Food and Agricultural Policy, University of Minnesota, St. Paul. Minnesota, (November, 1996).

Romano D. (2000), I sistemi locali di sviluppo rurale, in: *Secondo Rapporto sull'Agricoltura. L'agricoltura tra locale e globale. Distretti e filiere*, CNEL - Consiglio Nazionale dell'Economia e del Lavoro, Roma, pp. 221-293.

Rostow W.W. (1960), *The Stages of Economic Growth: A Non-Communist Manifesto*, Cambridge University Press, Cambridge.

Saraceno E. (1992), Alternative Readings of Spatial Differentiation: The Rural versus the Local Economy Approach in Italy, *European Review of Agricultural Economics*, **21**(3/4): 451-474.

Scott A. – Angel D.P. (1987), The U.S. Semiconductor Industry: A Locational Analysis, *Environment and Planning A*, **19**: 875-912.

Stöhr W. – Todling F. (1977), Spatial Equity: Some Anti-Theses to Current Regional Development Doctrine, *Papers of the Regional Science Association*, **38**: 33-53.

Tani P. (1987), La decomponibilità del processo produttivo, in: *Mercato e forze locali: il distretto industriale*, Becattini G. (ed.), Il Mulino, Bologna, pp. 69-92.

PRINCIPLES OF RURAL TOURISM DEVELOPMENT

William C. Gartner [*]

SUMMARY

Rural tourism development is almost as difficult to do right as it is to define what it means. In this paper a number of critical features necessary to successful rural development were examined. Some of the features mentioned such as transportation fall into the domain of the public sector. Others such as human resource training and communication technology may be the responsibility of both the public and private sectors.

A number of tourism trends were examined that relate to rural types of touristic experience. Most of them indicate growth for rural products are on the increase. However a warning sign can be seen as world growth in tourism has created awareness of its beneficial impacts. Many communities are now trying to create their own rural tourism product in order to capture a share of an expanding market. The increased competition requires that communities pay attention to the principles of tourism development if they are to be successful.

The majority of this paper has been devoted to examining ways to increase visitor flows to an area. The negative implications of that action were not discussed in detail but can be devastating for a community unable or unwilling to engage in tourism planning. It is recommended that this paper be viewed as only part of the rural tourism development process. Many more issues need to surface and be discussed before the complete tourism development process for rural destinations can be fully understood.

1. INTRODUCTION

Tourism in peripheral areas has been a focus of study for a number of years. The origins of this work can be traced to the late 1960's and early 1970's. Jafari (1988) refers to this period as the Advocacy platform. During this time most of the published works on tourism were supportive of the activity often touting its beneficial (mostly economic) impacts. In the mid 70's a spate of studies appeared that countered many of the favorable benefits of tourism development. This period has been labeled the Cautionary platform

---

[*] Department of Applied Economics, Tourism Center Director – University of Minnesota

(Jafari, 1988). It is during this time that rural based community studies began to assess some of the environmental and socio-cultural impacts resulting from unplanned or poorly planned tourism development. As a result of the criticisms brought by contributors to the Cautionary Platform new community development models began to appear. Terms such as eco-tourism, cultural tourism, green tourism etc began to appear which in some cases became major selling themes for the tourist trade. Studies proposing new models of tourism development were categorized as part of the Adaptancy platform (Jafari, 1988). Most of the new models called for less intrusive types of development, more sensitivity to local needs and a greater reliance on local capital for development. Since urban areas were already physically transformed most of the attention for these new types of tourism development was centered on rural or peripheral areas.

Demand for touristic use of rural areas has accelerated in recent years. Qualities inherent in a rural setting such as personal contact, authenticity, heritage, and individualism resonate with an increasingly urban based population (Long et al., 1994; Stokowski, 1996). Media attention on the "authenticity" of rural areas and, especially in the United States, a rural life that some see threatened by the expansion of large retailers (e.g. Wal-Mart) and global food service chains (e.g. McDonalds) and loss of the traditional rural economic base (i.e. agriculture) has led to the search for the "unspoiled" rural community. Unfortunately it is exactly the unspoiled nature of the experience that results in rapid transformation of the resource base to accommodate increasing amounts of visitors. Understanding and exploiting tourism for rural communities while trying to maintain a traditional lifestyle is a difficult process (Perry et al, 1986; OECD, 1994).

This paper will examine some of the necessary ingredients for successful rural tourism development. It will not delve too deeply into many of the problems experienced by community residents as a result of tourism development instead the reader is referred to Van der Stoep (2000), Gartner (1996) and Stokowski (2000) for more depth on this subject. In this paper some of the key ingredients for opening up rural areas to tourists will be examined. It is recognized that this approach naturally can lead to many of the problems cited by the authors who have studied tourism's impacts on host communities. However failure to understand the development process can lead to as many problems as can understanding how tourism works for rural areas but without a management plan in place to accommodate tourism's growth potential.

2.   WHAT IS RURAL?

Defining rural may seem to be an elementary exercise. Countries are fairly good at keeping accurate numbers of where people are living. Using a standard definition of the amount of people residing in a particular area is one criterion fairly commonly used to categorize communities as urban or rural. In the United States the Economic Research Service (ERS) classifies any community with less than 2500 permanent residents as rural. Communities with 2500 to 19,999 permanent residents are classified as less urbanized with any community over 20,000 classified as urban. The U.S. Census Bureau defines rural areas as all non-urban areas with urban defined as a community of over 50,000 permanent residents. Other federal agencies use other definitions for what is rural. For a more in depth review of the various classification schemes currently used to define rural the reader is referred to Flora et al (1992).

From a tourist's point of view travel to areas that have not been extensively developed for tourism can be considered rural. In that sense even urban areas that have a core untouched by tourism could be considered "rural". Conversely areas that have been heavily transformed for tourism development but have low levels of year round residents can also be considered rural. Using this definition Mediterranean resort areas such as Palma de Majorca would be classified as rural. Similarly areas in my home state of Minnesota that have become tourism magnets such as the Brainerd Lakes area with its up scale resorts and world class golf courses are considered rural even though during the high use summer months the population density exceeds what one would normally encounter in an urban community.

Long (1998) proposes a definition of rural that reflects lifestyles one is likely to encounter in a visit to a "rural" community; "Rural can be perceived as a place of safety, with solid values, surrounded by open space and natural beauty, where one is treated respectfully and friendly". In a functional sense rural can be considered a place where small scale enterprises dominate the economic scene, open space is abundant, contact with nature or "traditional societies" is offered, development is slow growing using local capital and the types of touristic activity offered varies but reflects local resource capabilities (Lane, 1994). Getz and Page (1997) argue that even local enterprises are capable of growing quickly and rural tourism is still a possibility even with rapid transformation of the physical plant.

Definitional difficulties aside it is still important to attempt to review some of the principles necessary to grow rural tourism. The level of physical plant transformation, issues of authenticity, maintenance of traditional lifestyles etc are still relevant but they will only be addressed in this paper from an awareness perspective. In other words what eventually happens in a rural community is only partially dependent on external forces and trends. The ultimate development objective should be based more on community values than other forces that may be driving the development process.

3. TRANSPORTATION

Tourism transportation systems are built on existing infrastructure. Very little transportation planning around the globe considers the relationship between tourism development and transportation systems. An historical review of tourism development will help put this into perspective.

Rural domestic tourism in the United States is heavily tied to highway infrastructure. It is estimated that U.S. domestic tourism is a multiple of 9 or 10 times the level of international tourism accounting for almost 500 million trips in 1999. Even though a great deal of attention has been paid to alliances and mergers within the airline industry it is the private automobile that accounts for over 80% of all tourism miles in the U.S. The highway system in the U.S. with the development of the interstate network occurring shortly after the end of World War II and still expanding today, coupled with an extensive system of secondary roads has made almost every rural area in the lower 48 states plus Hawaii easily accessible. However the highway system has not been utilized at the same level for international tourism as it has for domestic tourism. The most frequently visited destinations for international visitors to the U.S. have been Hawaii, Southern California, New York City and the consumption cities of Las Vegas and Orlando. Inhibiting the growth of international travel to the states rural areas has been the almost complete reliance on highways for accessibility. The lack of an extensive rail system and the vast distances between places make highways the only viable transportation link for many rural communities. The problem is further exacerbated by an airline system that prefers hub and spoke operations with very little transportation connectivity between international airports and rural communities in the states in which the airports are located. Additionally packaged tourism such as charter motorcoaches, which utilize existing highways, although important have not reached the same level of development as found in Europe.

European rural tourism has some advantages for the international market over that found in the U.S. Firstly the rail system in Europe is much more developed allowing for greater transportation options for tourists. Secondly the trade has engaged in a great deal of vertical and horizontal integration in the last decade offering many more packaged options to rural areas than one finds in the U.S.. Finally the amount of holiday time made available to workers throughout Europe is at least 3 times that in the U.S. and Asia allowing for a greater volume of trips, of both the short and long haul variety, to be consumed by European residents.

Developing nations face an even greater challenge to developing their rural tourism products than found for developed countries. Since they are developing countries the level of domestic tourism is usually quite limited requiring the focus be placed on international visitors. More often than not their transportation system is not at the same level, with respect to quality or quantity of the infrastructure, as that encountered in developed countries. To overcome these difficulties most developing country rural tourism relies on the trade to develop packages to bring people to their area. This sets up dependency relationships that often limit community control over the types of products offered and promoted. To understand this relationship and its implications it is necessary to review tourism distribution systems.

4. DISTRIBUTION SYSTEMS

Tourism products require different distribution systems than for conventional trade goods. The most important differences are:

1. An inverted distribution channel. Instead of product flowing from manufacturer to distributor to consumer the tourist must come to the source of product production,
2. Tourism products are produced and consumed at the same time. There is no shelf life for the tourism product and all consumption is individually unique,
3. Tourism, as a luxury good and one that can not be pre-tested, is heavily dependent on external forces, such as regional conflicts or internal political problems, which can have a major impact on use in the short term.

Because of the product differences noted above the tourism trade has developed an extensive distribution system for moving large amounts of tourists. Given where the product is located there may be as many as five separate business entities engaged in the distribution channel. A typical arrangement for a tourist wishing to travel to a remote region of the world may look like that depicted in Figure 1.

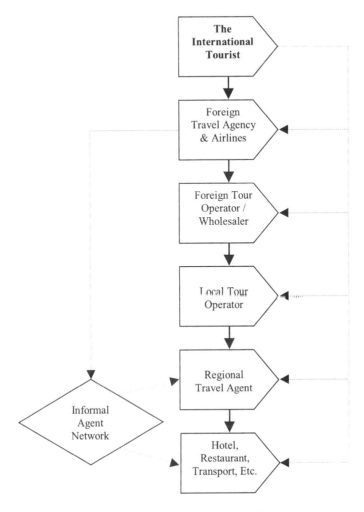

Figure 1. International Tourism Distribution Channel

Travel agents in developed countries will sell packages that have been developed and packaged for them. They do not control the product but act as the retail sales agent for it. Travel wholesalers in the developed country will arrange packages based on what they can either put together from their sources in the destination area or more likely will buy packages from wholesalers or agents in the destination. Travel agents in the destination will sell to wholesalers in the market or sell to the Free Independent Traveler (FIT) who has arrived in the developing country and is searching for travel products. Travel agents in the developing country will sell products assembled by in country tour operators or wholesalers. Most of the travel agents and wholesalers in developing countries will be located in urban centers. For travel packages to rural areas wholesalers and/or travel agents will purchase packages from rural based tour operators/agents that deal with local service providers (e.g. hotels). For each member in the distribution channel a fee for service must be charged to cover expenses which in many cases greatly increases cost for tourists. In addition the reliance on the distribution channel for local, rural area service providers sets up dependency relationships that are very hard to break. In many cases local service providers are threatened with loss of business if they try to circumvent key parts of the distribution channel.

The tourism trade has recognized the cost implications and dependency relationships inherent in the distribution channel and through consolidation has begun to squeeze cost out of the system. In Europe where the consolidation and integration wave is reaching its zenith many companies are trying to secure control over rural destinations by acquiring retail operations and wholesale packaging companies, purchasing airlines and other forms of transportation and in some cases purchasing the local service operations (i.e. hotels) (Cavlek, 2000). This may be good news for the companies that control many features of the tourism products but could be bad news for consumers as choices may become limited. It could also be very bad news for the destination if a critical mass of local service providers simply become part of a multi-national concern. How then do rural areas maintain their own identity and break away from the dependency relationships inherent in the distribution channel?

5. TECHNOLOGY

Rapidly changing communications technology may be one of the key ingredients for changing how rural areas engage in tourism development. The adoption of WEB based systems is rapidly changing distribution channel relationships that have existed for years. An essential feature of WEB based systems is the ability for producer to connect with consumer even though the two maybe thousands of miles apart. However embracing the new technology and thus eliminating many of the intermediaries in the distribution channel is not as easy as it may appear. First there is no evidence that tourists are using WEB based systems to search for potential destination options. There is a great deal of evidence that once a destination is chosen WEB based systems are used to learn more about the area and possibly to select services and products for consumption.

The number one touristic use of WEB based systems to date is for the purchase of airline tickets. Second would be to secure accommodations. Therefore it becomes critical that destination images be analyzed and changed when necessary to move destinations into tourists' perceptual map of holiday places. (For an extensive review of how images are formed and how rural areas may manipulate them to their advantage the reader is referred to Gartner, 1996). This is much easier to accomplish with developed countries that have engaged in positive image formation for many years. For developing countries, which are usually short of cash for positive image formation purposes, the trade is viewed as critical to creating favorable place images. This again reinforces the dependency relationships described above. Using new technology to overcome the dependency relationships would require that an area WEB site be constructed that fairly represents all local service providers and is able to direct business directly to them. Given that most developing county service providers are not using WEB based systems now, either individually or collectively, there is a technical training gap that first must be overcome before communication technology can be effectively used to deal with local problems inherent in the presently constructed distribution system. Even with the freedom of operation the new technology brings with it there are still distribution issues to be resolved. Registration with search engines and alliances with providers that are known by the market still require some dependency relationships and the problems that come with them.

The use of WEB based systems is, primarily, to attract the FIT tourist. Packaged/group tourism will still depend on the trade to assemble all or most of the pieces of a holiday package for those tourists who do not wish to create their own travel product. For example it is highly unlikely that Asian markets would be prime candidates for using Web based systems to learn more about rural destination choices as they exhibit group travel behavior and have low levels of internet access (<10% of Japanese consumers now use the WEB for product purchases). North American markets on the other hand are those most likely to utilize WEB based systems as they are more FIT prone than either the Asian or European markets and exhibit high levels of Internet connectivity (> 60% of U.S. households are connected to the Internet) (Buhalis, 2000). However the development and use of WEB based systems is viewed as an essential ingredient for successful rural tourism development for all markets. One trend noted is that all tourist markets are going through a rapid period of evolution. Even though Japanese tourists are still likely to rely on the trade for travel advice there is evidence that younger Japanese travelers are behaving in a more individualistic manner. Developing WEB based systems now means that as other markets evolve the rural destination will be ready to meet their trip planning needs.

6. HUMAN RESOURCES

The World Tourism Organization has identified human resources as one of the most important elements in providing quality tourism products. Education and training are essential to establishing a labor force ready to meet the challenges of an expanding, knowledgeable, and value conscious tourist market. Skills required for trade in tourism are different than for those required for trade in other commodities. Much of the current educational effort worldwide has been on developing customer service skills that are necessary for any retail service operation. However education and training is needed in areas such as marketing, planning, operations, technology and interpretation to name a few. In developed nations these types of skills are readily available in established curricula at the technical skills level as well as in secondary educational institutions. Even though the focus of training at these institutions is generally not devoted to tourism supplementary education by businesses and trade organizations plus on the job skill development have been the primary means of acquiring tourism specific skills. Developing countries on the other hand are often faced with educational systems that are based on an old liberal arts model. These types of institutions are often not ready to deal with the "new economy" and the types of skills required by employees in a service industry setting. The lack of appropriate vocational school education is a compounding factor.

Earlier in this paper it was mentioned how rapidly communications technology has evolved and how important it is to know how to use these systems if only to free oneself from the dependency relationships present in the distribution system. Perdue (2000) discusses how new technology is being used by Colorado ski resorts to provide better customer service and at the same time reduce their own labor requirements. Airlines are following essentially the same path by embracing ticketless travel and increasing self - service options that serve the dual purpose of lowering labor costs and increasing customer satisfaction. Although labor shortages are not usually a problem in developing nations, such as they are in the states, the new technology does allow a service provider to increase customer satisfaction by reducing the chance for human error. In a self-service operation most mistakes are usually borne by the consumer rather than the service provider. The operational changes discussed by Perdue are clear indications of where the service part of tourism is headed. Rural area service providers interested in reducing transactional costs of money and time for a more value conscious consumer would be wise to explore the use of this new technology. Of course to do so would require an upgrade of educational offerings currently in place in the home country.

7. DEVELOPMENT ORIENTATION

There are a number of approaches rural communities can take to achieve market awareness of their products. Apart from numerous studies on impacts related to certain types of rural developments such as gaming (see Long, 1998 and Stokowski, 1996) or the acceptance of tourism over time very little has been done to compare generic types of development orientation with respect to economic development achieved balanced against negative impacts created. One of the earliest and most comprehensive studies of this type was performed by Lew (1989). In that study he looked at three development models which he described as Unplanned, Thematic and Historic Preservation. The Unplanned option, which essentially results from the strategic location of a community as a rural service area attracting its customer base from surrounding communities, resulted in the highest level of economic impact. This makes sense as the reason for the community's economic base is the presence of large retailers with a wide market area. However long term socio-cultural impacts were viewed as potentially high for this type of development orientation as there was no sense of place created for local residents. Tourism, for the Unplanned development orientation, was not viewed as an essential ingredient for local community development as tourists would simply be attracted to the area for the same reasons as rural residents in the market area are, which is to purchase needed goods.

The Thematic development orientation results in a tourism centered type of community that uses a central theme to organize its community look. Normally these communities will impose design standards on local business to ensure the penetration of the theme. If the theme is not authentic, which for example occurs when a north woods location uses an alpine theme for its community look, the chances for socio-cultural impacts to occur is high. Countering that is the economic impact that can come from theming. Thematic development can attract numerous tourists who spend time in the area thus increasing economic returns to the community. However the total economic impact does not appear to be as high as for a service area (unplanned) community of similar size.

The third development orientation is referred to simply as Historic Preservation. Where an opportunity exists to develop the community based on its historic past the opportunity to lower socio-cultural impacts is present. Using what exists with respect to architecture, important sites or other markers of the communities historic past are the basis for the Historic Preservation development orientation. Historic Preservation communities may not achieve economic impact levels as high as for the Thematic or Unplanned communities but they do create a strong sense of place and consequently pride on the part of local community residents.

The choice of development orientations is important when discussing rural tourism development if only because the nature of physical development is such that once mistakes are made they will pose problems for years to come. Development is often unidirectional with costs of re-development so high that they generally only occur when markets have eroded. Much more work on development orientation and its consequences is required to guide future rural community tourism development

8. MARKET TRENDS

There are number of market trends that indicate both positive and negative signs for rural tourism development. Some such as consolidation have already been discussed. Others with positive indications for rural tourism development include:

1. growing interest in heritage, tradition, authenticity and rural life,
2. taking multiple holidays per year with a desire to take a second short break in a rural area,
3. increasing health consciousness giving a positive appeal to rural lifestyles and values such as fresh air, activity opportunities and stress-free situations,
4. market interest in high performance outdoor equipment from clothing to all terrain bikes and high tech climbing equipment,
5. search for solitude and relaxation in a quiet natural place, and
6. an aging but active population retiring earlier but living and travelling far into old age (Long and Lane, 2000)

Worldwide demand for products such as found in a rural setting are on the increase. Cultural, nature, and adventure based tourism demand is expected to increase by a rate of 10-30% in the next five years. However when one reviews the definitions for what constitutes, for example, a culturally based trip it is not clear how much of the increase is due to demand or simply confusion over what comprises culture (Richards, 2000). Home stays and visits to museums, essentially different products, are both included in the current popular definition of cultural tourism.

Although most trends for rural tourism are positive there are few warning signs on the horizon. Probably the most significant, apart from consolidation within the trade, is an oversupply of rural tourism attractions. Richards (2000) argues that the supply of cultural attractions worldwide is on the increase so much so that supply is outpacing demand. Even though word tourism arrivals continue to increase at approximately 4% annually the number of communities and rural businesses embracing tourism development is increasing at a faster rate. One notable sub-trend is that tourist receipts are increasing at a faster rate than international arrivals. This implies that tourists are "buying up" rather than traveling more. Rural destinations that engage in a tourism development process should be aware that tourists are not simply paying more money for the same products they have consumed in the past. Rather they are expecting increasing levels of product quality or an expanded menu of activity options. They are willing to pay for increased value but are rural communities willing and able to provide it? To do so will require a thorough review of how tourists access the destination, how local service providers access the market, quality and quantity of product offering and needed training for service providers.

9. CASE STUDY

Puno, Peru is located on the shores of Lake Titicaca, the worlds highest navigable lake, at 3,800+ meters above sea level. Puno is the base for tourists who wish to explore the lakes indigenous island communities. Length of stay in the Puno area is only 1.4 days which is hardly enough time to do more than visit two of the lakes rural communities. The majority (90%) of visitors to the area take a boat from the main marina in Puno and visit the floating reed islands constructed by the Uros indians. On these islands tourists can catch a glimpse of how people live in the reed houses, complete with solar units, but for the most part the visit consists of nothing more than a opportunity to acquire souvenirs. Average length of stay on the islands is less than 30 minutes and the opportunity for cultural exchange is hampered by the inability of local residents to communicate in any other language but Spanish.

The second most visited site on the Peruvian side of the lake is the island community of Taquile. Approximately 40% of the 83,000 international visitors to the Department of Puno will visit Taquile. Taquile offers overnight stays in local homes using a rotating system to distribute tourists. Approximately 4,000 tourists elect the home stay option annually. The other tourists observe how local residents dress and of course buy souvenirs. There are some indications that this culturally based tourism attraction is beginning to fragment. The rotating home stay system is being circumvented by some tour operators who have struck under the table deals with local residents to house their clients with particular families. This is causing some concern in the community. Secondly overcharging at local restaurants is becoming more common to the point where it is mentioned in a popular guide book as a problem. Both the circumventing of the rotating home stay system and overcharging in restaurants is indicative of a rural tourism product that was originally advertised as a cultural immersion experience becoming more of a cultural commodity.

Another island community not far from Taquile is Amantani. Amantani is visited by only 10% of the tourists who arrive in the Department of Puno. Reasons for the low level of visitation are directly tied to transportation. Even though Amantani is close to Taquile there are no boats operating, except by charter, between the two islands. Instead both islands are accessed from the marina in Puno. Consequently the level of tourism development on Amantani is so low that visitors to the island question whether other tourists have preceded them. The only visible sign of tourism is a community crafts outlet. Overnight stays based on the rotating home stay system are in place at Amantani but unless a tourist can speak Spanish or their stay is booked in advance by a tour operator accessing the home stay option poses significant problems. In twenty years of tourism on the island no island resident has ever become a guide. This is another indication of a strong dependency relationship existing between the trade and local service providers.

Suasis is a private resort development at the north end of the lake. The setting, as well as the design of the lodge, make it a unique property able to cater to a discerning clientele. Although it satisfies demand that is unmet elsewhere in the Department of Puno it suffers from lack of market awareness and access difficulties. There are two options to access the resort at the present time. The first is by boat which takes at least 4 hours from Puno and costs so much to effectively increase the price of lodging by 50%. The second is by ground and a short ride to the island by row boat. Distance from the airport is over 4 hours by land and from Puno over 5 hours. The lodge at Suasi represents a prime example of how even high quality products can suffer from lack of decent transportation options and limited market awareness.

The final lake attraction on the Peruvian side is a new cultural tourism attraction on the island community of Anapia. At the present time only 3 day 2 night packages are offered by one tour operator/travel agent in Puno. In fact it is the work that this one tour operator has completed which makes the attraction even possible. She has worked with and continues to do so with all community leaders to make certain quality standards are maintained. Of all the island offerings this is the only one that offers a true cultural immersion experience.

Lake Titicaca has numerous cultural based attractions that have been available for tourist visitation for years. Yet only two are on the tourist path. Part of the reason for this has to do with how the trade has organized its clients itineraries. Dependency relationships do exist between island communities and members of the trade. As discussed for Taquile some members of the trade are trying to circumvent a locally prescribed way to distribute overnight guests so that all community residents may benefit from this type of tourism. Lack of market awareness and the difficulty in accessing some communities poses additional problems for further tourism development Because of the dependency relationships that have developed over time it will be very difficult for destination communities to offer alternative tourist products even when they are of very high quality. The failure to adopt new communications technology, the lack of a skilled labor force, and limited and costly transportation options all serve to limit the extent of tourism development for island communities on the Peruvian side of Lake Titicaca.

10. REFERENCES

Buhalis D. (2000), Trends in Information Technology and Tourism, in: *Trends in Outdoor Recreation, Leisure, and Tourism*, Gartner W.C. – Lime D.L., eds., CABI, London, pp. 47-62.

Cavlek N. (2000), The Role of Tour Operators in the Travel Distribution System, in: *Trends in Recreation Leisure, and Tourism*, Gartner W.C. – Lime D.L., eds., CABI, London, pp. 325-334.

Flora J. – Spears L. – Flora L. – Weinberg M. (1992), *Rural Communities: Legacy and Change*, Westview Press, Boulder, CO, USA.

Gartner W.C. (1996), *Tourism Development: Principles, Processes and Policies*, John Wiley, New York.

Getz D. – Page S. (1997), Conclusions and Implications for Rural Business Development, in: *The Business of Rural Tourism: International Perspectives*, Getz, D. – Page S., eds., International Thompson Business Press, London, pp. 191-205.

Jafari J. (1988), *Retrospective and Prospective Views on Tourism as a Field of Study*, paper presented at the 1988 meeting of the Academy of Leisure Sciences, Indianapolis, Indiana.

Lane B. (1994), What is Rural Tourism, *Journal of Sustainable Tourism*, **2**: 7-21.

Lew A. (1989), Authenticity and Sense of Place in the Tourism Development Experience of Older Retail Districts, *Journal of Travel Research*, **27**(4): 22.

Long P. – Clark J. – Liston D. (1994), *Win, Lose or Draw? Gambling with America's Small Towns*, The Aspen Institute, Washington, D.C.

Long P. (1998), *Rural Tourism Foundation Information Piece*, University of Colorado, Boulder.

Long P. – Lane B. (2000), Rural Tourism Development in Trends, in: *Trends in Outdoor Recreation, Leisure, and Tourism*, Gartner W.C. – Lime D.L., eds., CABI, London, pp. 299-308 .

OECD (1994), Tourism Strategies and Rural Development, in: *Tourism Policy and International Tourism*, OECD, Paris, pp. 13-75.

Perdue R. (2000), Service Quality in Resort Settings; Trends in the Application of Information Technology System, in: *Trends in Outdoor Recreation, Leisure, and Tourism*, Gartner W.C. – Lime D.L., eds., CABI, London, pp. 357-364.

Perry R. – Dean K. – Brown B. (1986), *Counterurbanization: International Case Studies*, Geo Books, Norwich.

Richards G. (2000), Cultural Tourism: Challenges for Management and Marketing, in: *Trends in Outdoor Recreation, Leisure, and Tourism*, Gartner W.C. – Lime D.L., eds., CABI, London, pp. 187-196.

Stokowski P. (1996), *Riches and Regrets: Betting on Gambling in Two Colorado Mountain Towns*, University Press of Colorado, Niwot, Colorado.

Stokowski P. (2000), Assessing Social Impacts of Resource-based Recreation and Tourism, in: *Trends in Outdoor Recreation, Leisure, and Tourism*, Gartner W.C. – Lime D.L., eds., CABI, London, pp. 265-274.

van der Stoep G. (2000), Community Tourism Development, in: *Trends in Outdoor Recreation, Leisure, and Tourism*, Gartner W.C. – Lime D.L., eds., CABI, London, pp. 309-322.

ENVIRONMENT & MARKETS

MARKET-BASED MECHANISMS FOR ENVIRONMENTAL IMPROVEMENT

Jay S. Coggins and Paolo Rosato [*]

SUMMARY

Market-based mechanism for pollution control are becoming more popular both in the environmental economics literature and in the real-world policymaking. This paper contains a review of the ideas behind these mechanisms. A brief example illustrates the appeal of both effluent taxes and permit-trading schemes, and compares and contrasts the two. This is followed by a review of some of the market-based schemes in use around the world. The paper finishes with a summary of performance in the U.S. sulphur dioxide allowance-trading market.

1. INTRODUCTION

There may be few matters about which economists agree more thoroughly than that pollution can be controlled more cheaply by using market-based instruments than by using alternatives that do not rely upon market incentives. In this paper a market-based pollution-control instrument refers to any policy measure in which polluting firms are given some latitude in making their own abatement decision based upon market signals of some kind. The idea is at least 65 years old, dating to Pigou (1932), who explored the possibility of applying a tax to each unit of emissions. The advantages of such a tax, Pigou argued, are that polluters are able to decide on their own how much to pollute and how to achieve their chosen level of abatement, and that the tax raises revenue for the public treasury. No regulator can know as much about each firm's operation to impose the optimal distribution of abatement across polluting firms. More recently, Dales (1968) went a step further, proposing that a market be created for the pollution right itself. Under Dales' proposal a regulator would issue coupons or permits, each granting its owner the right to emit a given amount of same pollutant. If all polluters were allowed to trade the permits, Dales argued, the resulting equilibrium in the permit market would automatically

---

[*] Jay S. Coggins , Department of Applied Economics – University of Minnesota. Paolo Rosato, Department of Civil Engineering, University of Trieste.

achieve the cost-minimizing distribution of abatement across firms. The idea was powerful, but because his paper lacked a formal justification it cried out for a more formal theoretical foundation. This was provided by Montgomery (1972), who proved that a permit market will have an equilibrium and that the equilibrium is indeed efficient or cost-minimizing.

These two approaches (emissions taxes and permit markets) are the primary types of market-based pollution-control instruments, though there are others. While similar in that market signals help determine abatement patterns, it is important to keep in mind two key distinctions between them.

First, an emissions tax by its nature raises revenue for the government. A permit system need not, although a regulator could choose to sell or auction the permits to polluters, which would remove this distinction. Second, with a permit-trading system the regulator must choose an aggregate emissions level. The total number of permits issued determines the total amount of emissions. Permit trading takes place given this total.

Market forces are used in both cases, but even with permits the market itself is a bit unusual in that it is created expressly by the government. Only by conferring the right to pollute upon the owners of permits, and granting the associated ability to trade, does a market arise.

The alternative to a market-based control regime is a command-and-control (CAC) regime, which usually consists of direct emissions controls or a technology requirement. For example, under the 1977 U.S. Clean Air Act new electric generating plants were required to install scrubbers. Such controls tend to be relatively rigid, giving polluters little latitude in their choice of control method. Even a program in which all sources are required to reduce their emissions by the same proportional amount is quite inefficient because it is insensitive to differences in abatement costs across firms. In general, the least-cost distribution of abatement will be such that marginal abatement costs are equalized across sources.

What is more, CAC systems do little to encourage the development and adoption of new control technologies (Stavins, 1999). A firm that adopts a higher level of technology, achieving thereby a lower abatement cost, is unlikely to gain financially. The extra abatement is good for society, but the firm cannot trade its reduction with another firm that would prefer to pollute more than the standard.

In an ideal world, where the ingredients for a well-functioning market are in place, a market-based scheme can overcome these problems. The regulator needs to know surprisingly little about the cost structure of firms, once an overall emissions target has been set. Firms can send each other the needed signals to assure that the responsibility for abatement is spread optimally.

In a permit scheme the government can require that trades be made public, which has the advantage of presenting a more transparent mechanism to all.

Of course, the world is less than ideal, so there can be problems as well. The purpose of this paper is to present the main arguments in favour of market-based systems, to summarize the main programs now in use around the world, and to discuss some of the shortcomings or pitfalls of these programs. The next section contains an example that illustrates exactly how a permit-trading or a tax scheme can achieve cost savings. The third section contains a summary of working market-based schemes. It turns out that, as a generalization, permit markets are more common in the U.S. and that tax systems of one form or another are more common in Europe. In the fourth section the U.S. SO_2 market is dis-

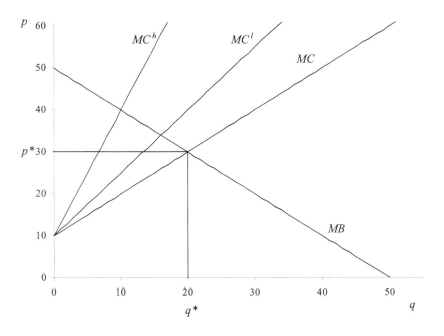

Figure 1. Example marginal abatement cost and marginal benefit

cussed at some length. Being the largest program of its kind in the world, and having been in place for over five years, this program seems to be of special interest.

2.   BASICS OF MARKET-BASED INSTRUMENTS

To illustrate the basic idea of a market-based instrument, consider a two-firm polluting industry whose emissions are perfectly mixed in a confined air shed. Both firms behave competitively, taking input and output prices as given. Initially, they each pollute $e^i=25$ where $i=h,l$ denotes the high- and low-cost firms respectively. The firms are different only in respect to their abatement cost functions, which are given by

$$C^h(q^h) = 10q^h + 3/2(q^h)^2 \quad \text{and} \quad C^l(q^l) = 10q^l + 3/4(q^l)^2$$

Note that if $q^i=0$ then i's abatement cost is zero. Marginal abatement costs are $MC^h = 10 + 3q^h$ and $MC^l = 10 + 3q^l/2$ respectively.
An environmental regulator has estimated the benefits to society resulting from abatement as

$$B(q) = 50\,q - 0.5\,q^2$$

where $q = q^h + q^l$ is aggregate abatement. This gives marginal benefits from abatement of $MB = 50 - q$. The following diagram illustrates the problem.

The regulator's goal is to maximize social welfare, so it decides to reduce emissions from the status quo of $e=50$ to 30, a reduction of 40%. This means that abatement will be $q=20$, where the marginal of abatement equals the marginal cost of abatement. Consider three alternative schemes that the regulator might choose to achieve this standard: proportional reduction (a CAC scheme); an emissions tax; and a permit system.

Proportional reduction. Under the proportional reduction scheme, each firm is required to cut back its emissions by 40%, from 25 to 15. Abatement for each firm is $q=10$. This leads to abatement costs of

$$C^h(10) = 100 + 3/2 \cdot 100 = 250$$

and

$$C^l(10) = 100 + 3/4 \cdot 100 = 175.$$

Total abatement costs are $C = 425$.

Emissions taxes. Under the tax scheme, the regulator must choose a tax so that when each firm sets its marginal cost equal to the tax, aggregate abatement will be $q=20$. From the diagram it is easy to see that the required tax is $t^*=30$. At this tax level, setting $MC^i(q^i) = 30$, we obtain:

$$q^l = 20/3 \quad \text{and} \quad q^h = 40/3$$

Abatement costs for the two firms are $C^h=133.33$ and $C^l=266.66$, for a total of $C=400$. Tax collections from the two firms are $T^h=550$ and $T^l=350$, for a total of $T=900$. The total of the two costs is $TC=1300$. These results are presented in Table 1.

Tradable permits. Under the permit-trading scheme, the regulator must choose a quantity of permits. It wishes to achieve abatement of $q=20$, so it gives out 30 permits, and let us assume that each firm receives $a^i=15$. Each firm then chooses q^i to minimize

$$TC^i = C^i(q^i) + p(e^i - a^i)$$

Together with the constraint that $a^h + a^l = 30$, solving the two cost-minimization problems yields

$$q^l = 20/3$$

and

$$q^h = 40/3$$

Table 1. Example: Comparison of CAC, taxes, permits.

	Proportional Reduction	Emissions Tax	Tradable Permits
High-cost firm			
Emissions	15	18.33	18.33
Abatement	10	6.67	6.67
Abatement cost	250	133.33	133.33
Tax paid	-	550	-
Permit cost	-	-	100
Low-cost firm			
Emissions	15	11.67	11.67
Abatement	10	13.33	13.33
Abatement cost	175	266.67	266.67
Tax paid	-	350	-
Permit cost	-	-	-100
Totals			
Emissions	30	30	30
Abatement	20	20	20
Abatement cost	425	400	400
Tax collected	-	900	-

which is the same as in the tax case. The resulting permit price is easily found to be $p=30$, which is the same as the tax. The low-cost firm achieves most of the abatement and sells the excess permits, in the amount 10/3, to the high-cost firm. Abatement costs are the same as in the tax case. The cost of the permit trade is (10/3) 30=100, which is a gain to the low-cost firm and a cost to the high-cost firm. Total abatement costs are once again $C=400$ and the net effect of the permit trade is zero. These results are also presented in Table 1.

The results of our simple example show that the market-based scheme yields savings of 25 in aggregate abatement costs. This is the main point of the argument. But there are other noteworthy features of the table as well. The tax collected in the tax regime is more than twice total abatement costs. This will not always be true, of course, but one can expect the emissions tax to be significant relative to abatement costs in most cases. We can see, then, why the industry is likely to favour a permit-trading scheme (so long as the permits are given out for free) over a tax scheme. The permit market has some equity effects among the firms, but as Montgomery (1972) showed the initial allocation has no effect on the ultimate distribution of emissions. Unless one of the firms is forced to leave the industry, the initial permit allocation can be divided among the firms in any way without changing the total abatement costs or the distribution of abatement. This leaves the regulator some leeway in achieving other goals, perhaps favouring one industry over another, through the initial allocation.

The comparison between the tax scheme and the permit-trading scheme is itself more interesting than this example shows. Indeed, the choice amongst market-based control instrument can be quite complicated and has attracted a large literature. If the marginal benefit or aggregate marginal cost curves are uncertain from the perspective of the regulator, then Weitzman (1974) showed that the choice between a price instrument (taxes) and a quantity instrument (permits) depends on the relative slopes of the marginal benefit and marginal cost curves. The goal of the regulator is to present polluters with an incentive scheme that most resembles the marginal benefit curve. Thus, if marginal benefits are steeper than marginal costs, permits are preferred in the sense that they lead to the lower expected deadweight losses. If the marginal cost curve is steeper, then an emissions tax is preferred.

Weitzman's paper has been influential, spawning a sizable literature. Kwerel (1977) explored the incentive properties of a Weitzman-like scheme and recommended that a hybrid price-quantity instrument is required in order to induce firms to behave optimally. Roberts and Spence (1976) developed a different combined price-quantity instrument that reduces deadweight losses below those experienced in Weitzman's optimal scheme. More recently, the properties of price and quantity instruments in a dynamic setting has received considerable attention (see Baldursson and von der Fehr 1999, Hoel and Karp 1999, and Rubio and Escriche 1999). In these papers the firms are naive but face a clever, dynamically optimizing regulator who foresees the future and changes the level of the policy (either the tax or the number of permits issued) in each period so as to incorporate the information it gleans from the behaviour of the industry in the previous period.

Moledina, *et al.* (1999) turn the tables and compare the outcomes of price and quantity instruments in a dynamic setting in which the regulator is naive and firms are strategic. They find that firms can gain significantly by strategizing against the regulator's dynamic adjustment rule. The strategic behaviour is complex because firms must take account of both the regulator's adjustment rule and each other. The findings indicate that the usual recommendation (that a price instrument is preferred if marginal benefits are flat) can be reversed.

The choice between permits and taxes is further complicated if one accounts for the macroeconomic or public-finance effects of regulation and taxes. A growing literature has investigated the possibility that emissions taxes can yield a "double dividend." In this literature, the first dividend is the gain from reducing emissions toward their optimal level. The second dividend, if it exists, is the reduction in deadweight losses as the revenue collected from polluters displaces other tax revenue and thus eliminates still more deadweight losses elsewhere in the economy (Goulder, *et al.* 1999).

Apart from the distinction between permits and taxes, the advantages of market incentives generally can be sensitive to a variety of market imperfections. Early on, Hahn (1984) argued that market power in the permit market can cause the attractive efficiency properties of a permit market to unravel. Innes, *et al.* (1991) showed that if one firm holds monopoly power in the *output* market then permit trading can be good as it induces the monopolist to increase its output, reducing the deadweight cost of monopoly. Uncertainty about a permit market - whether it will continue or change in the future - has been a significant hindrance to many markets in the past, as trading firms pull back from buying permits that the regulator may revoke at some time (Hahn and Hester 1989).

Some of the industries that face the most stringent environmental regulations, including the U.S. power sector, are also regulated economically. The combination of economic regulation and permit markets can lead to interesting and sometimes unexpected outcomes (Coggins and Smith, 1993; Bohi and Burtraw, 1991). Banking and borrowing of permits, which allow polluters to move their emissions optimally through time, can also be important in determining how well a permit market works and whether a true social optimum is achieved (Cronshaw and Brown Kruse, 1996; Rubin and Kling, 1997).

The main points to keep in mind, then, are the following. First, if markets are competitive and information is complete, and absent other complications such as no convex or discontinuous abatement costs or significant transactions costs, then market-based instruments can achieve the cost-minimizing distribution of emissions across firms. Second, in this circumstance both taxes and permits achieve this optimum. Third, the outcome of permits and taxes can be quite different, in regards to both tax revenue and social welfare. If markets aren't competitive, or in the presence of uncertainty, or if any number of imperfections is present, then one's enthusiasm for market-based instruments should be tempered somewhat.

Let us now turn to a summary of extant market-based environmental policies. In this discussion, the potential for permits or taxes to suffer from one or more of the problems highlighted above will be mentioned.

3.   A SAMPLING OF PERMIT-TRADING SCHEMES [1]

In a comprehensive survey of current and past policies that employ market-based instruments for the control of environmental quality, Stavins (forthcoming) divides the many types of instruments into four categories. *Pollution charges* or taxes involve a tax on each unit of emissions that a firm produces. *Tradable permits* involve establishing a desired level of aggregate emissions, issuing permits in that amount, and allowing firms to trade the permits with each other. *Market-barrier reductions* involve market creation (as with water markets), liability rules designed to provide firms with incentives to protect the environment, and information programs such as labelling requirements. *Subsidy reductions* appear in the form of reductions in subsidies that promote polluting practices.

All of these approaches have appeared in many forms, and Stavins has catalogued their incidence. In the present paper, only the first two types -taxes and permits- will be reviewed. The reader will get an idea of the scope of programs already in place, and also of the challenges that each approach presents. A pattern that will emerge, though it doesn't always hold, is that the U.S. tends to prefer permit systems while countries in Europe tend to prefer pollution taxes.

3.1.   Pollution taxes

Pollution taxes or effluent fees are used widely in Europe and to some extent in the rest of the world. In most cases there is some question as to whether the tax rates are high enough to have a significant effect on the level of emissions (Stavins, forthcoming, fn. 11; see also Blackman and Harrington 1999). Carbon taxes are levied in Denmark, Finland, Italy, the Netherlands, Norway, and Sweden. Except in Finland, the effect of these taxes in the Nordic countries has been lessened by various tax exemptions. Sulphur emissions or fuel sulphur content is taxed in Norway, Sweden, France, Denmark, Italy, and Galicia. Nitrogen oxides (NO_x) are taxed in France, Italy, Sweden, and Galicia. In Sweden, large plants pay a tax of $5 per kilogram on NO_x emissions, with the tax revenue neutral and the proceeds above administrative costs returned to polluters in proportion to energy

---

[1]   This section borrows heavily from Stavins (forthcoming).

produced. Water pollution is taxed in one form or another in the Netherlands, Germany, and France.

Of the transition economies, Poland stands out as having implemented pollution taxes at a rate sufficiently high to have a significant effect on emissions levels. For most pollutants, the Polish tax includes a low rate for emissions below a standard, and a higher rate for emissions above the standard. Though they are quite common, taxes in the rest of the region are too low to have had a significant effect on emissions. In the rest of the world, pollution taxes are also in place, but once again they are typically too low to be effective in reducing emissions. China taxes 29 water pollutants and 13 industrial gases, as well as solid and radioactive waste. The revenues collected help fund environmental projects. Pollution in wastewater from industrial sources is taxes in part of the Philippines. Malaysia taxes pollution from the palm oil industry. South Korea taxes emissions of ten air pollutants and fifteen water pollutants, but only for emissions levels above a regulatory standard. Japan taxes SO_2 emissions at a relatively low level. In South America, Colombia recently began a pilot program that taxes water pollution, with the tax based on biological oxygen demand (BOD) and total suspended solids (TSS). The city of Quito, Ecuador, has a tax on water pollution in which firms pay a tax on TSS when emissions exceed a national standard.

3.2.   User charges

User charges are a variant of pollution taxes, in which the tax is paid by those who use a product whose production involves pollution of some sort. The revenue raised by the U.S. tax on gasoline, for example, is used for highway construction and repair and is thus a user tax. Likewise, in Europe airline taxes are often used to finance noise abatement programs. In more than 100 jurisdictions in the U.S., households pay for the disposal of solid waste in proportion to volume. Switzerland has a similar system in place nationally, as do Bolivia, Venezuela, Jamaica, and Barbados. In many countries, a variety of products carry a tax based on the cost of disposing of the product. Batteries are taxed for this purpose in Belgium, Denmark, Italy, and Sweden. Tires are taxed in Denmark, Finland, and Sweden. Lubricant oils are taxed in France, Finland, and Italy. Surplus manure is taxed in Belgium and the Netherlands. Nuclear waste is taxed in Finland. In all of these cases, according to Stavins (forthcoming) the revenue from the tax is earmarked to remediation of the harm caused by the product (that is, to disposal), and the tax is also aimed at changing behaviour.

3.3.   Permit-trading systems

Permit-trading systems have grown in popularity in the U.S. in recent years, due in large part to the apparent success of the SO_2 allowance program. But such systems are becoming more common in other countries as well. Stavins (forthcoming) divides trading programs into two categories. The first is *credit programs*, in which an emissions standard is set and polluters may sell the rights accompanying any excess abatement to other firms. The second is *cap-and-trade programs*, in which permits are issued in an amount chosen by the regulator and are then traded among polluters. The U.S. Environmental Protection Agency (EPA) began devising credit programs under the original 1970 Clean Air Act. The first versions involved "netting" or "bubbles", in which multi-plant firms were allowed to assign their required abatement as they wished among their plants. Later, in 1976, under the "offset" program polluters were allowed to trade their emissions credits between firms. These programs have never been large, but according to Hahn and Hester (1989) they led to savings of more than $5 through 1990.

In the early 1980's the EPA developed a lead-trading program for U.S. gasoline refiners. The trading program accompanied a mandated reduction in lead content nationally to 10 percent of its previous level. Refiners who reduced the lead content in their gasoline were allowed to sell the right to sell leaded gasoline to other refiners. Banking of lead credits was first permitted in 1985, and during the life of the program more than 60 percent of lead in gasoline had been "traded". The program ended in 1987 when the target was reached. The EPA Office of Policy Analysis estimated in 1985 that the savings due to lead trading amounted to about $250 million per year.

Trade in rights to pollute water is still rare in the U.S. and elsewhere, though there are exceptions. A program to reduce pollution in Dillon reservoir in Colorado allowed point sources to pay for abatement by no point sources rather than control their own emissions further. A similar program is in place in Minnesota, where the Rohr malting plant recently paid for control measures on agricultural lands bordering the Minnesota River, thereby avoiding stricter controls on their own effluent. Canada has instituted two programs for trading emissions permits, the Pilot Emission Reduction Trading (PERT) and Greenhouse Gas Emission Reduction Trading (GERT) projects. PERT, an Ontario program, was established in 1996 and permits the registry of emission reduction credits for emissions below what is required by regulations. It applies to NO_x, VOC, CO_2, SO_2, and CO emissions. As of 1997, under PERT 14,000 tons of NO_x, 6,000 tons of SO_2, and over one million tons of CO_2 had been registered. The GERT project, established in 1997 and ended in 1999, applied to Alberta, British Columbia, Manitoba, Nova Scotia, Saskatchewan, and Quebec. The main aim of the program is to register willing trading participants. This program has not seen heavy use.

Under the international Framework Convention on Climate Change, carbon abatement activities can be conducted through "activities implemented jointly" (AIJ). An AIJ project involves investment by public or private entities of one country in carbon-abatement programs in another country, in order to satisfy part of the investing country's obligation under the Convention. Jepma (1999) reports that 133 AIJ projects had been approved by national authorities through September 1999.

Cap-and-trade programs, in which the trade is permits is an explicit ingredient, also come in many varieties. In the U.S. a market was established to comply with that country's obligation to reduce CFCs under the Montreal Protocol. The U.S. program requires each producer of CFCs to hold "allowances" at least equal to the amount of CFC material produced. Because different chemicals in the class of CFCs lead to different depletion levels, weights are assigned to each chemical and trades occur based on these weights. Through 1991, before the program ended, 80 trades had occurred. A similar program for CFC production quotas was instituted in EU countries between 1991 and 1994. Singapore and Canada have also experimented with trading programs for CFCs.

Other cap-and-trade programs include the RECLAIM program, which is used to reduce NO_x and SO_2 emissions in southern California through the use of tradable permits (Harrison, 1999), and the program in trading for total suspended air particulates in Santiago, Chile (Montero and Sanchez, 1999).

The U.S. program for trade in SO$_2$ allowances is the largest of the trading programs, and for this reason is dealt with in somewhat greater detail in the following section.

4.   THE U.S. SULPHUR DIOXIDE EXPERIENCE

The 1990 Clean Air Act Amendments overhauled U.S. air-quality regulations in many ways. Included in the law was Title IV, which instituted the first truly large-scale trading program in the country's history. Indeed, it is evidently the largest program of its kind in the world, with annual trading volume well above nine million tons in 1998. The unit of trade is an "allowance", which grants its bearer the right to emit one ton of SO$_2$. In 1998 the allowance price was close to $200, which means that the value of trade in the market was nearly $2 billion for that year alone.

The program was implemented in two phases. Under Phase I of the SO$_2$ program, which began in 1995, there were 110 so-called Table 1 plants (and 263 Table 1 units) [2]. These plants were listed in the legislation and were required to comply with the law during Phase I. In addition, during Phase I electric utilities had the opportunity to include additional units in the program to "substitute" or "compensate" for affected units. Table 2 shows how many units were included in the program from 1995 through 1998. All units in the program are required to hold allowances at least equal in number to their emissions. During Phase I each was granted allowances sufficient to emit SO$_2$ at a rate of 2.5 pounds per mmBtu. Those who reduced their emissions below 2.5 pounds of SO$_2$ per mmBtu, were free to sell their additional allowances to other utilities.

Table 2 also lists total allowances allocated and expended each year from 1995 to 1998. Total allowance allocation was 8.74 million in 1995, 8.30 in 1996, 7.15 in 1997, and 6.95 in 1998. The decline is due to a corresponding decline in the number of substitution and compensating units. Annual emissions are also reported in the table. Emissions by all units in the program were 5.30 million tons in 1995, 5.44 in 1996, 5.47 in 1997, and 5.29 in 1998.

Phase II of the program began on January 1 of 2000. During this phase all electricity-generating units with nameplate capacity of at least 25 megawatts are required to participate. This brings a total of more than 600 units into the program. Now, each unit is granted allowances sufficient to emit 1.2 pounds of SO$_2$ per mmBtu (down from 2.5 pounds, in Phase I). The number of trading utilities will increase, but perhaps not by as much as one would expect because the majority of utilities owned at least one unit that was affected in Phase I.

What has happened to the price of allowances during this period? Figure 2 shows how the market price index calculated monthly by Cantor-Fitzgerald as moved over time. One can see that it began at around $150 in late 1994, fell to $70 in early 1996, rose gradually through early 1998, and then rose steeply through mid-1999. By this time market participants had become familiar with the market, their plans for compliance in Phase II were set, and the demand for allowances was high as traders sought to hold surplus allowances to ease the transition into Phase II.

At the time the bill was signed into law, estimates of the marginal cost of SO$_2$ abatement ranged from $750 per ton upward. Clearly this estimate was far too high. Why were

---

[2]   A plant is a facility with a single stack. A unit is a single boiler. Most plants are made up of several units.

Table 2.	SO$_2$ Allowance Market Summary			
	1995	1996	1997	1998
Table 1 units	263	263	263	263
Subs. and Comp units	182	161	153	135
Opt-in units	0	7	7	10
Total number of affected units	445	431	423	408
Total alloc. (millions)	8.74	8.30	7.15	6.95
Total emissions (millions)	5.30	5.44	5.47	5.29
Trading volume (millions)	1.90	4.40	7.90	9.50

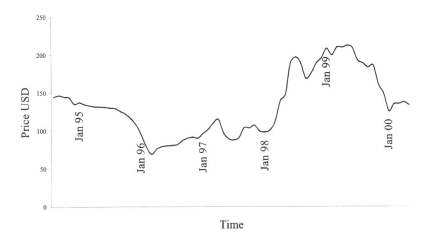

Figure 2. Cantor-Fitzgerald SO$_2$ allowance Market Price Index

the estimates wrong? Some have argued that the very possibility of allowance trading has struck fear into the various industries that supply alternative abatement methods (Burtraw, 1995). These include scrubber manufacturers, low-sulphur coal producers, and railways. Prices of scrubbers and low-sulphur coal, and rail rates, are all much lower than was anticipated. Carlson, *et al.* (1998) have estimated that the long-run cost savings due to the allowance-trading program could be as high as $800 million per year.

Perhaps the most striking feature of the diagram is the precipitous drop in the allowance price beginning in November 1999. In that month, the EPA filed suit against several utilities in the Midwest, the dirtiest region of the country. The suit alleged that the utilities had failed to comply with pre-1990 provisions of the Clean Air Act that require scrubbers on new sources. Several plants were upgraded since 1990, and the EPA has claimed that those revisions amounted to new sources under the old law. If the suit is successful, it is believed that many plants will be forced to install scrubbers. The effect on the allowance price has been two-fold. First, if the suit is successful, demand for allowances will fall and the supply will rise. Second, the EPA action has created some doubt about whether the market is as reliable as most had come to believe. Some utilities have pulled back from their plans to rely heavily on allowances for compliance and are looking at other alternatives such as switching to low-sulphur coal or relying more heavily on their gas-fired plants.

As of January 1999, utilities held 9.63 million unused allowances. This amounts to almost three years' worth of surplus allowances. Because allowances can be carried from one year to the next, or "banked," utilities are saving them for the future when they know their annual allowance allocations will shrink in number. The substitution provision itself appears to have increased the total number of banked allowances, though it was not intended to do so. Montero (1999) has found that utilities chose to include units in the substitution program that were already emitting less than 2.5 pounds of SO_2 per mmBtu. Thus, they were awarded more allowances than was required under their prior strategy. The excess allowances are now being banked for use in Phase II.

Another interesting feature of the data, though not reported here, is that about 25 percent of allowance trades are for "future vintage" allowances -allowances that have not yet been issued at the time of the trade [3]. Why would utilities be willing to pay not for an asset that won't come into existence until one or even two decades into the future? One conjecture is that they engage in this sort of transaction in order to manage inter-temporal risk. There is no active futures contract for allowances. Perhaps this is because utilities have found an alternative mechanism for sharing price risk over time. The fact that utilities have been willing to engage in trades for allowances of distant future vintage seems to indicate their considerable faith in the market over the long term. It will be interesting to see whether this changes in the next year as the EPA lawsuit wends its way through the courts.

5. CONCLUSIONS

Market-based instruments for the control of pollution seem to be on the rise. Both taxes and tradable permits are being used more frequently and in more places on the globe each year. The basic concept of harnessing market forces in order to achieve least-cost abatement patterns has been compelling to economists for a long time. The fact that politicians in many countries have become convinced as well in recent years represents a triumph for economic reasoning in the policy arena. Time will tell whether these instruments work as well in practice as we've been saying for years that they work in theory. Early indications are mixed, but there is reason to be optimistic. The Clinton administration has made it clear that the U.S. plans to insist that some type of trading provision be included in the implementation agreement for the Kyoto protocol. International trading, especially on the scale required for abatement of global carbon emissions, presents challenges that are in a different category from those laid out here.

---

[3] There have been trades in allowances that won't be issued until the year 2025.

6. REFERENCES

Baldursson F.M. – von der Fehr N.H.M.(1999), *Prices Quantities: The Irrelevance of Irreversibility*, mimeo, January.

Blackman A. – Harrington W. (1999), *The Use of Economic Incentives in Developing Countries: Lessons from the International Experience with Industrial Air Pollution*, Report 99-30, Resources for the Future, Washington, D.C..

Bohi D.R. – Burtraw D. (1991), Utility Investment Behaviour and the Emission Trading Market, *Resources and Energy*, **14**: 129-153.

Burtraw D. (1995), *Cost Savings Allowance Trades? Evaluating the SO$_2$ Emission Trading Program to Date*, Report 95-30, Resources for the Future, Washington, D.C..

Carlson C. – Burtraw D.- Cropper M. – Palmer K.L. (1998), *Sulphur-Dioxide Control by Electric Utilities: What are the Gains from Trade?* Report 98-44, Resources for the Future, Washington, D.C..

Coggins J.S. – Smith V.H. (1993), Some Welfare Effects of Emission Allowance Trading in a Twice-Regulated Industry, *Journal of Environmental Economics and Management*, **25**: 275-297.

Cronshaw M.B.- Brown Kruse J.(1996), Regulated Firms in Pollution Permit Markets with Banking, *Journal of Regulatory Economics*, **9**: 179-189.

Dales J. (1968), *Pollution, Property and Prices*, University Press of Toronto, Toronto.

Goulder L.H. – Parry I.W.H. – Williams R.C. – Burtraw D. (1999), The Cost-Effectiveness of Alternative Instruments for Environmental Protection in a Second-Best Setting, *Journal of Public Economics*, **71**: - .

Hahn R.W. (1984), Market Power and Transferable Property Rights, *Quarterly Journal of Economics*, **99**: 753-765.

Hahn R.W. – Hester G.L. (1989), Where Did All the Markets Go? An Analysis of EPA's Emissions Trading Program, *Yale Journal on Regulation*, **6**: 109-153.

Harrison D.jr. (1999), Turning Theory Into Practice for Emissions Trading In the Los Angeles Air Basin, in: *Pollution for Sale: Emissions Trading and Joint Implementation*, Sorrell S. – Skea J. (eds.), Edward Elgar, London, pp. - .

Hoel M. – Karp L. (1999), *Taxes and Quotas for a Stock Pollutant with Multiplicative Uncertainty*, mimeo, March.

Innes R. – Kling C. – Rubin J. (1991), Emission Permits Under Monopoly, *Natural Resources Modelling*, **5**: 321-343.

Jepma C. (1999), Planned and Ongoing AIJ Pilot Projects, *Joint Implementation Quarterly*, **5**: 14.

Kwerel E. (1977), To Tell the Truth: Imperfect Information and Optimal Pollution Control, *Review of Economic Studies*, **44**: 595-601.

Moledina A.A. – Coggins J.S. – Polasky S. (1999), *Dynamic Environmental Policy with Strategic Firms*, mimeo.

Montero J.P. (1999), Voluntary Compliance with Market-Based Environmental Policy: Evidence from the U.S. Acid Rain Program, *Journal of Political Economy*, **107**: 998-1033.

Montero J.P. – Sanchez J.M. (1999), *A Market-Based Environmental Policy Experiment in Chile*, Mimeo, Catholic University of Chile.

Montgomery D.W. (1972), Markets in Licenses and Efficient Pollution Control Programs, *Journal of Economic Theory*, **5**: 395-418.

Pigou A.C. (1932), *The Economics of Welfare*, 4th Edition, Macmillan and Co., London.

Roberts M.J. – Spence M. (1976), Effluent Charges and Licenses Under Uncertainty, *Journal of Public Economics*, **5**: 193-208.

Rubin J.D. – Kling C. (1997), Bankable Permits for the Control of Environmental Pollution, *Journal of Public Economics*, **64**: 101-115.

Rubio J. – Escriche L. (1999), *Strategic Pigouvian Taxation, Stock Externalities, and Polluting Non-Renewable Resources*, mimeo, April.

Stavins R.N. (forthcoming), Experience with Market-Based Environmental Policy Instruments, in: *The Handbook of Environmental Economics*, Maler K.G. – Vincent J. (eds.), North-Holland/Elsevier Science, Amsterdam.

Weitzman M.L. (1974), Prices vs. Quantities, *Review of Economic Studies*, **41**: 477-491.

EVALUATING SUSTAINABILITY
IN PLANNING AND DESIGN

Patrizia L. Lombardi [*]

SUMMARY

It has often been recognized that planning and design can play an important role in the achievement of sustainable development of cities. However, problems still exist with regards to both a clear understanding of sustainability in the built environment and a means of evaluating it within the context of urban planning and design.

This chapter will briefly compare different evaluation methods in urban planning, both *ex ante* approaches and monitoring, and their philosophical paradigms. Some significant limitations will be identified and discussed in the context of sustainability, such as the reductionism within many of the approaches and the lack of holism in the evaluation. The identified deficiencies will provide the motivation for the development of a new framework, which is able to integrate the different dimensions of sustainability in the built environment. The resulting framework will also have the potential to allow evaluation of the concept of sustainability over time.

1. INTRODUCTION

Sustainability is a difficult notion to define in substantive terms and may remain an academic idea, a 'fuzzy buzzword' (Palmer et al., 1997) or a 'paradox' (Merret, 1995), unless we develop a clearer understanding of which dynamics and mechanisms are required to transform the sustainability principles into practice.

Much of the early work on sustainable cities was focused on the ecological dimension of the problem, as reflected in the policy agendas of various local authorities. In the literature, this is reflected in the concept of 'metabolism' which aims at showing the demand for materials which give rise to resource depletion and pollution (Capello et al., 1999). On the other hand, the softer and more 'fuzzy' dimensions of urban sustainability (e.g. political, social, cultural, aesthetic, and so forth) are still poorly addressed, since

[*] Dipartimento Casa-Città, Politecnico di Torino. This paper is based on the author's Ph.D. research conducted at the University of Salford (UK) under the supervision of P.S. Brandon.

contemporary analytical tools do not sufficiently handle them. Consequently, the need still exists to incorporate sustainability principles and criteria in current decision making processes.

Devising strategies for the sustainable development of cities is difficult not just because the nature of a city is complex, but also because the concept is ambiguous, multi-dimensional and generally not easy to understand outside the single issue of environmental protection (BEQUEST, 1999). Mitchell (1996) suggests that effective urban sustainability strategies and sustainable development plans can best be identified by ensuring that decision makers and developers are adequately briefed on sustainability issues, local characteristics and community needs. This process can require the application of a suitable operational framework, an evaluation method or approach able to guide developers through the decision making process. However, such a structure for organizing the information required in decision making is not yet available or agreed upon among the different disciplines and fields of work.

Given the lack of an agreed upon structure which can help decision makers move towards greater sustainability, the overall objective of this paper is to find an integrating mechanism or framework. This framework should bring together the diversity of interests necessary to assess the impact of the built environment and urban design on urban sustainability.

This framework could be used by political and technical decision makers (public local control officers and/or planners or designers) to check a design or a plan for sustainability and to learn from it. It should be able to assist the process of devising sustainable planning strategies, ensuring that all sustainability aspects and quality of life issues are included and nested within each other.

The proposed framework is based on a Cosmonomic theory developed by a Dutch philosopher (Dooyeweerd, 1958) and adopted by the Swedish Multimodal system school of thinking (de Raadt, 1994). This is suggested as being more appropriate than other theories and conceptual paradigms for dealing with sustainability problems.

This paper is structured as follows. Section 2.0 provides an overview and a critique of the main evaluation approaches to sustainability. Section 3.0 illustrates the development of the framework and some examples related to the evaluation of sustainability in the built environment. Finally, Section 4.0 provides some conclusions and future directions for work.

2.   EVALUATION METHODS IN PLANNING AND DESIGN

The following section provides a critical view of the principal evaluation approaches and procedures (or methodologies) used in planning and design for sustainability. For the sake of clarity, this illustration will follow a classification of the methods based on the stage of the planning evaluation, i.e. *ex ante* and *ex post* or monitoring.

2.1.   *Ex ante* evaluation methods

Ex ante evaluation generally means a technical-scientific procedure for expressing a judgement, based on values, about the impacts of a policy or of an action on the physical (natural and/or built) environment, or for assessing the effects of these impacts on the community (Bentivegna, 1997). This implies the clarification of consequences of different choices and planning options, comparing the characteristics of various choice-possibilities in an explicit and systematic manner.

Methods, such as cost-benefit analysis and other 'formal' techniques, have been developed and widely applied in spatial planning during the last twenty-five years to support this task. Thus, they are also named "problem-solving" methods. All these methods are based on the assumption that the impacts of a policy proposal can be assessed for all relevant variables of the proposal. They differ in the way this assessment is done and in the way the results are presented. However, in all cases, the point of departure is the application of scientific logic for measuring the effects (Voogd, 1998).

Multicriteria methods have often been used as an alternative approach to cost-benefit analysis. Compared with cost-benefit analysis, in multicriteria analysis the measure of benefits is not related to the concept of willingness to pay but to the achievement level of the goals set (or preferences). In other words, the benefits are the outcomes from a project, which are evaluated positively in relation to one or more criteria (Nijkamp et al., 1999). Multicriteria analysis is also suggested for use in a communicative ideology in planning evaluation since it requires, on the part of the decision-makers, an explanation of the individual preferences assigned to the various objectives-criteria. Therefore, discussion and negotiations should be conducted where representatives of different groups, political parties and lobbies, as well as the promoters and executors of the proposed actions, may be represented (Fusco Girard and Nijkamp, 1997).

However, some handicaps do exist in relation to an applicability of current evaluation methods within the new framework of communicative ideology, as discussed in Voogd (1998) and Kakee (1998). In particular, the problem of 'how best to organize the form of discourse, to develop inclusionary arguments and to build interrelations' exists. This requires the application of a suitable operational framework, an evaluation methodology or an approach that can guide decision makers, and particularly planners, to understand the problems implied in a planning decision.

2.2. *Ex post* and monitoring approaches

Ex post and monitoring evaluations are strongly emphasized in the latest planning guides (DoE, 1992). This is reflected in the 1991 Planning and Compensation Act, which introduces the so-called 'plan-led system'. The aim is to make the planning system "simpler and more responsive, reducing costs for both the private sector and local authorities and making it easier for people to be involved in the planning process" (DoE, 1992, 1.10).

In particular, monitoring activity is strongly related to contingent and contextual aspects, and the methods tend to reflect the administrative character of this activity (Lichfield and Prat, 1998). A major difficulty with this activity is the selection of the right indicators able to represent the situation under study. An indicator or an index can be defined as a means devised to reduce a large quantity of data down to its simplest form, retaining essential meaning for the questions that are being asked about the data.

At the moment there is a great demand for indicators which are able to measure sustainable development as a prerequisite to promoting a sustainable society (OECD, 1997). This demand arose as a consequence of the UN Conference on Environment and Development (UNCED, 1992) which stated that 'indicators of sustainable development need to be developed to provide a solid basis for decision making at all levels and contribute to self regulating sustainability of integrated environmental and development systems'.

The European Community's Fifth Environmental Action Programme 'Toward Sustainability' (CEC, 1993), also notes that 'there is presently a serious lack of indicators and environmental assessment material'. This has added to the demand for effective sustainability indicators. However, according to Mitchell (1996), despite the considerable attention devoted to sustainability indicators, no set has emerged with universal appeal.

There are a number of limitations related to both current sets of urban sustainability indicators and their classification systems. All these indicators are quantitative in nature and are usually classified on the basis of a reductionist view of reality, in which one aspect is given undue emphasis to the detriment of others. Such lists emphasize the distinctness of the aspects, and give no indication of their inter-dependencies. These problems come from the empirical genesis of the lists: they are merely a compilation of people's ideas and have little theoretical, ontological foundation (Lombardi and Basden, 1997). Approaches based on reductionism are unable to handle all the elements and components of the system, both deterministic and nominal ones, demonstrating ignorance and imbalance. They may also be misleading, suggesting an unbalanced path in the future developments of a town.

2.3.   Main limitation of current evaluation approaches

Recent reviews of current evaluation methods in planning and design for sustainability show a lack of holism (Brandon et al., 1997; Litchfield et al., 1998; Mitchell, 1999). All the methods are limited, taking into consideration only a few of the multiple aspects required for developing sustainable solutions. The evaluation is mainly technical and economic, and there is not a mechanism or tool that is able to take into account all sustainability issues in a comprehensive manner.

A major limitation of existing assessment methods is the focus on the empirical measurements of specific effects, economic as well as environmental ones, rather than an identification of the multiple effects the project has on human and natural resources. This often leads to immediate and short term solutions rather than to a prevention of negative effects. A second problem is that very often the non-technical issues are not addressed, or insufficiently covered, in the evaluation.

Although multicriteria methods have often provided a guide for selecting suitable planning and design solutions in evaluation (Nijkamp, 1991), they lack content and a conceptual framework, or a theoretical guide that can help designers and decision makers to address the problem of sustainability in the built environment. The evaluation methods are many and there is no agreement among scholars on the theoretical framework to be used (Bentivegna, 1997).

A further problem is that experts use a specialized and codified vocabulary which is not common for all disciplines and stakeholders involved in the planning process. Yet, each discipline involved brings its own agenda, own classification system and own techniques to the problem. Often the disciplines are unwilling (or unable) to consider the views represented by others because there is not a common language or a systematic methodology which will allow a fruitful dialogue to take place (Brandon et al., 1997).

This chapter postulates a new scientific paradigm, named multi modal thinking, based on the philosophical work of Herman Dooyeweerd (1958). It was chosen mostly because it offers a pluralistic ontology of aspects that may guide the planning and design evaluation process, ensuring that all aspects of human life are present in the design. In particular, this has been used for both understanding sustainability in the built environment and establishing a more holistic classification system for sustainability aspects in the evaluation of the built environment at a local planning level.

3. DEVELOPMENT OF A HOLISTIC FRAMEWORK FOR DECISION MAKING

The challenge for political and technical actors (planners, designers and urban authorities) is to devise strategies and policies, urban plans and projects that can guide cities along a more sustainable development path. As previously suggested, there is a lack of a decision support system or a tool which is both comprehensive and holistic to harmonize the different aspects of sustainability in planning and design. In this section a new framework will be illustrated.

The proposed framework aims at guiding designers and planners, official public developers and decision makers through the process of understanding and evaluating sustainability in planning. The framework may also act as an educational tool, since it includes information on sustainability issues.

3.1. The theory underpinning the framework

The development of the scientific procedure that underlies the framework is supported by the theory of "The Cosmonomic Idea" of Herman Dooyeweerd, a Dutch philosopher, 1894–1975. This has been recently postulated in a number of studies related to cybernetics, information systems and organization learning basically because it offers an extremely useful check-list to guide systems development and usage, ensuring that not only one, but all aspects of human life, from the numerical to the credal be present in the design (de Raadt, 1994). In addition, it has been studied and developed by other contemporary authors, who have illustrated some of its benefits to understand and explain how social systems and institutions work (Hart, 1984; Clouser, 1991).

A specific feature of the theory is its ability to explain complexity without falling into reductionism and/or subjectivism. This suggested that the theory would be useful in structuring sustainability in the built environment, overcoming the problems of current assessment methods.

The "Cosmonomic Idea of Reality" acknowledges an external reality which is independent of the acting and knowing subject (hence the term, Cosmonomic). We are affected by it but also affect it and have views and desires concerning it. In particular, it claims there are two 'sides' to reality as we know it: a Law Side and an Entity Side. The Entity Side concerns things, systems, and in fact anything that does something, e.g. a person, a flower, a house, a government, a symphony, a town. The Law Side concerns modalities in which entities operate, e.g. physical, social, biotic, ethical, technical, etc..

The two sides can be seen as orthogonal: an entity crosses several modalities. In everyday living the entities stand to the fore, as it were, and the Law Side recedes into the background, but in science the Law Side comes to the fore while the entities recede. That is, when we analyze reality we should study the Law Side, not the behavior of entities. It is the Law Side that expresses the fundamental Meaning, and it is the Law Side that enables entities to 'exist'. However, existence is not denied. Rather, it is seen as essentially dependent and meaning-bound (Basden, 1996).

The theory of the Cosmonomic Idea shows some distinctive and useful features for this study. Specifically each modality has some essential characteristics which is irreducible to any other, and reality is rich and varied, since it is organized in such a way as to explicitly show different interrelations between irreducible parts.

3.2.   Toward a multimodal framework

The development of the framework has followed three main steps:

1. Definition and classification of sustainability aspects on the basis of a literature search and modalities provided by the Cosmonomic Idea theory.
2. Adoption of a scientific tool for guiding both the definition and the evaluation of each sustainability aspect by decision makers.
3. Identification of a number of current assessment methods and evaluation approaches within each sustainability aspect.

3.2.1.   Step 1: Definition and classification of sustainability aspects

The fifteen meaning-nuclei of the modalities are suggested to provide decision makers with a qualification system to classify relevant sustainability issues in an urban design or planning situation. Specific attention has been paid to a definition of each modal aspect with headings which may be more familiar to a larger number of stakeholders, and more specifically, to technical decision makers.

A literature review on sustainable development and its understanding by various members of the community was the basis of the development of this new vocabulary. This also takes into account the results of an assessment of the comprehension of the modalities by non-expert users, that was undertaken at different stages of this framework's development.

The list of modalities and their new definitions that reflect sustainability is provided in Table 1.

3.2.2.   Step 2: A scientific tool for guiding the identification of sustainability

As the definition of a sustainability aspect is a process which also includes non technical aspects, the process must be guided by a scientific tool. In this context the PICABUE approach (Mitchell et al., 1995), whose benefits in gauging the degree of common commitment of societal actors and representatives of different disciplines towards sustainable development have already been highlighted by a number of recent studies (Palmer et al., 1997; Curwell and Cooper, 1998). However, it may be useful to develop appropriate questions under each sustainability aspect, linking them to the four recognized principles underlying sustainable development:

• Futurity - a concern for future generations, i.e. not cheating on our children;

Table 1.   Modalities, nuclei of meaning and sustainability aspects

Modalities	Nuclei of meaning	Redefined modalities to reflect sustainability
Numerical	'How much' of things	Numerical accounting
Spatial	Continuous extension	Spaces, shape and extension
Kinematics	Movement	Transport and mobility
Physical	Energy, mass	Physical environment, mass and energy
Biological	Life function	Health, bio-diversity, eco-protection
Sensitive	Senses, feeling	People perceptions towards environment
Analytic	Discerning of entities	Analysis and formal knowledge
Historical	Formative power	Creativity and cultural development
Communicative	Informatory	Communications and the Media
Social	Social intercourse	Social climate and social cohesion
Economic	Frugality	Efficiency & Economic appraisal
Aesthetic	Harmony, beauty	Visual appeal and architectonic style
Juridical	Retribution, fairness	Rights and responsibilities
Ethical	Love, moral	Ethical issues
Credal	Faith, trustworthiness	Commitment, interest and vision

Source: Lombardi (1999)

- Social Equity - a concern for today's poor and disadvantaged, i.e. equal access to resources;
- Environment - preserving the eco-system, i.e. ensuring that human activities do not threaten the integrity of ecological systems;
- Public participation - a concern that individuals should have an opportunity to participate in decisions that affect them, i.e. ability to influence decisions.

Other approaches exist which aim at understanding and classifying sustainability issues, such as the Pentagon Prism Model developed by Nijkamp and Pepping (1998) within a comparative study of sustainability initiatives in order to identify critical success factors in energy policy. Yet PICABUE seems more appropriate in the context of this study. First, it is not limited to renewable energy policies or other sectorial and compartmentalized policy making as in the Pentagon model; rather, it is general and holistic. Second, it is founded on the sustainable principles that underlie official reports on Agenda 21, and therefore it is not too general or broad as is S.L.E.P.T. (Social, Legal, Economic, Political and Technical) system or the Flag and the Spider models (Bruinsma et al., 1999; Nijkamp et al., 1999). PICABUE has been used to classify quality issues and sustainability indicators within the context of a sustainability Information Technology (I.T.) model for cities (May, Mitchell et al., 1997). Third, it is very simple and easy to understand, as revealed by the studies on stakeholders' concern about sustainability in planning and construction. In this study, it has been useful to help link each aspect with the official understanding of sustainability principles. An example related to the economic aspect is shown in Table 2 (column 3).

3.2.3.   Step 3: Methods and approaches for evaluating each sustainability aspect

A number of current assessment methods and tools have been specified for each sustainability aspect. These evaluation methods belong to various technical fields and

Table 2.   An example of key questions for the (re)development of urban areas

Modalities	Sustainability aspects	Key-questions
Economical	**Efficiency & economic appraisal** It is concerned with wise use of limited resources. Efficiency is defined as the ability to achieve desirable goals by managing limited resources. It asks developers and designers to consider how to make best use of all the available resources.	Futurity: Has a long term financial appraisal been undertaken? Equity: What is the financial distribution for the stakeholders? Has employment of the local labour force in construction activities been considered? Environment: Is there an efficient environmental management system? Is there an exhaustive city-wide recycling programmes from which the development could benefit? Participation: How many of the stakeholders have committed themselves to the financial appraisal?

Source: Lombardi (1999)

scientific disciplines such as economics, different branches of engineering, structural technology, architecture, town planning, etc. An example of this list of assessment methods under each modality is illustrated in Table 3.

3.3.   The final framework

The final evaluation framework involves all the following: a technical and ecologically oriented assessment of the construction under development (a "green design") that illustrates the environmental compatibility of this development within the existing context; an assessment of the historical and cultural significance of the planning asset and of its social desirability; an analysis of the economic and juridical feasibility; a check of the visual appeal of this new (re)development and of its flexibility or adaptability, which may allow the meeting of some future user's needs; and an understanding of what interest or concern there is in the local agenda of the city.

The structure of this framework is a multi-layers tool-kit which makes use of current knowledge for pinpointing tools and problem solving methods for each situation related to sustainability in the built environment. It enables the linking of aspects for evaluation with questions for examining sustainability and, finally, with a number of problem solving assessment methods. However, in order to operationalize the framework, it needs further work and possibly the support of an I.T. tool, such as a knowledge based system which can evolve and cope with incomplete and uncertain information. It requires pragmatic testing, revision, implementation and a convenient user friendly interface.

4.   CONCLUSION

Decision making for sustainability, particularly in the field of planning or design, requires a framework which is able to structure the problem in order to understand and

Modalities	Examples of evaluation approaches
Numerical	Numerical indicators, Quantitative index
Spatial	Design approaches, Planning approaches, Geographic Information Systems, Computer Aided Design
Kinematics	Transport and traffic planning scheme, Transport evaluation tools, Infrastructure capacity
Physical	Strategic environmental analysis, Environmental impact analysis, Multicriteria methods
	Energy planning schemes, Physical indicators
Biological	Ecological footprint approach, Carrying capacity, Environmental impact analysis
Sensitive	Lynch's theoretical outlook approaches, I.T. tools, Virtual reality, Surveys and polls, Questionnaires techniques
Analytic	Analytical hierarchy process, Logic, scientific reasoning and deductive thinking
Historical	Design approaches, Technological analyses, Goals achievement matrix
Communicative	Monitoring and audit, Argumentative approaches, I.T. tools, Virtual reality
Social	Polls and surveys, Questionnaire techniques, Audit and monitoring
Economic	Life cycle costing, Cost-benefit analysis, Community Impact Evaluation, Multicriteria analysis
Aesthetic	Design approaches, Lynch approach to design, Polls and surveys, Workshops, meetings and consultation
Juridical	Public Committees, Public advisory boards, Public Planning Councils, European, National and Local Planning Laws and Regulations
Ethical	Community Impact Evaluation, Environmental impact analysis, Strategic impact evaluation, Contingent valuation
Credal	Strategic regional plan, Focus groups, Consultation

Table 3. Modalities and evaluation approaches

Source: Lombardi (1999)

evaluate the implications that the (re)development may have in relation to the existing situation.

In this chapter a new framework has been illustrated for understanding and evaluating sustainability in urban planning and design. The framework aims at enabling planners and decision makers to acquire commitment and consensus among stakeholders rather than arrive at a determinative policy agenda or evaluation outcome specified by themselves.

The framework developed in this study is based on the philosophical theory of "The Cosmonomic Idea". This is useful, not only because it recognizes different levels of information but also because it suggests an integration between decision makers and the affected parties, recognizing a multi-person approach to the problem.

Some major problems faced in decision making for sustainability are the amount of information required for an evaluation of this type, which is time consuming and certainly costly; the variety of vocabulary employed and required by each assessment method; the elements of uncertainty included in the available data; the difficult access to different data-bases; etc. The framework, as it has been developed in this study, does not overcome all of these problems directly, but it shows how it is possible to use current assessment methods within the framework. It does provide new opportunities for collaboration between disciplines, experts and lay people. It adds new dimensions that were traditionally uncovered in the evaluation (e.g. aesthetics). It links all the knowledge and the special contributions of science within the same structure, providing order, continuity and integration without falling into reductionism or lack of transparency. Thus, it can also act as a learning tool, answering current demands for higher education in the field of planning.

Further knowledge of sustainability, further development of the information on which the framework relies, and pragmatic testing in real world-wide contexts will certainly be required. Practical applications could also be improved if the model is linked to expert systems or G.I.S. (Geographic Information System). At present, research findings show that the framework is reliable as a model to be used for challenging planning towards greater sustainability in the built environment.

5. REFERENCES

Basden A. (1996), Towards an Understanding of Contextualized Technology, in: Proceedings of the International Conference of the Swedish Operations Research Society on *Managing the Technological Society: the Next Century's Challenge to O.R.*, University of Lulea, Sweden, October 1-3, pp. 17-32.

Bentivegna V. (1997), Limitations in Environmental Evaluations, in: *Evaluation in the Built Environment for Sustainability*, Brandon P. – Lombardi P.L. – Bentivegna V. (eds.), E&Fn Spon, Chapman&Hall, London, pp. 25-38.

Bequest – Building Environment Quality Evaluation for Sustainability Through Time Network, (1999), *Report 1998-99*, EC Environment And Climate Research Programme, Theme 4: Human Dimensions & Environmental Change, Directorate D – Rtd Actions: Environment – E.U. Dg12.

Brandon P.S. – Lombardi P.L. – Bentivegna V. (eds., 1997), *Evaluation in the Built Environment for Sustainability*, E&Fn Spon, Chapman&Hall, London.

Bruinsma F. – Nijkamp P. – Vreeker R. (1999), *Experts at Arm's Length of Public Policy-Makers. A Case Study on Utrecht* (forthcoming).

Capello R. – Nijkamp P. – Pepping, G. (1999), *Sustainable Cities and Energy Policies*, Springer Verlag, Berlin.

CEC – Commission of the European Community, European's Fifth Environmental Action Programme (1993), *Toward Sustainability*, CEC.

Clouser R. (1991), *The Myth of Religions Neutrality*, University of Notre Dame, London.

Curwell S. – Cooper I. (1998), The Implication of Urban Sustainability, *Building Research & Information*, **26**: 17-28.

De Raadt J.D.R. (1994), Expanding the Horizon of Information Systems Design, *System Research*, **2**: 185-199.

DoE – Department of the Environment (1992), *Assessing the Impact of Urban Policy*, Inner Cities Research Programme, Hmso, London.

Dooyeweerd H. (1958), *A New Critique of Theoretical Thought*, 4 Vols, Presbyterian and Reformed Publisher Company, Philadelphia, Pennsylvania.

Fusco Girard L. – Nijkamp P. (1997), *Le Valutazioni per lo Sviluppo Sostenibile della Città e del Territorio*, Franco Angeli, Milano.

Hart H. (1984), *Understanding Our World*, Univ. Press of America, Usa.

Kakee A. (1998), The Communicative Turn in Planning and Evaluation, in: *Evaluation in Planning. Facing the Challenge of Complexity*, Lichfield N. – Barbanente A. – Borri D. – Kakee A. – Prat A. (eds.), Kluwer Ac. Press, Dordrecht, pp. 97-111.

Lichfield N. – Barbanente A. – Borri D. – Kakee A. – Prat A. (eds., 1998), *Evaluation in Planning. Facing the Challenge of Complexity*, Kluwer Ac. Press, Dordrecht.

Lichfield N. – Prat A. (1998), Linking Ex Ante and Ex Post Evaluation in British Town Planning, in: *Evaluation in Planning. Facing the Challenge of Complexity*, Lichfield N. – Barbanente A. – Borri D. – Kakee A. – Prat A. (eds.), Kluwer Ac. Press, Dordrecht, pp. 283-298.

Lombardi P.L. – Basden A. (1997), Environmental Sustainability and Information Systems, *Systems Practice*, **10**: 473-489.

Lombardi P.L. (1998), Sustainability Indicators in Urban Planning Evaluation, in: *Evaluation in Planning. Facing the Challenge of Complexity*, Lichfield N. – Barbanente A. – Borri D. – Kakee A. – Prat A. (eds.), Kluwer Ac. Press, Dordrecht, pp. 177-192.

Lombardi P.L. (1999), *Understanding Sustainability in the Built Environment. A Framework for Evaluation in Urban Planning and Design*, Ph.D. Thesis, University of Salford, September 1999.

May A.D. – Mitchell G. – Kupiszewska D. (1997), The Development of the Leeds Quantifiable City Model, in: *Evaluation in the Built Environment for Sustainability*, Brandon P. – Lombardi P.L. – Bentivegna V. (eds.), E&Fn Spon, Chapman&Hall, London, pp. 39-52.

Merret S. (1995), Planning in the Age of Sustainability, *Scandinavia Housing & Planning Research*, **12**: 5-16.

Mitchell G. (1996), Problems and Fundamentals of Sustainable Development Indicators, *Sustainable Development*, **4**: 1-11.

Mitchell G. (1999), *A Geographical Perspective on the Development of Sustainable Urban Regions*, Earthscan (in press).

Mitchell G. – May A. – Mcdonald A. (1995), Picabue: A Methodological Framework for the Development of Indicators of Sustainable Development, *Int. Journal Sustainable Development World Ecol.*, **2**: 104-123.

Nijkamp P. (ed., 1991), *Urban Sustainability*, Gower, Aldershot.

Nijkamp P. – Pepping G. (1998), A Meta-Analytic Evaluation of Sustainable City Initiatives, *Urban Studies*, **35**: 1481-1500.

Nijkamp P. – Bal F. – Medda F. (1999), *A Survey of Methods for Sustainable City Planning and Cultural Heritage Management* (forthcoming).

OECD (1997), *Better Understanding Our Cities. The Role of Urban Indicators*, OECD, Paris.

Palmer J. – Cooper I. – Van Der Vost R. (1997), Mapping Out Fuzzy Buzzwords - Who Sits Where on Sustainability and Sustainable Development, *Sustainable Development*, **5**: 87-93.

UNCED - United Nations Conference on Environment and Development (1992), *Earth Summit '92 (Agenda 21)*, Regency Press, London.

Voogd H. (1998), *The Communicative Ideology and Ex Ante Planning Evaluation*, in: *Evaluation in Planning. Facing the Challenge of Complexity*, Lichfield N. – Barbanente A. – Borri D. – Kakee A. – Prat A. (eds.), Kluwer Ac. Press, Dordrecht, pp. 113-126.

TRANSFER OF DEVELOPMENT RIGHTS: AN INNOVATIVE APPROACH TO URBAN AND REGIONAL MANAGEMENT IN ITALY

Ezio Micelli [*]

SUMMARY

Planning intervenes have been used to regulate the numerous externalities that characterize cities and regions. Having recognized inefficiency of the authoritative command-and-control tools, some Italian local administrations have been trying to implement and manage urban plans through the use of market-based tools.

The transfer of development rights represent an innovative market-based tool that has created a great interest. Several significant elements emerge from analysis of the major case studies in Italy. For instance, markets for development rights do not replace the command-and-control tools traditionally used in planning. In reality, the success of the new markets seems to depend significantly on their integration with the latter.

Markets for development rights have not proven to be automatic devices led by an invisible hand: in order to reduce transaction costs, the *visible hand* of the administrations must establish the market rules and promote them.

1. INTRODUCTION

The issue of the effectiveness of planning has led economists and urban planners to debate and to experiment with new tools to manage urban and regional plans. The major shift concerns the use market-based tools in order to restore conditions of efficient resource allocation, instead of the traditional authoritative tools, among which zoning is the prime example.

The transfer of development rights in order to implement urban and regional plans has been used in the US and in several other developed countries (Renard, 1998). In Italy, a significant number of local municipalities has developed transferable development

---

[*] Department of Urban Planning, University of Venice.

Economic Studies on Food, Agriculture, and the Environment
Edited by Canavari *et al.*, Kluwer Academic/Plenum Publishers, 2002

rights (TDR) programs for urban and regional management while trying to avoid the traditional zoning procedures.

By analyzing the urban plans whose management is based on TDR programs, we can assess the possibility of passing these tools from economic theory to the actual practice in urban government.

This chapter considers the Italian experience and the relative pitfalls. The first section takes into consideration some of the theoretical aspects related to public intervention in urban planning. The second examines the Italian experiences with TDR programs. The third presents a critical evaluation of these experiences in light of economic theory.

2.   THE TOOLS FOR SPATIAL EXTERNALITIES REGULATION

In public economics, planning can be considered a way of regulating the externalities that affect urban and regional systems (Chung, 1994; Ferraro, 1990, p. 36 and ss.).

The planning technique of zoning can be defined as a device for regulating land use within a spatial area. It represents a tool with which a community can deal with the externalities raised by the physical and spatial interaction typical in a city or regional setting. Through the attributions of specific-use functions for land, zoning establishes land uses and the ways in which property can be exploited. Zoning attempts to avoid incompatible uses that can be mutually damaging and to integrate activities capable of generating positive externalities. In addition, zoning identifies the areas designated for use by the community – which in turn generate external economies – including public works and urban facilities that the market would produce, if at all, in a sub-optimal way (Chung, 1994, pp. 78-79).

It is also worthwhile noting that regulating land and building ownership and ensuring that certain land is designated for public facilities, the local administrations change position with respect to the economic agents. In the first case, the public entity regulates the interaction between the acts of production and consumption of economic agents. In the second, the local administration is directly involved in the process of forming externalities; by designating specific areas for urban infrastructures and facilities, it actually sets up projects that generate external economies.

In the field of urban planning, this act of local administrations is well-known primarily through the concepts of classical economics. Planning identifies the area designated for the city and determines the work to be done and/or facilities to be created. In doing so, it contributes to establishing land values and buildings. In other words, it determines the formation of the economic rents related to the areas' improved quality (Alonso, 1960; Camagni, 1992, pp. 279 and ss.).

Thus, externalities, public goods and urban rent indicate the connected phenomena that planning proposes to control through norms, standards and constraints. An economic interpretation of public intervention allows us to consider how traditional planning does not represent the only possible way of regulating urban and regional externalities.

Externalities can be regulated through the approach of market-based devices and through the direct normative approach of "command-and-control" regulation, or governance through the determination of standards and norms. In the former case, it is possible to intervene to correct market inefficiencies without resorting to legislative and normative tools, which are generally held to achieve less effective and efficient results (Turner, Pearce and Bateman, 1996, p. 188).

In addition to the difficulties public intervention encounters in regulating markets, economists have argued that market failures, represented by externalities and public goods, have been followed by the inefficiency of government based on the command-and-control approach (Petretto, 1987, pp. 159 and ss.; Wolf, 1987).

The weak efficiency of regional planning can then be attributed, at least in part, to the authoritative nature of the tools for implementing and managing plans and to the clear inequalities they induce. As a result, there is a growing interest in creating innovative planning tools – through real estate taxation and the creation of new markets – that do not replace the market, but are limited to intervening in it (Lanotte and Rossi, 1995; Stellin and Stanghellini, 1997).

The creation of a development rights market to manage urban growth has roused great interest in various developed countries[1]. According to Coase (1960), the establishment of a property rights market can replace direct forms of public intervention. The key concept around which his reasoning is developed is the property right. Coase's Theorem affirms that, if the property rights of any resource are clearly attributed, there is an automatic tendency to strive toward a socially optimal solution through negotiation between the parties, regardless of who holds the rights. The implications of this turn out to be highly significant for public decision-making and urban policies. If the theorem is correct, public administrations no longer need to regulate externalities if there is a possibility of establishing a specific property rights market; in such a case, the demand and supply autonomously and automatically re-establish conditions of efficient equilibrium.

The environmental permits programs are analogous to the development rights markets. As in the case of the rights market, trading environmental permits takes advantage of the market itself, modifying its signals with the aim of orienting economic agents' choices toward socially shared goals (Turner, Pearce and Bateman, 1996, pp. 235 and ss.). The first step towards establishing a market for environmental permits lies in determining a level of licencing (for example, pollution) to attribute to economic agents. Once this initial allocation has been made, the economic agents owning permits are free to market their rights. In the United States, various environmental policies have successfully adopted the tool of negotiable permits, insuring consistent benefits for companies without increasing pollution (Gastaldo, 1992).

The planning market-based tools under consideration are not without limits and objections. The limits of Coase's theorem are acknowledged by the author himself.. In particular, it may be impossible to establish an efficient market for property rights because of the high transaction costs associated with the negotiation between the parties involved or – still earlier in the process – because of the difficulty in precisely identifying which entities generate externalities and which, are affected by them (Pearce and Turner, 1991, pp. 77 and ss.; Frank, 1992, pp. 659 and ss.).

Regarding the use of environmental permits, various objections have also been raised. Among the most important, are whether it is proper to sell use of the environment (one can, for example, actually buy the right to pollute) and, once again, they are charac-

---

[1] The section of the journal *Urbanistica,* edited by Fusco Girard (1997), presents comparative cases in Italy, Spain and the United States, with specific exploration of case studies. Within the context of the OECD research project, Renard (1998) examines the experiences in France, New Zealand and the United States.

terized by high administrative costs. Though these criticisms significantly condition the effectiveness of similar approaches, the prospect of enhancing planning performance through tools based on market behaviours, which limit – to whatever extent possible – the use of command-and-control tools, appears nonetheless stimulating. It is not merely by chance that the International Conference on the Environment, held in Kyoto in 1997, explicitly indicated environmental permits as a tool for managing on an international scale the exhaust gases linked to the greenhouse effect. [2]

The prospect also appears of interest in urban planning in light of the experimentation carried out in recent years. This implies the possibility of going beyond traditional land use management through new markets in which development rights are exchanged.

3. DEVELOPMENT RIGHTS MARKETS TO MANAGE PLANS IN ITALY: STRATEGIES AND EXPERIENCES

In Italy, the market transfer of development rights-methodology has a wide range of applications, varying from the conversion of consolidated urban areas to the protection of environmental heritage[3]. What follows are the major points of reference and strategies that have guided administrative actions in setting new tools for plans' management. This is followed by an examination of the major problems confronting those managing development rights markets.

The experiences of plans based on the transfer of development rights in Italy follow a common scheme. Within a long-term planning framework, the public administration identifies the local areas designated for urban projects. These areas are then examined both in terms of their actual status (i. e. if they are used for urban purposes or agricultural ones) and in law (the norms of the existing plan).

The areas designated for urban transformation are classified on the basis of the characteristics identified. Every class of soils is give a building index that is applied, without distinction, to the areas designated to private and public use alike. Every class is then subdivided into sections, within which the property owners can negotiate the development rights on their own.

Owners of property designated for collective facilities and public infrastructures possess development rights that can only be used for the areas of the plan designated for private building. The owners of those areas use volumetric rights and "host" the rights of other property owners. Once the development rights have been established, the property owners of the areas designated for the city relinquish their areas to the administration at their opportunity cost (farmland prices).

This scheme aims at reaching several goals simultaneously. In the first place, the ownership of the land designated to be converted is only treated with reference to its status in fact and law, without respect to the choices made in the plan. The inequity of the

---

[2] Kyoto Protocol on Climate Change, United Nations, Framework Convention on Climate Change, 1-11 December 1997.

[3] Several important studies have already been conducted on this topic. In addition to those already cited, see Barbieri and Oliva (1995), Forte and Fusco Girard (1998), Micelli (1999), Pompei (1998) and Stanghellini (1993). As far as the cases in the United States are concerned, see essays by Hagman and Misczynski (1978) Jacobs (1997 and 1999), Johnston and Madison (1997) and Pruetz (1993), for the most complete summary of already completed projects.

zoning is thus mitigated by distributing the land value among all the property owners involved in the city's transformation. Moreover, the equalization principle makes land ownership of no consequence to the planner's choices; in the moment in which the property owners obtain the same building index – leaving the actual land designations aside – they are no longer interested in diverting public decisions toward private interests. Finally, the equalization of development rights allows the administration to purchase the land required for public use, at farm prices or for nothing at all, in agreement with the land owner, to whom a share of the property value – substantially related to development potential - is in any event recognized.

The transfer of development rights based on this general scheme has been applied within two different strategies. First, the TDR principle is applied to all the urban areas designated in the plan for urban transformation. The distribution and the transfer of development rights become pervasive tools for regulating the use of city land, whether it be for the areas that the plan designates for conversion from agricultural to urban use, or for those that are the object of significant urban renewal (such as abandoned areas). Examples of this first approach include Turin, Reggio Emilia, Piacenza, Casalecchio di Reno near Bologna, and Parma.

In the second strategy, the TDR principle is only applied to that portion of the areas under transformation which have been attributed the role of catalyst in the specific project or program. When the TDR principle is only applied to a portion of the areas designated for urban transformation, the scope of the plan includes two distinct regimes: that of traditional zoning and that of equalization regarding a certain land class. The most significant example of this second strategy is represented by the rehabilitation project for the city of Ravenna's Wharf and the concomitant development of the "green belt" surrounding the city (Crocioni, 1998). Similar experiments, however, are now underway in Padua and Venice.

In all the cases considered, the administration only attributed development rights to specific areas in the plan. Development rights are not in fact attributed to all the land designated for urban transformation in the case where the TDR principle was employed throughout the entire plan. They are only distributed to some of the areas designated for urban transformation where the administration had decided to adopt the TDR principle selectively or, rather, to manage specific parts of the plan. In both cases, nevertheless, it is the administration that establishes the land to which the transfer development rights can be attributed.

The areas involved in the TDR programs are thus subdivided into classes (Table 1). The land designated for urban transformation is actually recognized by a different statute. From an economic point of view, the owner of property inside the city is in a different position than a property owner of an area designated to be converted from agriculture to urban use. Analogously, a property owner whose area has already been designated by the plan for urban use (a new building, for example) is in a different situation - from a legal point of view – than a property owner of an area that was previously designated for public facilities.

The administrations recognize the different property owners' situations by intermeshing the legal and economic characteristics of the areas, thereby protecting the "acquired rights". Other criteria may also be added to these basic premises, linked to the area's specific economic characteristics or even to the administration's design goals. In

Table 1. A comparison of eight cases of equalization and transfer of development rights by project type, land classification, use of incentives and building indexes

Municipality	Gener-alized use	No. of classes	Land classification	Use of volumetric incentives	Building index sq.m/sq.m
Casalecchio di Reno	Yes	2	Marginal areas inside the city	No	0.23
			Peri-urban region		0.115
Reggio Emilia	Yes	3	Abandoned areas	No	0.40
			Converted settlements		0.25
			Green areas		0.10
Piacenza	Yes	6	Abandoned areas < 3 ha	No	0.50
			Abandoned areas > 3 ha		0.35
			Productive areas		0.30
			Mixed-use areas		0.30
			Military areas		0.25
			Open areas		0.10
Venice	No	1	Areas of environmental up-grading	Yes	0.44
Padua	No	2	Abandoned areas of the Urban	No	0.40
			Redevelopment Plan		0.50
			Abandoned areas with greater building density		
Ravenna	No	1	Green belt areas	Yes	0.10
Turin	Yes	4	Urban renewal zone	No	0.70
			Renewal areas for services		0.23
			Urban and river parks		0.05
			Natural park areas of the hill		0.03
Parma	Yes	3	Areas inside the urban center	No	0.50
			Areas outside the urban center		0.15
			Previously restricted areas of the urban center		0.25

Source: administrations' data.

Piacenza, for example, the very large abandoned areas were differentiated from the smaller ones taking into consideration the overall returns bound to land values.

Whereas the land classification is limited to grouping land with analogous characteristics, the attribution of the indexes establishes how much usable surface area (or cubic volume) can be developed. This step is crucial to the extent to which the land value is primarily a function of building capacity. The plans currently being implemented point out certain similarities and differences, from the point of view of the attributed indexes. The experiences completed in the Emilia-Romagna region show how the land owners and administration reached an understanding with regional indexes of about 0.1 sq.m/sq.m for the land converted from agricultural to urban use.

Higher values were determined for the areas that were abandoned, under-utilised and undergoing progressive abandonment. In virtue of the different rights acquired in terms of potential building volume, the areas included in these classes can reach indexes that are significantly much higher. In Reggio Emilia, abandoned areas were attributed an index of 0.4 sq.m/sq.m, while in Piacenza, analogous areas reached a building density of 0.5 sq.m/sq.m (see Table 1).

Building indexes can be defined by a unilateral decision made by the municipality or, more frequently, they are the result of a more or less structured orchestration between the administration, the property owners and interested social parties. In Parma, for example, the determination of the indexes and their possible forms of use were the object of a long phase of negotiation with property owners during the development of the City Plan, with the declared aim of maximizing the tool's effectiveness.

Once the general aspects are defined, it is necessary to establish the development rights market in an effective way. For the equalization mechanism to enter effectively in the operation, property owners have to transfer the development rights granted to them within the framework of the project provisions established by the Municipality.

The tool most often used for this is the *agreement* (convenzione) between property owners included within the same *urban section* (comparto). Here, the case of Ravenna is exemplary. The transfer of development rights agreement works on two distinct levels. On the more general level, the administration drew up a *basic agreement* that defines the essential points for any agreement between the administration and private land owners. For more specific projects, the property owners involved established the concrete form of the building rights negotiation autonomously within the framework set up by the basic agreement.

In most cases, the rights market is just beginning to come into use and manifests itself in the form of land trades for properties to which the administration has attributed various use. Nonetheless, in some cases there seems to be an autonomous building rights market. Ravenna again furnishes an interesting example. Transactions have recently included areas designated to become an urban park that were put up for auction only for their developing rights. In this way brokers and property owners acquired building rights without necessarily using them directly, waiting instead for interesting investment opportunities in those areas designated to receive the transfer development rights of the city's Wharf. In these transactions the development rights are separate from the lands to which they were originally bound and become immaterial *assets* that can be bought and re-sold, as has occurred in the most significant cases in the United States[4].

4. TDR AND TRADITIONAL PLANNING: THE COORDINATION OF *COMMAND-AND-CONTROL* AND MARKET-BASED TOOLS TO MANAGE PLANS

Two major problems historically emerged with respect to plans' implementation and management: first, the resistance of property owners to land use regulations meaning a significant reduction in the value of their holdings; second, the administrations' need to recover part of the positive externalities generated by the plan with the aim of financing its realization. If well managed, the development rights markets can contribute to the resolution of these two problems.

---

[4] The most advanced case is represented by the transfer development rights bank for Pinelands Park in New Jersey. It performs two functions: on one hand, it furnishes the necessary informative framework to the operators – property owners and brokers – regarding the value of the rights and the possibilities linked to marketing them; on the other, it purchases rights for property owners that are unable to find a buyer and sells them to brokers and property owners interested in increasing the buildable volume of their lots. For more information on the Pinelands, see Johnston and Madison (1997) and Micelli (1997).

The analysis of the TDR programs shows that newly created markets do not aim at replacing the plan's traditional management tools, but rather at integrating the former to make the latter more efficient. The integration of market-based and command-and-control tools is the general trait that marks TDR programs in Italy.

The attribution of development rights is actually made on the basis of a plan choice, and their marketing is organized in a significant way by the public. Furthermore, the rights can only be marketed within given sections, and their use is subordinated to adhesion with respect to the design proposals furnished by the administration. This is coherent, moreover, with an innovation that begins with the administrations' search for tools capable of ensuring better performance in terms of the plans' effectiveness without completely transforming the available tools.

Analysis of the Italian experience shows how the rights markets have been designed for regulating specific externalities. Is it possible to hypothesize new rights markets oriented not only to managing the two problems already discussed, but also the externalities that rise from the interaction between consumers and/or producers in the city? An analysis of transaction costs is crucial in answering this question. In reality, the negotiation around the externalities bound to the city form and function regarding a substantial number of subjects. As foreseen by Coase (1959, p. 26), in the presence of high transaction and operations costs, "... as a practical matter, the market may become too costly to operate. In these circumstances, it may be preferable to impose special regulations." Thus, the effectiveness of the rights markets is not ensured and the return to command-and-control tools currently remains the only possible solution.

The transfer development rights markets make it possible to find solutions to certain significant urban management problems such as the inequity bound to zoning, and the recovery of portions of value produced by the public and otherwise designated only to a few property owners. Other issues, such as land use designations and building density, that regulate the externalities tied to land use have been left to traditional tools of command-and-control. The use of the former tools for regulating other externalities could prove to be wrong for structural and not contingent reasons; the high transaction costs make it rational to employ command-and-control tools. In all probability, the elimination of every form of norm and standard in urban planning belongs to the utopia of certain ultra-liberal groups (Jacobs, 1997, p. 64), which go beyond the positions of Coase (Chung, 1994, p. 92).

In the future, it is probable that the success of the market-based tools will be tied to their capacity to be integrated with traditional urban tools (Renard, 1999, p. 16). The failure of certain projects, for example, demonstrates how the inflexibility of norms bound to forms and functions can heavily condition the take-off of the transfer of development rights. It also inhibits their marketing, failing in terms of both equity – if the transfer development rights are not the object of transaction, the land designated for public infrastructures and facilities is not compensated – and efficiency – the public and the private city remain simple provisions of the plan.

The administration's activity significantly conditions the form of the development rights market. It is the municipality that establishes the areas to which the building rights are to be assigned: in the case of pervasive equalization of the property rights, they go to all the areas of expansion and/or urban transformation; in the cases of partial equalization, only to a part of the latter. The public entity also organizes the land classification and subsequent attribution of building indexes.

Nevertheless, the administration's visible presence during the rights allocation phase is common to most of the cases where rights and environmental permits markets have been created. The initial distribution of the development rights actually has significant analogies with the initial distribution of environmental permits. When the administration gives out environmental permits (permits to pollute, for example), it usually makes an initial allocation of permits based on levels of pollution recorded in the past. This procedure follows a historical approach known as *grandfathering*: pollution rights are tied to past pollution levels (Turner, Pearce and Bateman, 1996, pp. 235-236). In the development rights programs as in the management of environmental permits, there are no automatic mechanisms for the initial allocation of rights. The administration necessarily becomes the entity that has to establish the rules. Usually, the rules held to be most equitable take into consideration the "rights acquired" in the past by the subjects to whom these rights or permits are attributed.

The exchange of rights has taken on different forms. On one hand, it can arrive at actual land transactions managed by the administration, which organizes and favors the exchange of public and private areas through, for example, a barter game. In this case, the rights market tends to disappear in favor of a technique of land re-composition oriented toward increasing the plan's equity and efficiency. On the other hand, where the economic operators learn the new rules of the game, and have faith in whoever has promoted them, it is possible for the development rights to become the object of a local market endowed with its own autonomy. The example of the development rights auctions held, for example, in Ravenna confirms the plausibility of this approach.

In any event, the rights exchange between property owners is always affected by significant transaction costs. Public administrators have usually found it useful to reduce these costs by decreasing the number of property owners involved in the urban sections or the minimal spatial unit subject to urban transformation. Thus, constraining the transactions seems to depend significantly on the lower number of economic agents involved. However, this leads to the possible formation of monopolies and/or monopsonies among the property owners within the section, which significantly reduces the chance for real market prices to form (Renard, 1999, p. 15; Jacobs, 1997, p. 63). Once again, the administration has to be ready to intervene, through the backing of a "bank", for example, to purchase the development rights and allow the property owners of the rights not to give in to potential situations of monopoly or monopsony (Heeter, 1975).

Thus, the institution of a previously non-existent market does not appear to be a risk-free operation. On the contrary, it requires the administrations entrusting the implementation of their plans to these kind of tools to make a significant effort toward innovation. Moreover, it would be misleading to maintain that the market of development rights can function immediately in a decentralized way. Such markets require an important effort in communication and training to the extent that the marketing of development rights is not an operation to be taken for granted (Renard, 1999, p. 14).

The administration's investment in training the property owners and the real estate operators – together with specific normative provisions, especially in the field of taxation – represents a ground for important experiments in reducing the transaction costs present in the development rights markets, with important implications for their success.

5. CONCLUSION

Planning can be understood as a device to regulate the externalities affecting cities and regions. Having recognized the inefficiency of the authoritative command-and-control tools, some administrations have been trying to implement and manage urban and regional plans through the use of tools that intervene in the market, orienting the behaviour of the agents toward socially shared goals.

The allocation and the transfer of development rights represent innovative tools of great interest. Several elements emerge from an analysis of the major case studies in Italy. First, markets for development rights do not replace the command-and-control tools traditionally used in planning. In reality, the success of the new markets seems to depend significantly on their integration with the latter.

Moreover, development rights markets have not proven to be automatic devices led by an invisible hand. In an analogous way to other markets for rights and environmental permits, the *visible hand* of the administration takes steps to establish the market rules and to promote its functioning, reducing transaction costs as much as possible. In a perhaps paradoxical way, the use of tools that intervene in the market seems to require significant managerial and administrative investment, on which the success of the initiative depends.

Future research could concern the crucial aspects for the success of municipal market-based tools, especially beginning with the *best practice* approach, that in recent years certain municipalities have been able to design and implement, with "a strong innovative quality in the planning tools servicing objectives of both efficiency and equity" (Camagni, 1999, p. 169).

6. REFERENCES

Alonso W. (1960), A Theory of the Urban Land Rent, *Papers of the Regional Sciences Association*, **6**: 147-157.
Barbieri C.A. – Oliva F. (1995), *Le prospettive perequative per un nuovo regime immobiliare*, Urbanistica Quaderni, 7.
Camagni R. (1992), *Economia urbana. Principi e modelli teorici*, NIS, Roma.
Camagni R. (1999), *Considerazioni sulla perequazione urbanistica: verso un modello percorribile e giudizioso*, in: *Le misure del piano*, Lombardi P. – Micelli E. (eds.), Angeli, Milano, pp. 158-169.
Chung L. (1994), The economics of land-use zoning: a literature review and analysis of the work of Coase, *Town planning review*, **65**(1): 77-98.
Crocioni G. (1998), *Il progetto della darsena di città nel nuovo Prg di Ravenna: un primo bilancio critico*, lecture prepared for the conference on La perequazione urbanistica: i temi e le esperienze, May 15, 1998, Department of Urban Planning, IUAV, Venice.
Coase R. (1959), The Federal Communications Commission, *Journal of Law and Economics*, **October**: 1-40.
Coase R. (1960), The Problem of Social Cost, *Journal of Law and Economics*, **October**: 1-45.
Dal Piaz A. – Forte F. (eds., 1995), *Piano urbanistico, interessi fondiari, regole perequative*, Clean, Napoli.
Ferraro G. (1990*), La città nell'incertezza e la retorica del piano*, Angeli, Milano.
Forte F. – Fusco Girard L. (1998), *Valutazioni per lo sviluppo sostenibile e perequazione urbanistica*, Clean, Napoli.
Frank R. (1992), *Microeconomia. Comportamento razionale, market, istituzioni*, McGraw-Hill, Milano.
Fusco Girard L. (ed., 1997) La perequazione urbanistica: le esperienze e le questioni, *Urbanistica*, **109**.
Gastaldo S. (1992), Les droits de polluer aux Etats Unis, *Economie et Statistique*, **October-November**: 35-41.
Heeter D. (1975), Six Basic Requirements for a Tdr System, in: *Transferable Development Rights*, Pas Report No. 304, American Society for Planning Officials, Chicago IL.
Hagman D. – Misczynski D. (eds., 1978), *Windfalls for Wipeouts*, Planners Press, Chicago IL.

Jacobs H. (1999), Regolazioni basate su meccanismi di mercato in un sistema di governo decentrato, in: *Urbanistica e fiscalità locale*, Curti F., ed., Maggioli, Rimini, pp. 135-150.

Jacobs H. (1997), Programmi di trasferimento di diretti edificatori in Usa: oggi e domani, *Urbanistica*, **109**: 62-64.

Johnston R. – Madison M. (1997), From Landmarks to Landscapes, *Journal of the American Planning Association*, **63**(3): 365-378.

Lanotte H. – Rossi D. (1995), Négocier les droits sur le sol, *Etudes foncières*, **68**: 19-26.

Micelli E. (1997), Il piano dei Pinelands nel New Jersey, *Urbanistica*, **109**: 65-67.

Micelli E. (1999), La perequazione urbanistica per le politiche dei suoli urbani: i modelli e le esperienze, *Estimo e territorio*, **4**: 13-20.

Micelli E. (1999), La négociation des droits de bâtir en Italie, *Etudes foncières*, **82**: 17-22.

Petretto A. (1987), *Manuale di economia pubblica*, Il Mulino, Bologna.

Pearce D. – Turner R.K. (1991), *Economia delle risorse naturali e dell'ambiente*, Il Mulino, Bologna.

Pompei S. (1998), *Il piano regolatore perequativo*, Hoepli, Milano.

Pruetz R. (1993), Putting Transfer of Development Rights to Work in California, Solano Press Books, Point Arena, CA.

Querrien M. (1999), Pour en finir avec les transferts de droits de construire, *Etudes foncières*, **83**: 31.

Renard V. (1998), *L'utilisation des permis négociables dans le domaine de la gestion des sols*, Atelier sur les systèmes de permis négociables nationaux pour la gestion de l'environnement: questions et défis, Direction de l'environnement, OCDE, Paris.

Renard V. (1999), Où en est le système des transferts de Cos?, *Etudes foncières*, **82**: 8-16.

Stanghellini S. (ed., 1993), *Per la riforma urbanistica del regime immobiliare*, Urbanistica Quaderni, 13.

Stellin G. – Stanghellini S. (1997), Politiche di riqualificazione delle aree metropolitane: domande di valutazione e contributo delle discipline economico-estimative, *Genio rurale*, **7/8**: 47-55.

Turner R.K. – Pearce D. – Bateman I. (1996), *Economia ambientale*, Il Mulino, Bologna.

Wolf C. (1987), A Theory of Non-market Failure: Framework for Implementation Analysis, *Journal of Law and Economics*, **October**: 107-139.

TRANSFORMATION OF RECREATIONAL ENVIRONMENTAL GOODS AND SERVICES PROVIDED BY AGRICULTURE AND FORESTRY INTO RECREATIONAL ENVIRONMENTAL PRODUCTS

Maurizio Merlo [*]

1. INTRODUCTION

In addition to traditional market commodities, agriculture, forestry and the related environment, produce a large set of Environmental Recreational Goods and Services (ERGSs). Examples include pleasant landscapes, rural lanes and footpaths, habitats for various kinds of flora and fauna, grounds for sports and other recreational activities. At the same time agriculture and forestry can produce Environmental Bads and Disservices (EBD). These ERGSs, and EBDs, are generally perceived by our societies as public goods and/or externalities of farming and forestry, as people cannot be excluded from using them and rivalry is not greatly felt; everyone can enjoy and/or suffer from them without any market transaction taking place.

Failure to internalise the external effects of public goods supplied by agriculture and forestry is typically due to free riding. It subsequently provides fuel to 'the tragedy of the commons' types of problems. The right signals are not perceived by producers/consumers; the provision of ERGSs is not sufficient to meet the demand, while the prevention of EBD is not adequate [1]. The consequence can be a loss of social welfare.

[*] Maurizio Merlo, Centro di Contabilità e Gestione Agraria, Forestale e Ambientale - Università di Padova.
The present paper reports results of a EU financed research on 'Niche Markets for Recreational and Environmental Services – RES' (FAIR – CT95-0743) undertaken in collaboration with the University of Hamburg (co-ordinator Prof. Udo Mantau), the University of Vienna (Prof. Walter Sekot) and IBN-DLO-Wageningen (Dr. Kees Van Vliet).
Part of the results of the above research, specifically focused on forestry, is also presented in Merlo M. et al. (2000). This version should appear as a chapter of a book to be edited by F. Brower. The author is indebted for the support given, among others, by E. Milocco, P. Virgilietti and R. Panting.

[1] The price support measures that are offered under the Common Agricultural Policy (CAP) during the past 30 years are often recalled for having systematically helped intensive farming, rather than extensive and environmentally friendly farming practices and their related ERGSs.

Economic Studies on Food, Agriculture, and the Environment
Edited by Canavari *et al.*, Kluwer Academic/Plenum Publishers, 2002

293

prevention of EBD is not adequate [1]. The consequence can be a loss of social welfare. Therefore, specific policy tools have been developed to achieve sufficient provision of ERGSs, and prevention of EBDs.

The main policy tools presented in Figure 1 include: (i) mandatory regulations; (ii) financial/economic interventions by the public sector; and (iii) market creation for public goods/bads and externalities. Complementary measures such as persuasion and communication are also considered to be essential for implementing the above set of policy tools.

1.1. Mandatory regulatory tools

Mandatory regulations, i.e. the 'stick' approach which may be softened by some form of compensation, have traditionally been applied to provide ERGSs, and prevent EBDs. Legally binding tools are often included within constitutions and laws throughout Europe. The framework of property rights, planning/programming, environmental standards and licences supported by codes of practice and indicators are also important. Forest laws, soil conservation, and water management are well established examples. The adoption of mandatory measures is now conceived as an essential part of any policy package aimed at conservation of natural resources.

Of course societies' changing ethical and cultural values, common understandings and consensus, have always been at the root of legal options. Obligations were conceived as a social commitment, and compensation was not generally considered. Now, however, mandatory tools tend to be applied as remunerated 'voluntary means with powers of compulsion'.

A negative aspect of mandatory tools, which is often overlooked, is the high administrative transaction costs of policy implementation, requiring monitoring, control and policing (Whitby and Saunders, 1996); this is particularly relevant when social consensus is not widespread and administration/services are not well established. In these cases economic instruments are considered the most viable alternative. However, recent evidence shows that consolidated policies and well established administrations reduce transaction costs (Falconer and Whitby, 1999). Another problem with mandatory obligations, particularly in less productive marginal areas like the Alps and the Mediterranean mountains, has been land abandonment due to lack of profitability.

In these contexts purchase and management of land by public bodies like State/Regional Forest Enterprises has also been widely used, e.g. in Italy, France and the UK, to guarantee the provision of ERGSs. The experience, in general, is positive whenever management is supported by effective administration. However, neither purchase nor management by public bodies, are now considered efficient tools. Costs are generally too high.

---

[1] The price support measures that are offered under the Common Agricultural Policy (CAP) during the past 30 years are often recalled for having systematically helped intensive farming, rather than extensive and environmentally friendly farming practices and their related ERGSs.

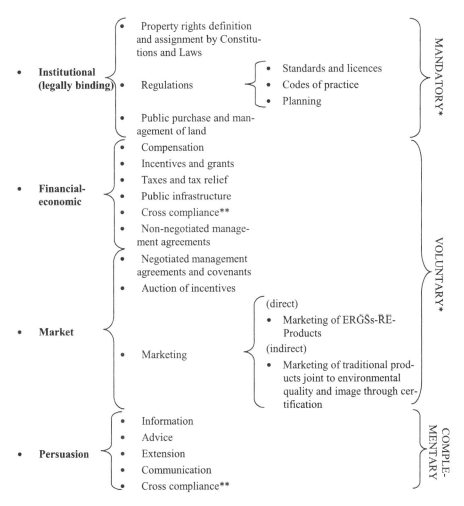

* In certain countries the difference between mandatory and voluntary measures remains vague. Certain mandatory measures should be understood as voluntary with powers of compulsion

** Cross compliance can be considered strongly persuasive, where the provision of ERGSs is conditional to the other financial measures (including income support).

Figure 1. Policy tools aimed at achieving provision of ERGSs and prevention of EBDs

Doubts are cast on the efficiency of this approach [2]. Semi-public bodies, as shown in Figure 1, are now more widely used and advocated, like Trusts, Consortia, Common

2 For instance in Italy costs have been estimated at around 200 Euro/ha for State forests; 150 per regional forests; 90 for communal and common property forests and 40 for private forests. Of course it must be recognized that public forests generally provide a larger share of ERGSs (Centro Contabilità e Gestione Agraria, Forestale e Ambientale, 1998).

Property, and various Non Governmental Organisations (NGOs), provided, of course, that these bodies can act without state support.

1.2.   Financial-economic voluntary tools

Financial-economic tools, following Pigou's (1920) internalisation approach, are based on the 'carrot' rather than on the 'stick' approach; these are positive instruments aimed at convincing farmers/landowners to implement certain measures in exchange for various advantages. According to OECD (1996), this is a 'state pays approach'. The Keynesian side must not be neglected since it is often applied to create jobs and activate depressed rural economies.

It has been the 1992 Common Agricultural Policy (CAP) to apply financial means extensively particularly in terms of incentives through Regulations 2078/92 and 2080/92, respectively aimed at the agri-environment and afforestation measures. The use of these financial tools has been confirmed by the Integrated Rural Development Incentives of Agenda 2000, and formulated under Regulation 1257/99 (specificare in nota di cosa si tratta). [3]

Financial-economic incentives vary according to their economic nature and admini-stration. One type is compensation to meet cost increases and/or income losses to main-tain certain types of land uses and the related ERGSs, as well as prevention of EBDs. Incentives and investment grants are also widely adopted in various countries. Unlike compensation, the payments include something more – i.e. a surplus to stimulate partici-pation in a programme including non-productive environmental investments. It is inter-esting to note that this extended view is fully accepted by Article 24 of Regulation 1257/99 of Agenda 2000 in Chapter VI (agri-environmental measures).

Fiscal measures might also contribute to the maintainance of traditional farm-ing/forestry systems and the related ERGSs, or prevention of undesirable land uses (EBDs). Tax exempts are widely adopted in Less Favoured Areas and forests in countries like Italy and France. Taxes, perhaps the most commonly envisaged instrument to control land use, and therefore environment quality, follow the much advocated, and little ap-plied, Polluter Pays Principle.

Cross-compliance can be seen as an indirect financial instrument. It was widely de-bated after the Berlin Summit in March 1999. The concession of existing payments, such as income support, is taken as conditional to the adoption of environmentally friendly techniques: i.e. eco-conditionality. In other words cross compliance could be considered a type of strong persuasion supported by various financial/economic payments, thus ex-plaining its appearance in Figure 1.

All Agenda 2000 'ordinary' payments fully paid by the EU, according to Regulation 1259/1999 [4], should be conditioned on the application of environmentally 'good farming practice'. Unfortunately this condition is completely left to Member States. It is worth noting how agri-environmental measures of Regulation 1257/1999, merely financed by

---

[3] Council Regulation (EC) No 1257/1999 of 17 May 1999, on support for rural development from the Euro-pean Agricultural Guidance and Guarantee Fund (EAGGF). It establishes the framework for EU support for sustainable rural development.

[4] Council Regulation (EC) No 1259/1999 of 17 May 1999, establishing common rules for direct support schemes under the common agricultural policy. It introduces constraints to payments related to environmen-tal protection requirements and principles of modulation.

the EU at 50%, are under EU direct control. The limited guidance offered by the European Commission in formulating such Codes of Good Agricultural Practices is probably one of the major shortcoming of the reform under Agenda 2000.

Payments, incentives and tax relief, have also been granted throughout Europe by various states involving, for instance, millions of hectares for reforestation in Central and Southern Europe. Timber production objectives were often declared, while environmental objectives were thought to be implicit.

Certainly it should be acknowledged that policy failure is an intrinsic risk of financial mechanisms as shown by several cases around the world (Paveri and Merlo, 1998). For instance, the various shortcomings of post-war reforestation policy in Southern Italy should be acknowledged, and may start to become recognisable in other countries. Whitby and Saunders (1996), with reference to agri-environmental policies, have shown British cases where landowners and farmers have been overcompensated by 'standard payments', undermining an ethical commitment to stewardship. Meanwhile, where public goods and positive externalities were in greater demand, appropriate incentives were not made available. Of course, this kind of reflection on the intrinsic danger of intervention policies can be extended to the overall CAP. [5]

1.3. Market tools

Marketing ERGSs by their transformation and development into Recreational Environmental (RE) Products is a relatively new policy tool, which is currently envisaged by the EU [6] and various Member States as a means to increase the provision of ERGSs, prevent EBDs, and to create additional income for the farming industry. The approach, based on the 'beneficiary pays' principle of the OECD (1996), is, to a certain extent, the one theorised by Coase (1960). Coase asserted that the best way to cope with public goods and externalities is through free market arrangements amongst the concerned parties, bypassing the various pitfalls inherent in regulations and internalisation through state payments.

As shown in Figure 1, management agreements providing payments subject to negotiations between land owners/farmers and the responsible public authorities can be considered the first step toward market creation (Bishop and Philips, 1993). It is still the 'state pays approach' (OECD, 1996), however mitigated by the negotiation process. The provision of standard payments should avoid possible excess payments, resulting in owners/enterprises' rents (Whitby and Saunders, 1996). A more extended view of management agreements, requiring registration of contracts, is given by the so-called covenants, or *servitutes praediorum* of Roman law, legally attached to the land. From the community's point of view they represent a stronger commitment to the provision of ERGSs, or prevention of EBDs.

Auction of incentives is another market-led tool based on competitive bidding directly submitted by landowners wishing to provide ERGSs under specific programmes financed by the public sector (Hamsvoort and Latacz-Lohman, 1996). Bids represent the

---

[5] Support has been given above all to the high productivity areas, which have been over compensated, while the low productivity areas (naturally disadvantaged) have been neglected. According to 1992 CAP reform premises, 20% of the farmers were draining 80% of the subsidies.

[6] E.g. Chapter IX, Art. 33 of Regulation 1257/99 where farm diversification, rural tourism and environmental conservation are considered as a means to achieve rural development.

amount of money for which landowners/entrepreneurs are willing to start the scheme and/or accept its restrictions.

However, it is with the marketing of ERGSs, more or less transformed into real RE-Products, that the market approach reaches its full extent and potentials. Cases are now surfacing all over Europe under different situations, institutions, managerial skills, and RE-Products typology.

In order to document this development, a survey has been undertaken and is reported in this paper. This paper first offers a theoretical background based on public and private goods theory, to be followed by a description of surveyed ERGSs and RE-products. The results of the transformation of ERGSs into RE-Products from the surveyed countries, as well as the underlying economic and institutional mechanisms that were utilized, are outlined in the subsequent section. Paths of transformation/development are shown with reference to the various areas conditions and case studies. The paper concludes with policy and management implications, stressing that the market approach to agri-environment issues must be seen in the context of appropriate policy mix and area packages of measures, including regulatory and financial means as well as information and communication.

2. THEORETICAL BACKGROUND FOR TRANSFORMATION OF ERGs INTO RE- PRODUCTS

Following the work of Samuelson (1954) and Musgrave (1959), goods can be defined in terms of rivalry and excludability, two terms much discussed, though sometimes misunderstood (Randall, 1987). Pure public goods are nonrival and nonexcludable in consumption, therefore fully available to the public. On the contrary pure private goods are fully rival and excludable.

ERGSs provided by agriculture and forestry generally appear and are perceived by our society, as public goods and/or externalities, implying low excludability and rivalry. The degree of excludability can, however, vary according to the property rights of each particular ERGS and are sometimes as different across Europe as are the local customs and the transaction costs to enforce exclusion. The degree of rivalry is equally variable according to each particular ERGS (e.g. a landscape visible from a public road does not generally have a high level of rivalry while the access to a nature reserve can create congestion and rivalry).

The two concepts of rivalry and excludability are shown in Figure 2, where excludability ranges from 0 to 1 along the horizontal axis, and rivalry from 0 to 1 along the vertical axis. Examples of similar presentations can be found in Buchanan (1967, p. 188), perhaps the first author to propose the diagram, and Brosio (1986) who added a third dimension from local to global goods.

2.1. The continuum from public to private goods: mixed cases

Pure public goods and pure private goods are 'polar' cases, as termed by Samuelson (1955). The actual situation is, in fact, much more differentiated and varied as argued by various authors criticising and trying to complete Samuelson's model. Buchanan (1965) introduced the new category of 'club' or collectively shared goods, which are offered by an organisation to its members, defraying the cost of the good from member payments

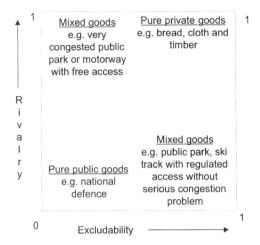

Figure 2. Public, private and mixed impure public goods

(McGuire, 1987, p. 414). These goods, with various degrees of rivalry and excludability, are defined as mixed impure public goods. In addition, Buchanan added a dynamic view, making clear that the group of users can be enlarged or restricted according to property rights (Buchanan, 1967, pp. 190-191). It was also shown that goods and services could move around the diagram according to rivalry and excludability - the allocation of hunting rights and their enforcement (p. 191) being a given example. Other examples of mixed goods were previously given by Tiebout (1956) who made reference to local public goods like hospitals and other services that are also characterised by various degrees of rivalry and excludability.

2.2. The marketability of ERGSs and RE-Products

The theories of local public goods (Tiebout, 1956) and club goods (Buchanan, 1965), have assumed momentum in the last decades regarding the environment and the related ERGSs and RE-Products, such as those provided by agriculture, forestry and the rural environment in general. Some researchers and policy-makers have started to question the 'State Pays Approach' (OECD, 1996), resulting, however, in increased attention to the 'Provider Gets Principle' (PGP) (Blochliger, 1994), while the 'Beneficiary Pays Approach' based on market transaction has remained much neglected (OECD, 1996). Nevertheless, many studies around Europe found a high willingness to pay for ERGSs (Dubgaard et al, 1994). Hanley (1995), in a review of case studies from various countries, found a high level of use of the 'State Pays Approach', in practice a 'Provider Gets Principle' justified in terms of payments to those providing positive externalities and public goods, but very little use of the 'Beneficiary Pays Approach'. On the contrary, extensive

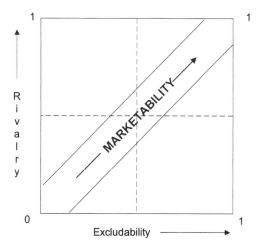

Figure 3. The marketability arrow: from public to private goods (Source: modified from Mantau, 1995)

research on countryside stewardship policies around the EU analysing some 351 policies in 8 member states shows that 12 % of the available ERGSs were already marketed, while 27% were potentially marketable (Gatto and Merlo, 1999).

The marketability concept already shown by Ferro *et al.* (1995) has been formally introduced by Mantau (1995) in the excludability/rivalry diagram depicted by an arrow from public to private goods (Figure 3), which is called a 'marketability arrow'. The feasibility of this hypothesis, that is the applicability of ERGSs marketing tools, has been tested in this chapter surveying 98 case studies in different countries of Europe.

3. THE LIST OF CASE STUDIES (ERGSs) AND THE TYPOLOGY OF RE-PRODUCTS

The 98 case studies investigated are grouped in Table 1 under the headings of each country: Austria (21 cases), Germany (28), Italy (29) and the Netherlands (20). The RE-Products supplied by, or linked to, agriculture and forestry have been classified in three broad categories of recreational, environmental and traditional products. Occasionally, it was not possible to assign a case study to a precise category, because of the complex structure of RE-Products. Therefore the mixed categories of environmental/recreational, traditional/recreational and traditional/environmental products have been included in addition to pure recreational and pure environmental as shown in Table 1.

Remarkably, the majority of RE-Products had a recreational component and fell, to a great extent, in the categories including recreational goods and services. (71 products out

Table 1.	Typology of RE-Products				
	Austria	Germany	Italy	The Nether-lands	Total
Individual category					
Recreational	9	10	16	7	42
Traditional	-	-	-	-	-
Environmental	6	4	1	5	16
Mixed categories					
Environmental/Recreational	3	10	8	4	25
Traditional/Recreational	1	3	-	-	4
Traditional/Environmental	-	1	4	4	9
Not classifiable	2	-	-	-	2
Total	21	28	29	20	98

of 98). These were mainly 'structured' [7] activities giving added value to the ERGSs and taking advantage of the agricultural and forest environment. RE-Products with an environmental component (50 out of 98) included case studies that were strictly dependent on the place where they occurred. Sponsorship provides a good example: the real product is an 'environmental image' promoting the business undertaken by the sponsor. Traditional products with a recreational or environmental component (13 cases out of 98) were marketable products normally produced by agricultural and forest enterprises (like timber and meat), that were transformed to include a certain remuneration for the quality of the environment and for possible recreational activities taking place where they are produced.

4. EMPIRICAL EVIDENCE OF TRANSFORMATION/DEVELOPMENT: INSTITUTIONAL CHANGES AND MANAGEMENT AND MARKETING TECHNIQUES

4.1. Quantification of excludability and rivalry

Throughout the survey, questions on excludability and rivalry dealt primarily with the modifications before and after transformation of the ERGs into RE-Products. The following evaluation procedure was undertaken:

i)   a scale of 5 grades was adopted, both for excludability and rivalry (0=none; 2.5=very low; 5=low; 7.5=high; 10=very high), in order to plot ERGSs and RE-Products in the public-private goods diagram;

ii) rivalry and excludability were evaluated independently, before and after transformation/development;

iii) excludability and rivalry were equal to zero when the RE-Product was newly established and the existing ERGS had free access without rivalry/congestion;

---

[7] That is, activities where "structures" like footpaths, sports and other facilities help the "consumption and usage" of the rural environment.

iv) after transformation/development of the ERGSs into RE-Products, excludability was evaluated according to the combined effects of actions undertaken in terms of institutional changes and management/marketing techniques; [8]

v) after transformation/development, rivalry was valued at the highest level, scoring 10, when transformation was related to a fully rival good (e.g. house lease and product certification) where it is compulsory to pay a price for the RE-Product and the attached ERGSs. [9]

The valuation of rivalry could be biased by subjective judgement. The following criteria have been adopted taking into account that, for club goods, rivalry by definition can not achieve a score of 10:

a) rivalry = 2.5, means that there is a club good without any problems of rivalry among users (internal and external);

b) rivalry = 5, is the standard club good, where there can be accidental/temporary congestion problems at some periods of the year, the week, or the day;

c) rivalry = 7.5, when there is a declared problem of congestion (high internal rivalry) and/or a demand greater than supply (high external rivalry).

4.2.   The state of ERGSs (and potential RE-Products) before transformation

In most cases (Tables 2 and 3) ERGSs, before transformation (and the resulting RE-Products) had the characteristics of public goods, with no excludability and no rivalry. This was sometimes due to the physical constraints to forbid use or access to the ERGSs. There were only a few existing pure private goods which did not require transformation, and in these cases the development of RE-Products dealt mainly with management and marketing.

The few existing pure private goods and services were mainly found in Austria, where in 9 out of 19 cases (Table 2) the ERGSs were already subject to very high exclusion before transformation (e.g. use of forest roads, spring water and fishing). In some cases this was due to a precise definition and assignment of property rights; for example in Austrian legislation water rights are privately owned. In 70 cases out of 95, however, the ERGSs that were to be exploited through RE-Products were non-excludable, or at least excludability was not clearly defined. This applied also to traditional products, such as those transformed into RE-Products, exploiting their environmental image and origin. In general, before transformation the image was that of a public good, since everyone could claim that a certain product came from a certain area where the desired agriculture practices were employed.

---

[8] Whenever the RE-Products 'price' was the only possible way to use the ERGS excludability was valued at 10. However, if paying the price for the ERGS could be avoided, a reduction in excludability was accounted (e.g. voluntary contribution for a picnic area) and excludability was valued around 5.

[9] For club goods and services two types of rivalry were distinguished as either (a) internal rivalry among people already entitled to use the good (i.e. club members) where congestion could be a problem, or (b) an external rivalry among potential users. For example, for admittance to a park, there can be rivalry to obtain the entrance ticket if the number of admitted persons is regulated and demand is greater than supply. Once tickets have been issued, internal rivalry can exist among the admitted persons, due to congestion within the park.

Table 2.   Excludability before transformation *

Excludability	Austria	Germany	Italy	The Nether-lands	Total
No (0)	7	25	23	15	70
Very low (2.5)	-	-	-	-	-
Low (5.0)	3	-	2	1	6
High (7.5)	-	-	1	-	1
Very high (10)	9	3	2	4	18
Total	19	28	28	20	95

* Not applicable to 3 case studies

Table 3.   Rivalry before transformation/development *

Rivalry	Austria	Germany	Italy	The Nether-lands	Total
No (0)	14	22	14	12	62
Very low (2.5)	1	1	3	2	7
Low (5.0)	1	2	9	1	13
High (7.5)	-	2	2	4	8
Very high (10)	3	1	-	1	5
Total	19	28	28	20	95

* Not applicable to 3 case studies

The concept of rivalry overlaps with that of congestion. A good is rival when the consumption by one individual detracts *all* the opportunities of consumption of the same unit of the good to other consumers. Congestion can be considered as 'the situation in which one individual's consumption *reduces* the quality of service available to others' (Cornes and Sandler, 1986). Randall's (1987) congestable goods are those 'that can be enjoyed by many individuals but are subject to a capacity constraint and for which the fixed cost of provision far exceeds the marginal cost of adding additional users until the capacity constraint is approached'. From this definition, most club goods can be regarded as congestable. It is clear that congestion, diminishing the possibility of consumption to other people, causes an increase in rivalry. Rivalry, however, can exist independently from congestion, if a certain good is divisible and consumption by an individual excludes any other.

As far as possible, rivalry due to congestion (internal and linked to indivisibility amongst people entitled to use a certain good) and to the good's intrinsic nature (external) were evaluated separately and summed up in the final assignment of rivalry level. It should be underlined that before transformation it was the rivalry concerning the ERGSs, and not the potential RE-Products, that was calculated, although the two identify themselves to a lesser or larger degree according to the level of their consumption complementarity.

4.3.   The state of RE-Products (previously ERGSs) after transformation

The situation of RE-Products (previously ERGSs) after transformation can be very different, with various levels of excludability, as shown in Table 4. In some cases, pay-

Table 4.   Excludability after transformation/development *

Excludability	Austria	Germany	Italy	The Nether-lands	Total
No (0)	-	-	-	-	-
Very low (2.5)	-	1	2	-	3
Low (5.0)	1	2	-	-	3
High (7.5)	-	-	-	1	1
Very high (10)	18	25	26	19	88
Total	19	28	28	20	95

* Not applicable to 3 case studies

Table 5.   Adoption of measures to prevent free riding *

Measures to prevent free riding	Austria	Germany	Italy	The Nether-lands	Total
No	7	16	9	11	43
Yes	13	3	17	7	40
Total	20	19	26	18	83

* Not applicable to 15 case studies

ment for the use of the RE-Product is voluntary, and not compulsorily linked to the re-lated ERGSs.

Measures were taken to prevent free riding and to enforce excludability, or to regu-late access to the RE-Products in 42 cases out of 83 (Table 5). These measures have been adopted in more than half of the Austrian and Italian cases. Only in Germany were meas-ures to prevent free riding rare. The adoption of these measures seems to depend upon RE-Products typology. Products like bike access to forest roads or mushroom picking need more control than Christmas market [10] or Cabin Leasing. However, the costs of these measures were considered relevant only in a few cases (Table 6). This can be due to the nature of the goods paid for by the consumers, which are usually familiar market products (e.g. car parks, guided visits, accommodation). In addition, in many cases (58) the issue of costs to prevent free riding was not even considered in the case studies analysed. This was particularly evident in Germany and the Netherlands where the issue rarely arose in the course of the survey.

Rivalry after transformation into RE-Products generally increases (Table 7). Reduc-tion is, however, also evident in some cases. Therefore, it is difficult to identify a clear relationship. Rivalry depends, to a certain extent, upon congestion, since too many users decrease the enjoyment of the good/service. RE-Products can therefore have different and opposite effects on the congestion of ERGSs:

i)   a promotion of access and use, due to attracting a larger number of users, in-creasing congestion;

ii)   a more regulated use of the ERGSs, due to zoning and introducing entrance fees or permits, reducing, or at least controlling, congestion.

---

[10] Christmas markets relate to forest enterprises selling Christmas trees together with other farm related Christmas decorations, goods, etc.

Table 6.   Relevance of costs to prevent free riding (cases where measures have been taken) *

Costs to prevent free riding	Austria	Germany	Italy	The Nether-lands	Total
No (0)	6	-	2	2	10
Very low/low (2.5-5)	6	1	11	2	20
High (7.5)	-	-	-	-	-
Very high (10)	1	2	4	3	10
Total	13	3	17	7	40

* Not applicable to 58 case studies [non sarebbe meglio includere una riga con i casi non applicabili?]

Table 7.   Rivalry after transformation *

Rivalry	Austria	Germany	Italy	The Nether-lands	Total
No (0)	-	-	-	-	-
Very low (2.5)	4	6	4	1	15
Low (5.0)	6	5	5	6	22
High (7.5)	1	2	13	3	19
Very high (10)	8	15	6	10	39
Total	19	28	28	20	95

* Not applicable to 3 case studies

It is interesting to note that, when rivalry was originally low, it tended to increase after transformation/development, while it increased very little or decreased when the level was already high – as in three Dutch cases. A low level of rivalry before transformation indicates that a certain ERGS is a typical public good with no congestion problems and so the aim of transformation/development is to increase use. Where there is already a high level of rivalry, the product transformation/development consists mostly of an increase in regulation and production efficiency. Finally, when there is a serious problem of congestion, transformation/development can often decrease congestion through market regulation, i.e. exclusion through pricing, as shown by the experience of well consolidated RE-Products.

In summarising excludability/rivalry aspects of transformation/development, it is clear that excludability has been successfully achieved in most cases - 88 out of 95 (Table 4). There are fewer cases that achieved a high level of rivalry - 39 cases out of 95 (Table 7). Overall about half the cases have shown full transformation into pure private goods when considering both rivalry and excludability [11]. Rivalry was the criterion most difficult to achieve. In general terms, development mainly led to the creation of club or local goods, rather than private goods, due to the intrinsic characteristics of the goods and services offered (e.g. car parks, picnic sites and cross-country skiing).

---

[11] In Italy, however, only a quarter of RE-Products were completely transformed into pure private goods with full excludability and rivalry.

4.4.   Mechanisms of transformation/development

The transformation of ERGS into RE-Products can be divided into two stages: the first concerning institutional factors (legal status and property rights, planning and permissions, contractual arrangements, etc); the second concerning development, mainly linked to management and marketing, the provision of complementary/additional goods and services, as well as promotion and information.

Institutional factor changes include:

i)   legislative changes concerning state and regional laws, often representing the first step towards transformation, creating the base for a RE-Products market niche (e.g. the Italian law on mushroom picking obliging pickers to buy a permit, or the EU Regulations 2081/1992 and 2082/1992 on product origin certification obliging producers to get approval and follow a code of practice);

ii) planning changes, concerning regional and local action, which can again initiate the transformation of an ERGS into a RE-Product. Land use and environmental planning, including for example zoning and the designation of parks and protected areas, create new opportunities for RE-Products;

iii) administrative changes, at an even lower level, implying local planning standards, licences and other regulations, again making possible the first step towards the establishment of a RE-Products on the basis of existing ERGSs.

Therefore, institutional changes, or at least compliance wih a certain administrative procedure, often represent the first step towards the transformation process, triggering the following developments made possible through management and marketing. Notable differences do exist among the four countries based on existing property rights and, above all, RE-Products typology. Only a few German cases required a true legal change, while in several Italian cases, such a change was required (Table 8). This is certainly related to products typology prevalent in German and Dutch cases. The development of products like farm hospitality, Christmas markets, ecomeat, etc, require clearly defined property rights. In Italy, and to a certain extent in Austria, more complex RE-Products have been developed, such as access to private land, cross country skiing, biking in forest roads, protected names (PDO or PGI) for traditional/regional food, in a context where, at least in Italy, property rights are affected by an heritage of communal rights, quite obviously more difficult to modify and regulate.

Management and market developments were generally based on the relationship between the ERGSs and the RE-Products actually sold in the market (a relationship often associated with consumption complementarity). In order to appreciate the environment (the ERGSs) one must use appropriate infrastructures and equipment: RE-Products. This complementarity can also be seen as the RE-Products adding value to the ERGSs. The relationship of complementarity and the additional value of the equipment and infrastructure therefore represents a final step for creating a market value for ERGSs through the RE-Products.

Usually, both institutional and management/market approaches are needed to achieve remuneration of ERGSs through RE-Products. The institutional approach builds the base for transformation (e.g. new regulations are introduced for using the ERGSs). However, it is management and marketing which make up the core for the development. In fact, the availability of additional goods and services is necessary to create a market for the

Table 8. Occurrence of legal changes

	Austria	Germany	Italy	The Nether-lands	Total
No	12	24	10	12	58
Yes	9	4	19	8	40
Total	21	28	29	20	98

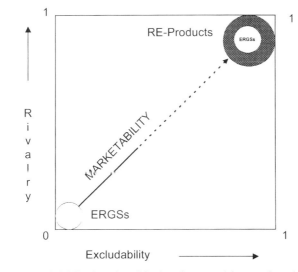

Excludability

—— Excludability through modification of property rights: transformation
- - - - Excludability through provision of additional services (management and marketing techniques): development

Figure 4. Transformation/development paths from ERGSs to RE-Products: the two components of the 'marketability arrow'

ERGSs through the RE-Products. Figure 4 exemplifies the transformation process, showing that the ERGSs for which institutional status is slightly changed (continuous line) are
enveloped within conventional complementary market goods such as car parks, tourist
and sport facilities (dotted line), and traditional farm/forest products, whose quality and
image is linked to the environment and the landscape (ERGSs) where they are produced.

The possible level of complementarity between ERGSs and RE-Products is variable
as shown by Figures 5 and 6 (e.g. there is relatively high complementarity between the
pure environment and the footpaths to gain access to the environment, while complementarity is lower with restaurants and shops. At the limit, when complementarity is very high
the ERGSs and the RE-Products overlap almost completely, as shown by Figure 5). The
case studies have also shown that it is almost always the RE-Products that are paid for in
the market, not the ERGSs.

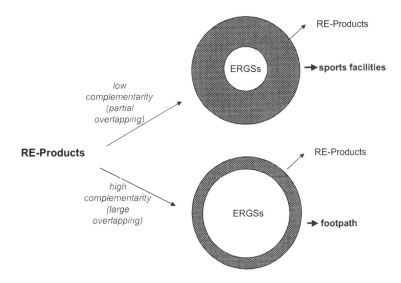

Figure 5. Degree of complementarity between ERGSs and RE-Products

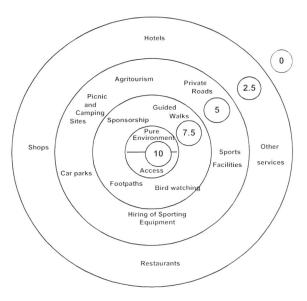

Figure 6.   Complementarity between the pure environment (ERGSs) and the means to make use of it
(RE-Products)*
Source: elaborated from Mantau (1995).
* The numbers 0, 2.5, 5, 7.5 and 10 indicate the level of complementarity between the RE-
Product and the related ERGS, where 0=non complementarity and 10=very high complementarity.

In addition to the 98 cases surveyed, some 20 cases of failures were also recorded. Causes of failure varied. However, in general it was mainly a case of property rights violation raising protest by the public. In other cases it was lack of demand and/or poor management or poor capability to communicate with consumers.

5. LINKS BETWEEN RE-PRODUCTS AND REGIONAL/LOCAL CONDITIONS DURING THE TRANSFORMATION/DEVELOPMENT

5.1. Reference to country case studies

It appears that the transformation/development involves both institutional and management/market based approaches. The case studies examined indicate that the two approaches can be applied alone or simultaneously, thus reinforcing the marketability of the RE-Product. Both approaches, however, aim at increasing excludability, while rivalry may or may not be increased.

Tables 2 and 3 (before transformation) and Tables 4 and 7 (after transformation), show the transformation/development for the four countries surveyed. The cases demonstrate that changes in the transformation/development 'paths' can be very difficult in terms of excludability and rivalry. However, the 'marketability' arrow of Figure 3 remains a common feature. It is also evident that transformation and development can either be applied separately or together according to the different case studies.

The interdependency between institutional and management/marketing approaches is particularly evident for Italy. In Germany and Austria management/marketing means alone are already frequently applied and sufficient, while initial support and gearing by institutional changes is less evident and less needed. This also applies to the Netherlands where, however, paths often move from an high rivalry/congestion to a less evident one, similarly to Italy, meaning that marketing of RE-Products is also aimed at reducing pressure on the environment. As a consequence Austria and Germany show transformation/development paths more similar to the 'marketability arrow' of Figure 3, compared to Italy and the Netherlands, whose paths are more dispersed in the diagram. The differences in paths can be linked to Austrian and German property rights for ERGSs, which are better defined and assigned by consolidated legislation, particularly in forestry (representing the largest number of case studies). Traditionally, in Austria and Germany, forest owners are also more business minded both for timber and tourism services linked to farming and forestry. The lack of business tradition in Italy can also be viewed in terms of 'late economic development' and inheritance law, which has left a legacy of fragmentation. Property rights are not always well defined and assigned, including communal rights on rural land. The present situation, characterised by a rather affluent and environmentally sensitive society, makes it difficulty to apply unpalatable land policies that were easily applicable centuries ago. Therefore, other more socially acceptable means have to be applied. Meanwhile, in the Netherlands, it is striking to note the high rivalry/congestion of ERGSs before transformation/development into RE-Products with private and/or club connotations. This helps to explain the complicated mix of institutional / management / marketing approaches as shown by the Italian and Dutch case studies. These two coun-

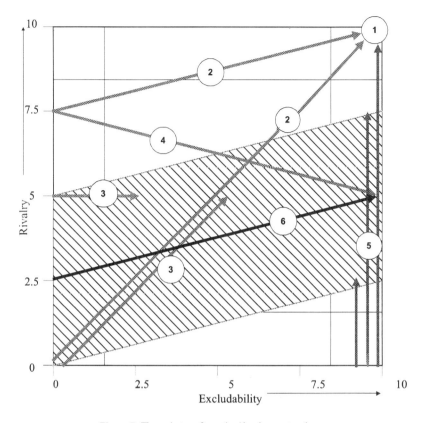

Figure 7. The main transformation/development paths

tries also show a common high pressure on rural resources determined by high population density, however allowing more opportunities to farmers and land owners.

5.2. Towards a synthesis of transformation/development paths

The different case studies can be grouped to establish six significant paths of transformation/development of ERGSs into RE-Products as illustrated by Figure 7. To simplify the diagram, only paths which consist of two or more cases are shown. These paths are detailed below:

1. Re-launch of an already private RE-Product. In these cases there was not any variation in the excludability/rivalry criteria.
2. From a public ERGS (pure or rival to various degrees) to a pure private RE-Product.
3. From a public ERGS (pure or rival to various degrees) to a club RE-Product with partial excludability.

4. From a public (reasonably rival in nature) or a club ERGS with partial excludability to a club good with a decrease in rivalry.

5. From an excludable club ERGS with low rivalry to a club or private RE-Product with partial and high rivalry.

6. From a public ERGS (pure or rival to a certain extent) to a club RE-Product with high excludability. Path six, the most evident among case studies, consists of several paths that are all within the shaded area, given various excludability/rivalry levels before and after transformation/development.

The simplification of Figure 7 allows the following common features to be pointed out:

i) Transformation/development paths tend to move from left to right and from down to up. This is the consequence of the fact that the main purpose of transformation/development actions is to obtain private goods where remuneration is easier;

ii) Some areas of Figure 7, characterised by high rivalry and low excludability do not have significant paths, indicating that these positions are difficult to deal with, meaning 'free riding' and 'tragedy of the commons' can easily take place;

iii) The key issue for transformation/development is the increase in excludability, while varying changes in rivalry result from this modification. Particularly in path 4 the goal is rivalry reduction, obtained through an increase in excludability;

iv) Transformation/development leads usually to the creation of private or club goods, depending on the products' intrinsic characteristics. Path 3 is quite unusual; it corresponds to the use of voluntary contribution/donation as a means of payment, and it could be considered as an incomplete transformation/development. The RE-Product remains in the central undefined area of mixed goods potentially club/private goods;

v) Path 4, including three Dutch cases, goes between goods with low excludability and high rivalry to club goods with reduced rivalry. More path 4's could have been expected, as it seems a logical further development of path 3 - more common with 7 cases observed in all countries. The likely reason for this is that the change from voluntary donations to compulsory payments is often unpalatable to the public;

vi) Finally it must be underlined that the most common case of transformation/development is illustrated by path 6, grouping 32 cases where transformation/development mainly aims at the creation of club/local goods with full excludability and various levels of rivalry.

6. CONCLUSIONS

The following points are important in references to the potential markets open to ERGSs and the related RE-Products:

i) Recreational products, requiring various types of facilities and structures in addition to the pure environment, can be more easily transformed and developed.

This is a good opportunity for several ERGSs, particularly those located within less intensive, nature-oriented farming and forestry where tourism is more important (e.g. on coastal and mountainous areas). Meanwhile, the environment alone cannot be easily marketed; however, traditional quality products labelled according to the environment/geographical area they come from (such as PDO), the well established *appellation d'origine contrôlée (AOC)* of wines and cheeses, production process certification in general, and sponsorship of a certain environment, represent viable opportunities that can be exploited in order to capture consumers' willingness to pay for the environment;

ii) Both public and private land ownership give opportunities for development of RE-Products. The role played by landowners associations, consortia, the traditional common properties of forests, and the modern trusts and environmental societies are remarkable. In the mean time, specialised contractors (which are often more efficient parties for transformation/development than individual landowners) must not be neglected, because they can more easily obtain the adequate land base necessary for RE-Products development;

iii) Institutional changes are certainly needed for triggering transformation of ERGSs into RE-Products. However, various case studies have shown that dramatic changes of property rights are not feasible as they are unpalatable to the general public, and in any case they are not necessary, as management and marketing approaches aimed at achieving 'exclusiveness through additional services' are often sufficient. Generally speaking both institutional and management/marketing changes are required to achieve transformation as outlined in Figure 4. According to Italian case studies, one-third of the path, measured according to the already mentioned scale of excludability/rivalry [12], was explained by institutional changes often granted by local authorities (through permission, licences, etc.) and two-thirds by management and marketing means;

iv) A positive relationship was also found (particularly in Italy) between the different ERGSs, the related RE-Products and other local market products. The term 'area product mix' can be used to describe the situation where RE-Products related to farming and forestry have been more successful due to the regional system to which they belong. Marketing of RE-Products is easier in an area where various infrastructures and other products are already available;

v) Finally, the most evident shortcoming of any transformation/development strategy (which was particularly clear in Italy, but also applies to other countries), is shown by the fact that those producing the positive ERGSs are not always those involved with the management/marketing of the RE-Products, but rather appear to be the contrary. The result is market distortions and inefficiency. Compensation amongst economic actors at the area level must be therefore considered and represents the most challenging task for market based environmental policies.

Concerning the various policy tools for promoting ERGSs, and RE-Products, all have both advantages and disadvantages. However, it is clear that they cannot be seen, and conceived, separately; this is a common feature of environmental policies aimed at in-

---

[12] For Italian cases, during the interview and the analysis of policy documents necessary to fill the questionnaire, it was also asked to separate the role of institutional means (legal changes including property rights, permissions granted by local authorities, etc.) from the managerial and economics aspects.

creasing the provision of ERGSs and RE-Products (Glück, 1998). Mandatory, legally binding instruments, new and old, should be applied together with economic-financial and market-led tools. The sustainable provision of the various ERGSs, RE-Products and the prevention of EBDs, involves jointly devised and applied policy tools that are integrated and 'transparent', including procedure for persuasion and communication as outlined in Figure 1. After all the various ERGSs have different value-types: from use-value (direct and indirect), to option, to existence and non-use value. It seems to be rather obvious that an effective and efficient agri-environmental policy needs various tools and measures.

It also seems important to stress that the policy mix should be devised area by area according to the 'subsidiarity principle', developing specific packages of policy measures and tools. The concept is not new, as it is for instance demanded by the EU 'Structural Funds'. It represents, however, a general demand related to compensation amongst local forest actors, taxation, devolution, and autonomous status (Rojas Briales, 1995), of each area able to guarantee local responsibility over farming, forestry and more generally over the environmental resources.

7.   REFERENCES

Bishop K.D. – Philips A.P. (1993), Seven steps to market: the development of the market led approach to countryside conservation and recreation, *Journal of rural studies*, **9**(4): 315-318.
Blochliger H. (1994), Main results to the study, in: *Main contributions of amenities to rural development*, OECD, Paris.
Brosio G. (1986), *Economia e finanza pubblica*, NIS (Nuova Italiana Scientifica), Roma.
Buchanan J.M. (1965), An economic theory of clubs, *Economica*, **2**: 1-14.
Buchanan J.M. (1967), *The demand and supply of public goods*, Rand McNally & Company, Skokie, IL, USA.
Centro Contabilità e Gestione Agraria, Forestale e Ambientale – Università di Padova (1998), unpublished data.
Coase R. (1960), The problem of social cost, *Journal of Law and Economics*, **3**: 144-171.
Cornes R. – Sandler T. (1986), *The theory of externalities, public goods and club goods*, Cambridge University Press, Cambridge.
Dubgaard A. – Bateman I. – Merlo M. (eds., 1994), *Identification and Valuation of Public Benefits from Farming and Countryside Stewardship*, Wissenschaftsverlag Vauk, Kiel.
Falconer K. – Whitby M. (1999), The Invisible Costs of Scheme Implementation and Administration, in: *Countryside Stewardship: Farmers, Policies and Markets*, van Huylenbroeck G. – Whitby M. (eds.), Pergamon, Amsterdam, pp. 67-87.
Ferro O. – Merlo M. – Povellato A. (1995), Valuation and Remuneration of Countryside Stewardship performed by Agricultural and Forestry, in: *Proceedings of the XXII International Conference of Agricultural Economists, Harare, Zimbabwe*, Peters G.H. – Hedley D.D. (eds.), Dartmouth, London, pp. 415-435.
Gatto P. – Merlo M. (1999), The economic nature of stewardship: complementarity and trade-offs with food and fibre production, in: *Countryside Stewardship: Farmers, Policies and Markets*, van Huylenbroeck G. – Whitby M. (eds.), Pergamon, Amsterdam, pp. 21-46.
Glück P. (1998), The task of research in evaluation of multifunctional forestry, *II International Forest Policy Forum*, Centre Tecnològic Forestal de Catalunya, Solsona, vol. IV, pp. 215-227.
Hanley N. (1995), Synthesis report to the rural development programme, in: *Rural amenities and rural development: Empirical evidence*, Hanley N. (ed.), OECD, Paris.
Mantau U. (1995), Forest policy means to support forest outputs, in: *Forest policy analysis - Methodological and empirical aspects*, Solberg B. – Pelli P. (eds.), EFI Proceedings 2, Joensuu, pp. 131-145.
McGuire M. (1987), Public goods, in: *The new Palgrave: a dictionary of economics*, Eatwell J. – Milgate M. – Newman P. (eds.), Mc Millan Press, London.

Merlo M. – Milocco E. – Panting R. – Virgilietti P. (2000), Transformation of Environmental Recreational Goods and Services provided by Forestry into Recreational Environmental Products, *Forest Policy and Economics*, **1**(2): 127-138.

Musgrave R.A. (1959), *The theory of public finance. A study in public economy*, McGraw–Hill, New York.

OECD (1996), *Amenities for rural development*, Paris.

Paveri M. – Merlo M. (1998), Formation and implementation of forest policies: a focus on the policy tools mix, *FAO XI World Forestry Congress*, Rome, vol. 5, pp. 233-254.

Pigou A.C. (1920), *The Economics of Welfare*, MacMillan, London.

Randall A. (1987), *Resource economics - An Economic Approach to Natural Resource and Environmental Policy*, John Wiley & Son, New York.

Rojas Briales E. (1995), *Una política forestal para el estado de las autonomías*, Editorial Aedos, Madrid.

Samuelson P. (1954), The pure theory of public expenditure, *Review of economics and statistics*, **36**: 387-389.

Samuelson P. (1955), A diagrammatic exposition of a theory of public expenditure, *Review of economics and statistics*, **37**: 350-356.

Tiebout C.M. (1956), A Pure Theory of Local Expenditures, *Journal of Political Economy*, **5**: 416-424.

van der Hamsvoort C.P. – Latacz-Lohman U. (1996), *Auctions as a Mechanism for Allocating Conservation Contracts amongst Farmers*, Agricultural Economics Research Institute (LEI-DLO), The Hague and Wye College, University of London, The Hague.

Whitby M. – Saunders C. (1996), Estimating Conservation Goods in Britain, *Land Economics*, **72**(3): 313-325.

INTEGRATING PREDICTION MODELS, EVALUATION METHODOLOGIES AND GIS: THE ASSESSMENT OF AGRICULTURE-ENVIRONMENT RELATIONSHIPS IN THE TRASIMENO WATERSHED

Paolo Abbozzo, Antonio Boggia, and Francesco Pennacchi *

SUMMARY

This study is part of a broader research project whose aim is to assess the impacts of the introduction of environmentally sound agricultural techniques in the area of the Trasimeno Lake watershed, in Umbria.

In this chapter the methodological path used in this study is explained. It is composed of different steps, each one of them with a specific methodological core, but all directed to get the same final result.

The first step is the economic and environmental assessment of the production practices at a single crop and farm level, both for the present situation, and for the alternatives (in terms of crops, but also practices) coming from field experimentation. To develop this step the Planetor computer program has been used. The output data from Planetor, together with other indicators, has been incorporated into a multicriteria matrix to get a comparative assessment of the different alternatives.

The second step is the economic and environmental assessment of different scenarios at the whole watershed level. To develop this step a combination of linear programming and multiobjective analysis has been used, to generate the optimal combination of all the different factors.

The third step is the assessment of land vulnerability and environmental risk of the watershed, developed through the implementation of regional environmental data in a Geographic Information System (GIS). The result is a number of vulnerability and risk maps.

---

* Department of Economics and Appraisal Sciences - University of Perugia. Paolo Abbozzo: section 1. Francesco Pennacchi: sections 2 and 5. A. Boggia: sections 3, 4, 6, and 7.

1. INTRODUCTION

This paper is part of a broader research project co-ordinated by the *Center for Teaching and Research Applied to Agricultural and Veterinary* Sciences of the University of Perugia, entitled "Experimentation and assessment of environmentally sound production techniques for the Trasimeno Lake agriculture", funded by the European Union through the Regional government of Umbria.

The project was developed during 1998-99 and finished in December 1999. The aim of the project was to assess the economic and environmental impact of the change to more sustainable farming techniques in the Trasimeno Lake watershed area.

To get the final results, the main steps have been:

1. to assess both from the economic and the environmental point of view the existing agricultural system and production techniques;
2. to develop, with field experimentation also, alternative production processes, with regard both to the crops and to the farming techniques;
3. to assess both from the economic and the environmental point of view the alternative agricultural system and production techniques.

This paper will focus only on the assessment methodological path and on the tools used for this purpose. Considering the limits of space and time it would be impossible to give a thorough report on the whole research, including the activities of all the scientists involved: agronomists, soil scientists, fertilizer and pesticide chemists, irrigation engineers, crops pests protection researchers, etc.

This part of the study is related to the evaluation of the economic results and the environmental impact of the most important crops cultivated in the watershed, taking into account the different farming techniques and the soil types and position. The evaluation was applied both for single crops and at the farm level. The results of this step were the starting point and the database for the next step: the assessment of different scenarios at the whole watershed level.

2. AGRICULTURAL, RURAL, AND ENVIRONMENTAL POLICIES

The agricultural and rural policy of the EU has increased its attention to the environment in the last few years. However, the discussions on the strategic objectives of sustainable agriculture, the criteria to take into account, the actions to develop, and the methodological tools to use for the evaluation of the above, are still not developed. These policies follow, so far, the general concept of sustainability as proposed by the Brundtland Commission in 1987, without any effort to better focus the contents of this definition or to understand what path of sustainability (weak or strong) it is possible to reach.

Discussions on the types of environmental impacts caused by agriculture are more developed. Water pollution and consumption, land use, climate changes, biodiversity and landscape are the most common problem areas linked to the intensification of agriculture and high input farming systems. But, again, attention to select specific criteria and methods to assess these problems are not well developed. Very often the need to develop criteria and methods is highlighted, but the task is assigned to working groups, which are not always effective.

The EU has already decided to enhance the role of environmental variables in the agricultural systems. Given this choice, it is now time that the policy become operative. Most probably, it would not be possible to do this at the central level; too many differences exist in a big system such as the EU. Following the guidelines from the EU policies and documents, local institutions could play an important role. In Italy, in particular, regional governments are the most suitable to get good results, because they are in charge of agriculture management, and also because the portion of land would be of the right size for full control.

First of all, the Regions should develop a regional model for environmental management. This model should clearly indicate the sustainability objectives to reach and the timeline for reaching them, as well as the procedures to achieve the objectives.

This means that a link between evaluation and decision making becomes more important, so that models and methodologies for evaluation can be developed together with decision making. Thus the policy objectives can be incorporated, and evaluations can be a support for decisions.

The Region of Umbria moved in this direction when, having the need to evaluate the level of sustainability of agriculture in the Trasimeno Lake watershed, decision makers decided to promote the above mentioned project

3. STEP 1: ECONOMIC AND ENVIRONMENTAL EVALUATION AT CROP AND FARM LEVEL

The tool used to develop this part of the study is Planetor. It is a computer program that integrates the economic and environmental assessment of farms and is able to give separate economic and environmental assessment of farm results. Planetor uniquely combines different kinds of ecological models, together with an economic analysis.

The Center for Farm Financial Management of the University of Minnesota is the author of the program. There is an Italian version also, developed by the University of Padova, ESAV and the Center for Farm Financial Management.

In this study the Italian version was used. However, Planetor works on the basis of specific databases that have to be built referring to the studied area. This makes it possible to get specific results for different environmental characteristics.

So, the first important thing to do was to develop the databases for the Trasimeno watershed. To develop databases normally requires a number of analyses and large data and information collection effort.

The developed databases are:

1. *Crop database*: information has been collected using farm questionnaires, available literature, field experimentation results, and discussions with extension technicians and practitioners.
2. *Livestock database*: in this case, the main source of data were also farm questionnaires, available literature, and discussions with extension personnel.

3. *Soil database*: all the types of soil existing in the watershed have been included in the database. Given the very high variability of soils in the watershed, soil scientists have made several field verifications.

4. *Pesticide database*: on the basis of the information collected, most of the ingredients and commercial products used in the area were not included in the existing database. So, all the pesticides used were added to the database.

5. *Fertilizer database*: even for fertilizer, it was necessary to add some more types, not included in the list.

6. *Machinery operations database*: this database has been developed using interviews with farmers and practitioners, as well as a literature search, simply converting the U.S. database in some cases and processing available data in other cases.

7. *Climate database*: this database is divided in two parts; the first one containing data for a RUSLE equation, for soil erosion assessment and the second one containing data for an NLEAP model, for nitrogen leaching assessment. Both parts have been implemented using available climate data for the area.

On the basis of diffusion in the area, the economic weight, and the production trend in the near future, the following crops have been selected to be evaluated: sugar beet, corn, sunflower, hard and soft wheat, barley, tomato, green pepper and cantaloupe.

For each crop, different farming techniques, on different types of soil, with a different slope, have been evaluated. The result is a number of combinations to assess with Planetor for each crop, as shown in Table 1.

The number of combinations varies depending upon how many farming techniques, besides the high input, low input and organic, have been developed from the results of the field experimentation.

The low input farming technique is the result of the adoption of the measure A.1 from EU Reg. 2078/92. It consists of a 40% reduction of the nitrogen fertilization. The farmers receive a subsidy to compensate for the yield reduction.

To provide an example of the results obtained for each crop, the output for corn is presented in Table 2. For this crop, besides the 18 combinations generally assessed, there are four more, resulting from the field experimentation.

The indicators obtained from the analysis are not only the Planetor outputs, but also some achieved through further processing, or from other data.

The meaning of each of the indicators in the columns is:

1. *Erosion index*: the ratio between the quantity of soil (ton/Ha) annual losses and the maximum annual soil loss tolerance.

2. *Pesticide leaching*: an index function of the risk of pesticide leaching. It may vary from 1 (no risk) to 1000 (high risk).

3. *Pesticide runoff*: an index function of the risk of pesticide runoff. It may vary from 1 (no risk) to 1000 (high risk).

4. *Nitrogen leaching*: the amount of nitrogen leaching per year (kg/Ha).

5. *Nitrogen leaching/yield ratio:* the kg/Ha per year of nitrogen leaching per each ton of product obtained from the crop.

6. *Phosphorus runoff*: an index function of the risk of phosphorus runoff. High levels indicate high risk.

7. *Water use:* the quantity of water used for irrigation (cubic meters).

Table 1.   Combinations to assess for each crop

Clay soil	Hilly fields	High input farming technique
Medium textured soil	Flat fields	Low input farming technique (Reg. 2078/92)
Sandy soil		Organic farming technique

Table 2.   Corn results

Farming technique	Soil position	Soil type	Erosion index	Pesticide leaching index	Pesticide run-off index	Nitrogen leaching kg/ha	NL/Yield kg/t	Phosphorus index	Water use mm/ha	Gross income 000 ITL/ha	Return over direct exp. 000 ITL/ha
Low input	Hilly	Clay	3.3	560	990	6	0.8	14.5	240	4,006	1,977
Low input	Hilly	Medium	2.8	725	948	8	1.0	14.0	240	4,006	1,977
Low input	Hilly	Sandy	0.5	725	917	15	1.9	9.5	280	4,006	1,922
Low input	Flat	Clay	0.2	560	990	6	0.8	9.5	240	4,006	1,977
Low input	Flat	Medium	-	725	917	10	1.3	7.5	240	4,006	1,977
Low input	Flat	Sandy	0.1	945	917	39	4.9	9.0	280	4,006	1,922
High input	Hilly	Clay	0.9	725	994	22	1.1	13.0	300	3,908	1,188
High input	Hilly	Medium	0.8	828	968	39	1.7	12.5	300	3,908	1,188
High input	Hilly	Sandy	0.1	828	948	86	6.0	12.5	350	3,908	1,123
High input	Flat	Clay	-	725	994	20	0.9	12.5	350	3,908	1,188
High input	Flat	Medium	-	828	948	37	1.5	10.5	350	3,908	1,188
High input	Flat	Sandy	-	966	948	88	6.2	12.0	380	3,908	1,123
Organic	Hilly	Clay	9.9			-	-	15.0		3,047	2,088
Organic	Hilly	Medium	8.5			1	0.2	14.5		3,047	2,088
Organic	Hilly	Sandy	1.4			6	1.0	3.5		3,047	2,088
Organic	Flat	Clay	0.5			1	0.2	4.0		3,047	2,088
Organic	Flat	Medium	-			3	0.5	3.5		3,047	2,088
Organic	Flat	Sandy	0.4			17	2.9	3.5		3,047	2,088
P.W.C.C.*	Flat	Sandy	0.1	888	832	65	5.4	12.0	420	3,908	1,156
Class 300 Seeds	Flat	Sandy	0.1	966	948	40	5.4	12.0	245	2,940	978
Reduced irrigation	Hilly	Clay	2.8	725	994	15	8.4	16.0	80	2,182	572
No irrigation	Hilly	Clay	3.2	725	994	12	8.8	17.5		1,858	612

* Precision Weed Chemical Control

8. *Gross income*: the value, in Italian lire, of the yield.
9. *Return over direct expenses*: the result of the difference between the gross income and the direct expenses.

At the farm level, following the information coming from the field survey, 24 farm models have been built, taking into account the size, the rotation, livestock (if applicable), and the soil type and position.

The output obtained is the same as the crop level, of course with regard to the whole farm, and not only the single crop.

4. STEP 2: COMPARISON BETWEEN RESULTS: A MULTICRITERIA ASSESSMENT

At this point, every different alternative (crop farming technique) evaluated with Planetor is represented by the same indicators set. The different values of the same indicator for different alternatives assume the meaning of the behavior of each alternative in meeting the objectives represented by the environmental and economic indicators.

The large number of evaluations developed with Planetor, means a large number of alternatives to compare. This increases the need for an understandable methodology capable of guiding the comparative analysis of a finite number of alternatives, given a number of evaluation criteria (indicators). This is the typical field of action for multicriteria methodologies, particularly multiattribute analysis.

The methodology adopted is the Electre method. It is based on a pairwise comparison of alternatives. For each pair of alternatives a dominance relationship must be found. To get this result, a concordance index and a discordance index are used.

Concordance is the level where alternative i is better than alternative i'. This index is defined as the sum of the weights of criteria included in the concordance set $C_{ii'}$, that is the set of criteria where alternative i is at least at the same level as alternative i'. Concordance index is:

$$c_{ii'} = \sum_{j \in C_{ii'}} w_j$$

Discordance is the level where alternative i is worse than alternative i'. A discordance set $D_{ii'}$ is the set of criteria where alternative i is worse than alternative i'. For every criterion of this set the difference between the values of both the alternatives is calculated. The largest of these differences is defined as the discordance index. Discordance index is:

$$d_{ii'} = \max_{j \in D_{ii'}} \left| x_{ji} - x_{ji'} \right|$$

The result is a ranking of alternatives. The first alternative is the one that reaches the most objectives simultaneously.

In this particular analysis, since the indicators used are both economic and environmental, the ranking has the capability to assess alternatives with widely varying objectives. The greater this capability is, the better is the assessment of the alternative approaches to the condition of economic and environmental sustainability.

Two different types of analysis have been carried out:

(1)   Crop: for each crop the different farming techniques have been compared, taking into account the type and position of soil. Depending on the field experimentation, for each crop a different number of alternatives has been evaluated, as follows:

- Soft wheat   19
- Hard wheat   20
- Barley   18
- Corn   22

Table 3. Ranking for some crops

Farming techniques	Soil position	Soil type	Sugar beet rank	Hard wheat rank	Soft wheat rank	Sunflower rank	Corn rank	Barley rank
Organic	Hilly	Sandy		9th	5th	4th	3rd	10th
Organic	Flat	Clay		5th	3rd	2nd	2nd	7th
Organic	Flat	Medium		4th	2nd	1st	1st	6th
Organic	Flat	Sandy		3rd	1st	3rd	6th	9th
Class 300 seeds	Flat	Sandy					20th	
High input	Hilly	Clay	10th	11th	11th	15th	14th	15th
High input	Hilly	Medium	9th	17th	17th	18th	13th	17th
High input	Hilly	Sandy	11th	19th	14th	17th	19th	18th
High input	Flat	Clay	7th	8th	9th	9th	15th	14th
High input	Flat	Medium	8th	15th	8th	12th	16th	16th
High input	Flat	Sandy	12th	20th	15th	19th	22th	11th
P.W.C.C.*	Hilly	Clay				16th		
P.W.C.C.*	Flat	Sandy					21th	
Low input	Hilly	Clay	4th	14th	19th	11th	9th	8th
Low input	Hilly	Medium	3rd	13th	18th	13th	10th	5th
Low input	Hilly	Sandy	5th	12th	13th	10th	11th	4th
Low input	Flat	Clay	2nd	1st	12th	7th	7th	3rd
Low input	Flat	Medium	1st	2nd	10th	8th	8th	2nd
Low input	Flat	Sandy	6th	16th	16th	14th	12th	1st
Reduced irrigation	Hilly	Clay					17th	
Minimum tillage	Hilly	Clay		18th				
No tillage	Hilly	Clay		10th	7th			

* Precision Weed Chemical Control

- Sunflower   19
- Sugar beet   12
- Cantaloupe   12
- Green pepper   18
- Tomato   12

Table 3 shows the rankings obtained for some of the crops studied. This is only to make it easier to understand the kind of results coming from this analysis, but it would take too long to explain and analyze the data, given the aim of this paper, that is to present a methodological approach.

(2)   Soil type and position: for each type of soil crossed with each position taken into account, all the crops and all the farming techniques of each crop are evaluated. The output is a ranking of crops and techniques more suitable for each soil and slope. Six classes of soil type/position have been considered, and for each of them the following number of crop/farming techniques have been evaluated:

- Clay flat   24
- Medium flat   24
- Sandy flat   26
- Clay hilly   31
- Medium hilly   24
- Sandy hilly   24

Table 4 shows the obtained rankings.

Table 4.   Ranking for soil type and position

Crops	Farming techniques	Medium Flat Rank	Medium Hilly Rank	Sandy Flat Rank	Sandy Hilly Rank	Clay Flat Rank	Clay Hilly Rank
		Soils					
Sugar beet	High input	24th	22nd	26th	22nd	24th	31st
Sugar beet	Low input	20th	20th	22nd	17th	19th	24th
Hard Wheat	Organic	4th	2nd	1st	2nd	2nd	2nd
Hard Wheat	High input	12th	9th	8th	9th	7th	8th
Hard Wheat	Low input	8th	7th	7th	8th	6th	9th
Hard Wheat	Minimum tillage						11th
Hard Wheat	No tillage						10th
Soft Wheat	Organic	2nd	1st	2nd	1st	1st	1st
Soft Wheat	High input	21st	16th	16th	12th	17th	18th
Soft Wheat	Low input	11th	8th	9th	7th	11th	21st
Soft Wheat	No tillage						15th
Sunflower	Organic	3rd	5th	4th	4th	4th	4th
Sunflower	High input	22nd	17th	20th	16th	15th	26th
Sunflower	P.W.C.C.*						28th
Sunflower	Low input	10th	10th	10th	10th	9th	7th
Corn	No irrigation						19th
Corn	Organic	1st	3rd	3rd	3rd	3rd	3rd
Corn	Class 300 seeds			19th			
Corn	High input	23rd	12th	23rd	20th	23rd	13th
Corn	P.W.C.C.*			25th			
Corn	Low input	14th	11th	13th	11th	10th	5th
Corn	Reduced irrigation						14th
Cantaloupe	High input	16th	18th	17th	18th	20th	27th
Cantaloupe	Low input	13th	19th	15th	19th	18th	23rd
Barley	Organic	5th	6th	5th	5th	5th	17th
Barley	High input	17th	15th	11th	15th	14th	22nd
Barley	Low input	7th	4th	6th	6th	8th	6th
Green pepper	Organic	6th	14th	12th	13th	12th	12th
Green pepper	High input	9th	13th	14th	14th	13th	16th
Green pepper	Low input	15th	21st	18th	21st	16th	25th
Tomato	Organic						30th
Tomato	High input	18th	23rd	24th	23rd	22nd	29th
Tomato	Low input	19th	24th	21st	24th	21st	

* Precision Weed Chemical Control

The second analysis looks like a first link with the evaluation at the watershed level. In fact, it gives important information on the crops and the farming systems to develop in every soil type and position in the area.

5.   STEP 3: THE SCENARIOS AT WATERSHED LEVEL ASSESSMENT

On the basis of the economic and environmental output data available after the evaluation with Planetor, it has been possible to make assessments of alternative scenarios at the whole watershed level. The simulation of different scenarios and the assessment have been carried out using linear programming in a first phase, and a multiobjective weighed goal-programming model later. Figure 1 shows the simulations path.

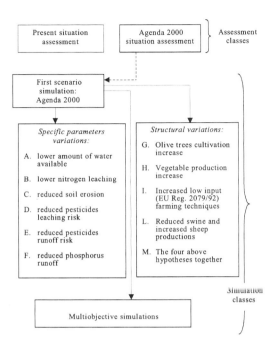

Figure 1. Watershed level assessment path

Table 5. Present situation and Agenda 2000 assessment results

Parameters	Present	Agenda 2000	% variation
Total return over direct expenses (000.000 Itl)	27,199	25,627	- 5.8
Water consumption (000 cubic meters)	9,284	9,284	0
Total nitrogen application (kg)	25,373	22,747	- 10.3
Total nitrogen leaching (kg)	2,975	2,502	- 15.9
Soil erosion (t)	315,981	288,510	- 8.7
Pesticide leaching (index)	18.2	17.9	- 1.7
Pesticide runoff (index)	19.01	16.7	- 11.9
Total phosphorus application (kg)	12,292	10,772	- 12.4
Phosphorus runoff (index)	1,654	1,581	- 4.4

The starting point was the present situation. However, since this will change very soon due to the Common Agricultural Policy reform with Agenda 2000, the present situation has been corrected, in economic terms, with the compensations and prices coming from the new regulations, but also in technical terms, following the consequences of the economic changes. In this way, even environmental indicators have changed. Table 5 shows the values of the most important parameters arising from the assessment of the present situation and the variations in the Agenda 2000.

All the environmental impacts decrease, but there is a cost for this change, because the return over direct expenses also decreases.

In the first phase, starting from the Agenda 2000 scenario, simulations have been done taking into account two types of variations. The first one includes environmental impact indicators or production factors variations, individually. In this way, using linear programming, it has been possible to find the optimal combination of crops, livestock and farming techniques to maximize the income, given the change of a certain parameter. The changes considered were:

- lower amount of water available;
- lower nitrogen leaching;
- reduced soil erosion;
- reduced risk of pesticides leaching;
- reduced risk of pesticides runoff;
- reduced risk of phosphorus runoff.

The other type of variation concerns structural factors, such as:

- olive trees and grape cultivation increases;
- vegetables production increases;
- increased low input (EU Reg. 2078/92) farming techniques;
- reduced swine and increase of sheep productions;
- the four above hypotheses together.

After this phase, the multiobjective model followed. The target is to get higher income values at watershed level and, at the same time, to lower environmental negative externalities. The model ran with different targets, so that different scenarios were assessed, some more economic oriented, others more environmental oriented. The choice of the best alternative should be defined depending on the weight given by the decision-makers to economic and environmental impacts.

This part of the study showed the important role of assessment models like the multiobjective ones in a highly complex situation such as the definition and evaluation of sustainability, and where it is possible for decision-makers to interact with the models.

6. LAND VULNERABILITY AND ENVIRONMENTAL RISK FROM AGRICULTURE: A GIS-BASED SYSTEM

This part of the study was carried out to try to represent on the territory the results of all the quantitative analyses developed.

First of all, the target was the creation of a map of land vulnerability in the Trasimeno watershed. To work on this, several different GIS computer programs have been used, depending on the operation to be done. Mostly ArcInfo was utilized for the digitalization phase and for the setting of the basic maps, Idrisi for the spatial analyses and the production of the thematic maps, and Arcview for the final phase of maps editing.

The map of land vulnerability is the product of the overlay and interaction of two separate maps: the map of soil erosion risk, and the map of water pollution risk. Both of these maps have been developed from the overlay of maps, each one representing a specific parameter. For example, the soil erosion map comes from the overlay of maps representing one of the RUSLE equation terms. So, to get the final map, the RUSLE equation is applied, in a spatial context.

For water pollution, a multiparametric model using parameters like the distance from surface water, the underground water level, the soil, type, etc., has been used.

The synthesis map of land vulnerability is the result of the overlay of the erosion and the water pollution maps. Of course, the concept of vulnerability is more complex than this suggest but these are two of the major environmental problems considered in agriculture.

With a cross classification procedure, the GIS software combined the classes of erosion and pollution, to get 9 classes of vulnerability for the Trasimeno watershed territory.

The last step in this path, only outlined and tested for a small part of the watershed, is to overlay on this map the exact position on the territory of each crop at the single field level. To do this, the crop grown has been defined for each cadastre field, from the application that farmers have to submit every year to get price compensation from the EU.

This information, linked and integrated with the other data and information coming from the other steps, can guarantee a strong information system, easy to update, and can also be the basis for specific and precise analyses and assessment of the agricultural systems in the watershed.

If one knows from the results of the previous steps that a certain crop produces better economic and environmental results, then it is clear with this system where in the territory more of that crop should be grown.

It is also possible to use the system for a continuous monitoring of where the crops are grown every year and where they should be moved the following year to lower the potential of environmental damage.

7. CONCLUSIONS

The methodological path developed in the "Trasimeno project" is shown (Figure 2).

The social function of agriculture and forestry is increasing, particularly with regard to environmental protection. The complexity of the assessment processes in this sector is also increasing due to the enlargement of the field of action of farms, following the multifunctionality, diversification and rural development policies of EU.

In environmental assessment, models can play a very important role; however, despite the great deal of progress made in the last few years, there is still a lot of work to be done both at the research and at the operative level.

In the field of environmental analysis however, models are the most important aid to decision making. However, decisions in environmental policy have a high level of uncertainty, irreversibility and complexity. These three characteristics are very difficult to manage, but models, especially through the use of computer programs, can be extremely helpful.

The whole evaluation process can be consistent and effective only if developed not at private property level, but in a common property and social context. Moving in this direction, while keeping in mind that in the environmental field, evaluations show a strong multidimensional characterization, the integration of results from models with multicriteria methodologies should play an important role.

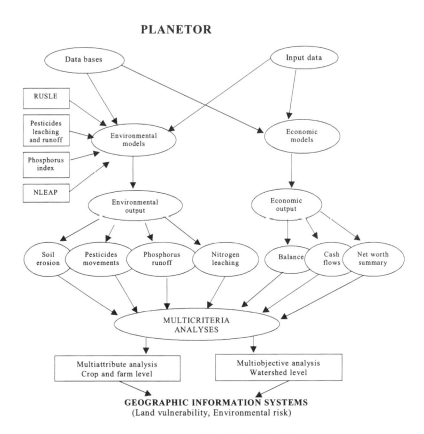

Figure 2. Integrated methodological path

From the practical point of view, the strongest advantage of the multicriteria (both multiattribute and multiobjective) methodologies is the capability to address problems from many standpoints.

For a complete methodological evaluation path, that is useful for decision makers and that includes realistic environmental policies, the Geographic Information Systems (GIS) is important. It can give a spatial representation of a territory using the results of quantitative evaluations.

An integrated system should include the following three key steps:

Simulation and assessment models ➔ multicriteria analysis ➔ GIS

This system can help assure good results in the evaluation processes, and aid in the decision making process.

8. REFERENCES

Boggia A. – Pennacchi F. (1999), *Sviluppo agricolo sostenibile del bacino del Lago Trasimeno*, Regione dell'Umbria, Assessorato Agricoltura e foreste.

Boggia A. – Abbozzo P. (1999), *Assessing Sustainability in Agriculture: a Multicriteria Approach*, Proceedings of the 6th Joint Conference on Agriculture, Food and the Environment, Minneapolis, MN (USA), august 31-september 2, 1998.

Boggia A. – Klair K. (1995), *Planetor: the potential to adapt it to Italy*, Department of Agricultural and Applied Economics, University of Minnesota, Staff paper n. P95/2, St. Paul, Minnesota , USA.

Center for Farm Financial Management (1994*), Planetor 2.0, Software rel. 0.65*, Minnesota Extension Service, University of Minnesota.

Janssen R. (1994), *Multiobjective decision support for enviromental management*, Kluwer Academic Publishers.

Keeney D. (1990), Sustainable Agriculture: definitions and concepts, *J. Prod. Agric.* **3**(3): 281-285.

THE APPLICATION OF REG. 2078/1992
IN THE PROVINCE OF VICENZA

Tiziano Tempesta and Mara Thiene [*]

SUMMARY

A survey was carried out in Vicenza province (Italy) in order to verify the environmental impact of the 1992 reform of the Common Agricultural Policy (CAP), including a series of accompanying agro-environmental measures.

The level of the acts implementation was defined as well as farmers attitude and motivation towards agro-environmental measures.

Results seem to highlight that agro-environmental measures could not be so important in land environmental improvement, especially in the plain as long as more than half of the land receiving subsidies is concentrated in the mountain areas.

Among all the measures, not all of them have received equal attention. Most applications have regarded the measures related to organic farming methods or the reduction in chemical inputs and some improvements to the landscape-environment.

Farmers age seem to be an important parameter in applying Regulation.

1. INTRODUCTION

The 1992 reform of the Common Agricultural Policy (CAP) includes a series of accompanying agro-environmental measures, that define the environmental nature of the new policy (INEA, 1997).

This study is in two parts. First, the level of implementation of the acts outlined in Reg. 2078/92 in the province of Vicenza (Italy) was verified. This gave a sufficiently objective picture of the conduct of farmers towards the implementation of interventions aimed at environmental improvement. In the second part, through a direct survey of farmers on the Vicenza plain, further knowledge was gained concerning their attitudes toward

---

[*] Tiziano Tempesta, Department of Economics and Policies on Agriculture, Agri-food, and Environment, University of Milan, has written sections 3 and 4. Mara Thiene, Department of Land and Agro-Forestry Systems (TeSAF) University of Padua, has written sections 1 and 2.

Economic Studies on Food, Agriculture, and the Environment
Edited by Canavari *et al.*, Kluwer Academic/Plenum Publishers, 2002

329

these acts. In this way, the motivations that influenced the farmers' participation can be better understood.

2.   THE LEVEL OF IMPLEMENTATION OF E.U. REG.2078/92

Reg. 2078/92 poses the following basic objectives (INEA, 1997):

- strive for production quality and contemporarily protect the environment through reduction in the use of production factors and conversion to organic farming methods;
- encourage the farmer to protect the environment and the agricultural system;
- diversify production activities and enhance the multifunctional nature of the primary sector.

These objectives, endorsed by the EC regulation, were subsequently adopted by every member state and have found practical application in the multiyear programs drawn up by the Regional Bodies in relation to the different territorial and economic situations. Each Region has therefore defined, within a general plan, a series of detailed measures, with incentives provided to the farmers for their realization.

The Veneto Region, in its multiyear program, has laid down the following measures (Tempesta, 1997):

- A1-A2: considerable reduction in the use of chemical fertilizers and/or pesticides (A1) and maintenance of reductions already made (A2);
- A3: introduction or maintenance of organic farming methods;
- B1: extensive plant production;
- B2: conversion of arable land to meadows and pasture;
- C: reduction in cattle density per unit surface area of forage;
- D1: use of other production methods compatible with environmental needs;
- D2: rearing of local species that are in danger of extinction;
- E: management of abandoned farmlands and forests;
- F: set-aside of arable land;
- G: land management for public access for recreational activities.

The incentives vary depending on the type of measure, but also in relation to the territorial context and environmental sensitivity of the area where the farm is located (i.e. the highest incentives are reserved for areas of environmental importance).

The highest subsidies are for conservation of natural spaces (D1-A, which is one of the D1 measures), which receive up to 3,500,000 ITL in hill or mountain areas, while for those on the plain the subsidy is more than 4,500,000 ITL per hectare.

Between 1993 and 1997 there were 540 grant applications under Reg. 2078/92 that included many of the communes in the province of Vicenza (73.5%). However, the level of implementing the agro-environmental measures was not homogeneous within the province. Farmers in the mountain communes expressed the greatest interest, by registering on average more than 25 applications per commune; overall these make up more than 28% of participants. In contrast, there have been no more than 15 participants per commune on the plain, with one exception.

Table 1.   Applications for financing and surface area by year of presentation of the application

Year of presentation	Applications (no.)	%	Surface area (ha)	%
1993	58	10.7	253.65	4.7
1994	72	13.3	720.35	13.5
1995	61	11.3	743.83	13.9
1996	147	27.2	1,833.41	34.3
1997	193	35.7	1,789.83	33.5
Data not available	9	1.7	7.82	0.1
Total	540	100.0	5,348.89	100.0

Source: authors' survey data

The participation level has increased progressively over time, from around fifty applications in 1993 to over 150 in 1997. However, as has been observed in other studies (Tempesta and Finco, 1997, Thiene, 1999), this increase has not been gradual, as applications almost tripled in 1996 (Table 1), and the land area involved has increased seven-fold in the last two years of financing. This trend could have been determined by many factors, including the limited information given out by the technical services during the early years of the agro-environmental measures, the lack of willingness of the farmers to accept the innovations deriving from the new common agricultural policy and, last but not least, the complicated bureaucracy involved with the necessary paperwork.

Of the farm owners who have adopted Reg. 2078/92, the largest age group is of individuals who are over 40 years old (almost 59.5%) (Table 2). This is fairly surprising when compared with the data on the average age of those active in the farming sector in the Population Census (ISTAT, 1990). In the age group under 35, that represents 25% of those in farming in the Vicenza Province, only 17% have adopted the agro-environmental measures [1]. It would normally be expected that the individuals with a greater interest in new agricultural policy instruments in the environmental sector would be younger. They should, in fact, be more open to innovation and have a greater ability for implementing and using new financing to obtain alternative forms of income.

Half of the applications for financing have been for measure A1/A2 for the reduction in chemical fertilizers and/or pesticides or maintaining previous reductions (Table 3). Measure D1-A is next (21.9%) in terms of popularity, and has generally used for planting of hedges and tree rows. Only 14.8% of the recipients applied for the measure relating to the rearing of local species in danger of extinction (D2). Interest in organic farming methods (A3) has been fairly low (8.3%), and there has been even less interest in the other measures.

---

[1] Comparison should be made with some caution, as the data relating to the age of those engaged in farming refer to 1991.

Table 2. Farmers who have adopted Reg. 2078/92 in the Province of Vicenza by age group

Age groups	Farmers who have adopted Reg.2078/92		Active in farming in province of Vicenza		Adoption rate
	No.(A)	%	No.(B)	%	% (A/B)
< 24	13	2.7	974	7.5	1.3
25-34	68	14.3	2,244	17.4	3.0
35-44	112	23.5	2,246	17.4	5.0
Over 45	284	59.5	7,437	57.6	3.8
Sub-total	477	100.0	12,901	100.0	3.7
Unknown	63	11.7			
Total	540		12,901		4.2

Source: authors' survey data

Table 3. Applications made and surface area accepted for financing by type of measure

Measure	Applications (no.)	%	UBA (cattle units)	Surface area (ha)	%
A1-A2	289	53.5	0	5,034.6	94.1
A3	45	8.3	0	230.6	4.3
B1	1	0.2	0	2.5	0.0
C1	1	0.2	102	0.0	0.0
D1-A	118	21.9	0	68.6	1.3
D2	80	14.8	1,278	1.8	0.0
E1	6	1.1	0	10.9	0.2
Total	540	100.0	1,380	5,348.9	100.0

Source: authors' survey data

Preference for each measure has not been constant over the five years. Although measure A1/A2 has been the one used most on average, in the first year of implementing the agro-environmental measures more than 40% of the applications were for measure D2, followed by the biological control methods with 27%, while reducing chemical use was less than 20%. Measure A1/A2 increased in the subsequent years, reaching 80% of all applications in 1995. Although this share has declined in recent years, it has remained above 50%.

The rearing of species in danger of extinction has registered a net decrease, from 40% to around 5%, and interest in organic farming methods has also declined. Only measure D1-A has increased constantly over the five years, mainly for the planting of hedges, tree rows and stands.

Adoption of the measures provided for in Reg. 2078/92 differ on the basis of age of the applicant (Figure 1); the younger farmers show a clear preference for measures A1/A2 and A3, which cover organic farming methods or the reduction in chemical inputs, both pesticides and fertilizers. As age increases interest turns to other measures, in particular, attention is concentrated on the possibility of rearing local species in danger of extinction and on planting hedges and tree rows or making other improvements to the landscape-environment. The young tend to show a greater sensitivity and openness to more innovative farming practices and forms of farm management, and are generally more flexible.

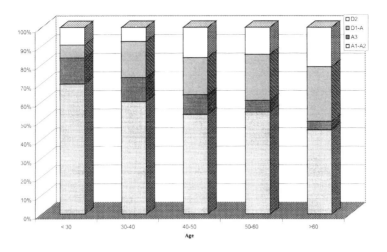

Figure 1. Applications made according to type of measure and age of farmer

It seems that the older farmers, if given the choice between different measures, tend to prefer crop practices or production choices more in line with past traditions, such as planting unproductive hedges and tree rows or rearing endangered species.

Almost 70% of the land being financed is concentrated in homogeneous territorial area no.1, that corresponds to the mountain areas, where adoption has almost exclusively consisted of measure A1/A2. On the plain, attention shifts to other measures, such as the adoption of biological pest control (A3) and the planting of hedges and tree rows (D1-A).

In total, a considerable amount in subsidies has been paid out, equal to more than 3,260,000 ITL. This amount does not, however, appear to have a homogeneous distribution over the territory (Table 4). Forty-seven percent of the total financing, over one and a half billion, has been allocated to the mountain areas, and applied almost exclusively to measure A1/A2 in the meadows or pastures. The technical rules of Reg. 2078/92 for measure A1/A2 in mountain areas provides for a simplified nitrogen balance, that must consider estimates of both nitrogen removal after mowing and of nitrogen supply by animal manure. Chemical nitrogen fertilization is forbidden.

It is of great economic benefit for the farmers to adopt this measure, since the subsidy they receive per hectare is fairly high (around 525,000 ITL /ha for meadows and pastures and 233,000 ITL/ha for alpine pasture), and the practices imposed by the technical rules are not very different from standard farming methods.

Measure A3 is mainly concentrated at the foot of the mountains and the hills of the Colli Berici, where the communes with a tradition for vine-growing are located. There is, in fact, a clear link between the adoption of measure A3 and intensive crops.

Table 4.   Subsidies paid out through measures and territorial areas.

| Measure | Homogeneous Territorial areas | | | | | |
	1	2	3	4	5	Total
A1-A2	1,444,257,900	200,776,500	104,835,200	385,922,300	282,241,900	2,418,033,800
	(94.0)	(66.2)	(78.5)	(60.4)	(42.9)	(74.0)
A3	31,231,500	12,937,600	17,812,700	72,616,200	73,385,800	207,983,800
	(2.0)	(4.3)	(13.3)	(11.4)	(11.2)	(6.4)
B1					715,100	715,100
	(0.0)	(0.0)	(0.0)	(0.0)	(0.1)	(0.0)
C1					24,226,600	24,226,600
	(0.0)	(0.0)	(0.0)	(0.0)	(3.7)	(0.7)
D1-A	2,364,500	22,842,700	9,276,500	73,320,600	187,862,300	295,666,600
	(0.2)	(7.5)	(6.9)	(11.5)	(28.6)	(9.0)
D2	57,979,800	64,763,400	1,669,200	106,283,300	86,725,100	313,420,800
	(3.8)	(21.4)	(1.2)	(16.6)	(13.2)	(9.7)
E1		2,002,000		405,200	2,637,500	5,044,700
	(0.0)	(0.7)	(0.0)	(0.1)	(0.4)	(0.2)
Total	1,535,833,700	303,322,200	133,593,600	638,547,600	657,794,300	3,269,091,400
	(100.0)	(100.0)	(100.0	(100.0)	(100.0)	(100.0)
%	47.0	9.3	4.1	19.5	20.1	100.0

Source: IRA data.

3.   FARMERS'ATTITUDES TOWARDS E.U. REG.2078/92

In the survey of farmers, 60 questionnaire responses provided information on the characteristics of the farms and motivations for not adhering to Reg. 2078/92.

Only 10% of the farmers surveyed declared that they had adopted Reg. 2078/92. Of these, only one farmer had chosen measure A3, while all the others had preferred measure D1-A. These are mainly livestock farms, usually with one young full-time person working.

The reasons given by the remaining respondents for not adopting the Regulation was (Figure 2), in 20% of cases, excessive bureaucracy (particularly in the medium-big farms), lack of interest (18.5%), and lack of information (16.7%, especially in the small farms). Among the other reasons given are the paltry amounts of the subsidy and the imposition of constraints, while the excessively high costs of adherence and lack of requisites were less common.

To try to more fully understand the factors that in some way affected adhesion to Reg. 2078/92 on the part of the farms surveyed, a stepwise logistical regression analysis was used to determine the probability of adhesion to the measures of the regulation (Table 5).

The probability that a farm would adhere to Reg. 2078/92 depends on the use of contractors for harvesting operations, on the percentage of land used for orchards and permanent meadows and, lastly, in an inverse proportion, on the age of the farmer. This means that if young people manage a farm there is a higher probability of adopting more innovative farming practices and a more flexible farm management approach.

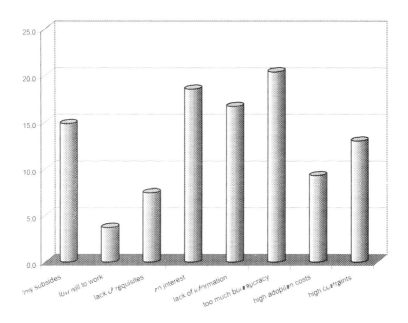

Figure 2. Reasons for not adhering to Reg.2078/92

Table 5.   Interpretative model of the probability of a farmer adopting the measures of Reg. 2078/92

$Pr = 1/1(1+exp(17.49-Y))$
With $Y = +18.123X_1+0.5909X_2+0.0479X_3-0.1132X_4$

Where:
- Pr   =   probability of having adhered to a Reg.2078/92 measure
- X_1 =   dummy use of contractors for harvesting operations
- X_2 =   percentage of farm land growing orchards on the commune orchard surface area.
- X_3 =   percentage of farm land with permanent meadow on the commune permanent meadow surface area.
- X_4 =   age of the farm owner.

Chi-square   =   22.327
Significance =   0.0002
% of replies correctly interpreted by the model = 93.33

Table 6 reports some possible scenarios referring to the probability of adopting Reg. 2078/92, varying the parameters selected by the model. First, the farms that do not use contractors for harvesting, essentially large production units without livestock, have

Table 6. Possible scenarios referring to the probability of taking up Reg. 2078/92, by varying the parameters selected by the model

Probability of adoption (%)		Use of contractors for harvesting	% of agricultural land planted as orchards	% of agricultural land planted as meadows
Age of the farmer				
30 years	65 years			
0.00	0.00	0	0	0
5.94	0.12	1	0	0
93.99	22.93	0	40	0
100.00	100.00	0	60	0
0.00	0.00	0	0	60
7.76	0.16	1	0	60

shown no propensity for adopting the agro-environmental measures. Likewise, the presence of orchards plays an extremely important role on the plain; on farms with more than 60% of the agricultural land planted as orchards, the probability of adoption is 100%. It is also interesting that the probability would, in this case, be clearly affected by the age of the farm manager. For example, the probability that a farm with 40% of the agricultural land used as orchards would adhere is 22.9% if the manager is 69 years old, but rises to 93% if he is 30. The presence of meadows seems to play a fairly minor role, as the probability increases from 5.9% on a farm without meadows to 7.6% where this is the only crop.

4. CONCLUSION

It can be assumed that the measures included in Reg. 2078/92 can, at least in theory, play an important role in improving the environmental quality of the territory. This appears to be even truer in the areas on the plain where modern production methods have impoverished the landscape and environment.

However, not all the measures have received equal attention from the farmers; there has been a small amount of interest in some measures and others have not been considered at all. Most applications have been for measures related to the reduction and/or maintenance of the use of chemical fertilizers and pesticides, and there has been reasonable interest in measure D1-A for planting hedges and trees.

Seventy percent of the land receiving subsidies is concentrated in the mountain areas. Paradoxically, the major part of the subsidies paid out has been in areas with fewer problems of environmental upgrading and where pollution was low. The remainder of the territory has therefore benefited in a marginal way from the positive effects coming from adhesion to the regulation (Franceschetti and Provoli, 1999).

On the plain, the interventions are concentrated at the foot of the mountains and on the hills they are concentrated in the meadows and pasture, both areas relatively intact from the environmental point of view, compared to the others.

Crops that already have reduced chemical inputs (e.g. integrated pest control in vineyards or orchards) can gain a notable advantage by adherence to the regulation, given that it doesn't involve any lowering of crop yield against a reduction in production costs.

Obviously the fact that different parts of the territory are inherently suitable for the various crops means that the subsidies also tend to be lower in some areas in relation to others.

The survey of the farmers has demonstrated a generally modest implementation of the agro-environmental measures, as only 10% of those interviewed declared they had done so. Analysis of the reasons for not adhering has been interesting, the main ones being the excessive bureaucracy, the absence of interest and lack of information, reasons that represent an opportunity to step in to alter the farmers' opinion.

5. REFERENCES

Finco A. – Tempesta T. (1997), L'applicazione del Regolamento CEE n.2078/92 in provincia di Venezia, *Studi di Economia e Diritto*, **1-2**: 104-138.

Franceschetti G. – Provoli A. (1998), Il Regolamento CEE n.2078/92 verso una nuova ruralità, *Genio Rurale*, **12**: 13-24.

INEA (1997), *L'applicazione del regolamento (CEE) n.2078/92 in Italia. Campagna 1996.* Osservatorio sulle Politiche Strutturali.

ISTAT (1990), *Censimento dell'Agricoltura.*

Regolamento CEE n.2078/92 Metodi di produzione agricola compatibili con le esigenze di protezione dell'ambiente e con la salvaguardia dello spazio rurale, Gazzetta Ufficiale delle Comunità Europee, 30 luglio 1992.

Tempesta T. (1997), Aree ambientalmente sensibili e misure di accompagnamento della PAC: un'analisi nelle provincie di Padova e Venezia, in: *XXXIV Convegno SIDEA*, Trevisan G. (ed.), Torino 18-19 Settembre, pp. 315-326.

Thiene M. (1999), Le misure di accompagnamento della PAC e i fondi strutturali in un'area ambientalmente sensibile: il Parco regionale dei Colli Euganei, *Agricoltura delle Venezie*, **5-6**: 53-72 and **7**: 47-72.

APPLICATION OF AGRO-ENVIRONMENTAL PROGRAMS IN EMILIA-ROMAGNA REGION

Guido Maria Bazzani, Alessandro Ragazzoni, and Davide Viaggi *

SUMMARY

The recent regulation 1257/99/EC provides for the funding of new agro-environmental measures, representing the continuation of those introduced by the 1992 CAP reform. Past application of agro-environmental policies raised a large debate about the best way of carrying out the interventions. The definition of payment levels and the setting up of suitable monitoring procedures are two major issues on which a large degree of disagreement still remains.

The objective of this study is the measurement of actual costs for implementing agro-environmental measures and the setting up and testing of a model for the quantification of optimal parameters for agro-environmental policies.

The model is tested on a specific data set related to the application of non-productive measures included in regulation 2078/92/EC and continued in regulation 1257/99/EC in Emilia-Romagna (Italy).

The results show large area for improving the efficiency of application. The work shows that quantifying the payment on the basis of the average compliance cost is not an obvious solution, and, on the contrary, it is normally unable to satisfy efficiency objectives. Instead, many instruments can improve efficiency. In particular, instruments in which payments are conditioned on environmental effects or are proportional to environmental improvements allow important savings for the public administration, through the concentration of participation on farms offering the cheapest environmental improvement. Among such instruments, the use of auctions for contracts appears to be particularly promising for future improvement of policy mechanisms.

* Guido Maria Bazzani, CNR Land and Agri-System Management Research Centre (Ge-STA), Bologna. Alessandro Ragazzoni and Davide Viaggi, Department of Agricultural Economics and Engineering, University of Bologna. The paper is a common result of the work of the authors. In particular, G. M. Bazzani has written section 3, A. Ragazzoni has written section 2 and D. Viaggi section 4 and 5. The introduction and the conclusions are a common work of the authors. The paper arises from research funded by the Emilia-Romagna Region and co-ordinated by C.S.A., Bologna.

Economic Studies on Food, Agriculture, and the Environment
Edited by Canavari *et al.*, Kluwer Academic/Plenum Publishers, 2002

339

1. INTRODUCTION

In March of 1999, the EU countries reached an agreement on the reform of the Common Agricultural Policy (CAP) presented in the document called Agenda 2000. The following chapter arises from what happened in Berlin and proposes a few reflections about the new CAP measures along with the potential effects on farm choices, with specific attention to rural development and agro-environmental measures.

The recent regulation 1257/99/EC provides for the funding of new agro-environmental measures, representing the continuation of those introduced by the 1992 CAP reform. Past application of agro-environmental policies raised a large debate about the best way of carrying out the interventions. The definition of payment levels and the setting up of suitable monitoring procedures are two major issues on which a large degree of disagreement still remains.

The objective of this study is the measurement of actual costs for implementing agro-environmental measures and the setting up and testing of a model for the improvement of policy performance through the use of auctions for contracts. In relation to such an innovative instrument, optimal parameters for agro-environmental policies are quantified.

The model is tested on a specific data set related to the application of non-productive measures included in regulation 2078/92/EC and continued in regulation 1257/99/EC in Emilia-Romagna (Italy).

2. THE APPLICATION OF AGRO-ENVIRONMENTAL MEASURES IN ITALY

2.1. Legislative aspects

The legislation concerns rural development, involving new general lines designed to be a coherent and supportable policy for the future rural European zones. A great number of agricultural functions have been recognized, including the combination of several important measures of support for rural zones, in particular the following three:

- **reinforcing agricultural and forest sectors**: among the principal measures there are those aimed at the modernization of farms and at the transformation and commercialization of quality agricultural products (for example, labeling Protected Designation of Origin, Protected Geographical Indication, and Specification Awards). There are financial interventions both for young entering farmers and for the advanced retirement of the entrepreneurs. The essential role of forestry in rural development has been recognized;
- improving the **capacity to compete of rural zones**: the principal aim of these measures is to find alternative sources of income and occupation for the farmers who live in particular rural zones, by favoring new activities, also including the services;
- **protecting natural environment and rural European patrimony**: the agro-environmental measures seek to promote methods of production compatible with environment. They represent a further recognition of the multi-functional role of agriculture.

Each member state of the EU has to follow an unique regulation to realize his own operative programs relative to rural development; this regulation gathers together several norms that characterized the 1992's Reform. The scopes of interventions are indicated in the Regulation 1257/99/EC of the 17 May, "on the support to rural development from the European Agricultural Guidance and Guarantee Fund (EAGGF)". The operative indications are in the Regulation 1750/99/EC of the 23 July, "dispositions of application of the Reg. 1257/99/EC of the Council, on support to the rural development".

For this purpose it is important to underline the pre-eminent function that the single Regions have in carrying out the regulations. In fact, it is their task to predispose the **Plan of rural development (PSR-Piano di sviluppo rurale)**, in adhesion to the mentioned EU regulations; within six months from presentation of the PSR, the EU has to release the approval judgement and, if that's positive, the Regions must become operative.

The reform of the rural development policy has been realized following some principal directions:

1) the simplification of instruments; the Regulation 1257/99/EC comprehends and co-ordinates the interventions that, from 1992, are inserted in distinguished regulations. Inside the Regulation, there are nine reasons for intervention:

- investments in agricultural holdings
- setting up of young farmers
- training
- early retirement
- less-favored areas and areas with environmental restrictions
- agro–environment
- improving processing and marketing of agricultural products
- forestry
- promoting the adaptation and development of rural areas

2) the reinforcement of the principles of subsidiarity and decentralization of decisions and management; the principal novelties introduced by regulations are:

- the European Commission gets a more centralized role in defining intervention's strategies and the objectives of plans and programs. In contrast, the formulation and the management of programs are more and more decentralized at territorial level. In the Italian case this involves an increased responsibility for regions;
- the scope of the decision making power of the Committees of surveillance for programs is modified, the European Commission assumes a more consultative role;
- the principle that programming and management of the EU interventions are founded on a strict arrangement between the European Commission, member state, regional/local authorities and economical and social parts;

3) the reinforcement of the instruments of control and evaluation is necessary for a greater efficiency and efficacy of expenses; it is opportune for the EU to provide greater control to increase the achievement of the objective because of the decentralizing in the programming and management of the interventions.

The concept of checking and evaluation is not only connected to the effective realization on the farms of the measures indicated by the PSR; it is also connected to the achievement, at the territorial level, of the proposed objective. For example, if an agro-environmental measure indicates, among the general objective, a 20% decrease of nitrogen distributed, it is essential to verify the congruence with the actual operations.

The European Commission, by the Regulation 1257/99/EC, seeks to unify within a unique juridical picture all the instruments of intervention used until now by the structural policy. The list of instruments can be more properly placed in three great categories:

- GROUP A: modernization measures
- GROUP B: farm and economic diversification measures
- GROUP C: accompanying measures to sustain incomes

For group A, the modernization measures include:

- investments in agricultural holdings (art. 4-7);
- setting up of young farmers (art. 8);
- training (art. 9);
- improving processing and marketing of agricultural products (art. 25-28).

These are instruments that were already included in the Regulations 950/97/EC and 951/97/EC. The investments in farms and in the enterprises of transformation and commercialization will only get the EU subsidies if they satisfy the following requisites:

- demonstration of returns;
- respect of minimum standards in the **theme** of environment, hygiene and animals welfare;
- existence of "normal outlets" in **markets** for interested products.

The public participation in the investments, in the final text of the Regulation, has been modulated, indicating a maximum share of financing in percentage terms on the total expense.

The group B of measures, relative to the farm and economical diversification, is totally included in the article 33 of Regulation.

These kinds of interventions are more directly aimed to promote, in rural areas, the integration between agriculture and the other activities. This is supposed to have a potentially larger impact on the local economy.

In relation to the potential beneficiaries, article 33 affirms that the subsidies can be awarded for the activities carried out in rural areas and not for agricultural activities only.

In the group C, the measures of accompaniment and the subsidies for disadvantaged zones include:

- early retirement (art. 10-12);
- less favored areas and areas with environmental restrictions (art. 13-21);
- agro-environment (art. 22-24);
- forestry (art. 29-32).

An element that joins all these measures is the financial competence of the EAGGF-Guarantee section in all the European Union regions, independently from the objective zones instituted by the structural policy. For the agro-environmental measures, the Regulation extends the objectives, including some other innovative activities such as:

- the management and **maintenance** of pasture systems at low intensity;

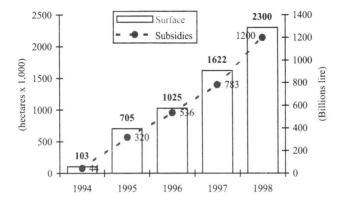

Figure 1 - Trend of the adoption to the Regulation 2078/92/EC in Italy

- the preservation of the **traditional** characteristic of the rural landscape;
- the covering of the fields with stubble crops for green manuring;
- the costs provided for the environmental planning in farm management (for example, rules ISO 14000).

The Emilia-Romagna Region proposed in November 1999, its own Regional Plan of Rural Development (PRSR), realizing it was in conformity with three principal sections of intervention:

- Priority section 1 ⇨ Sustain the farms capacity to compete
- Priority section 2 ⇨ Subsection 2a: *Agro-environment*
  Subsection 2b: *Environment and forestry*
- Priority section 3 ⇨ Local integrated development

Each priority section contains the measures of intervention indicated by the Regulation 1257/99/EC and the rules of application of the Regulation 1750/99/EC. For the aims of the research we will concentrate on the environmental section and, in particular, we will describe the non-productive measures that can be realized in the farms. When going over the principal agro-environmental activities that are more important for farmers of Emilian plain, it is interesting to evaluate whether there is continuity with the past and, in particular, with the Regulations 2078/92/EC and 2080/92/EC.

2.2. Applied aspects

The growing interest in the agro-environmental programs in the last period of adoption has been encouraging. We noticed a remarkable acceleration in the last year's adoption (Figure 1); in Italy, the surface area covered went from about 100,000 hectares in

Table 1 - Regional adoption to the Regulation 2078/92/EC in 1997

Area	Cultivated surface area (hectares)	Regional program (hectares)	Incidence on the total cultivated area	Adherence 1997 (hectares)	Incidence on the total cultivated area	Incidence on the Regional Program
Piemonte	1,099,684	299,913	27.3%	297,517	27.1%	99.2%
Valle d'Aosta	92,023	27,245	29.6%	42,967	46.7%	157.7%
Lombardia	1,082,247	232,593	21.5%	118,645	11.0%	51.0%
Bolzano	260,475	76,849	29.5%	157,143	60.3%	204.5%
Trento	139,325	94,156	67.6%	49,243	35.3%	52.3%
Veneto	870,948	103,600	11.9%	51,865	6.0%	50.1%
Friuli	252,288	39,060	15.5%	19,515	7.7%	50.0%
Liguria	75,505	6,495	8.6%	1,917	2.5%	29.5%
Emilia-Romagna	1,201,672	162,080	13.5%	72,881	6.1%	45.0%
Toscana	913,362	40,807	4.5%	197,055	21.6%	482.9%
Umbria	399,050	20,760	5.2%	31,786	8.0%	153.1%
Marche	528,530	90,055	17.0%	30,127	5.7%	33.5%
Lazio	781,618	217,370	27.8%	87,026	11.1%	40.0%
Abruzzo	491,709	45,830	9.3%	2,471	0.5%	5.4%
Molise	237,389	3,713	1.6%	6,736	2.8%	181.4%
Campania	612,497	103,491	16.9%	1,780	0.3%	1.7%
Abruzzo	1,402,776	51,548	3.7%	66,819	4.8%	129.6%
Basilicata	582,673	49,158	8.4%	80,939	13.9%	164.7%
Calabria	623,404	30,415	4.9%	23,606	3.8%	77.6%
Sicilia	1,525,000	70,298	4.6%	160,700	10.5%	228.6%
Sardegna	1,336,344	62,058	4.6%	120,865	9.0%	194.8%
TOTAL	**14,508,519**	**1,827,494**	**12.6%**	**1,621,603**	**11.2%**	**88.7%**

Source: Povellato A., Lo Piparo G. (1999), "Cresce l'interesse in Italia per le misure agro - ambientali", Informatore Agrario, n. 1.

1994 to about 2.3 million hectares in 1998; the distribution of subsidies increased from 44 billion to 1,200 billion ITL.

How did the farmers of the single regions behave?

The adoption is very differentiated both in absolute value and in relation to the dimension of agro-environmental programs evaluated at the beginning of the projects. In most northern Regions programmed, for the first years of application, an adoption had to cover more than the 20% of the cultivable area available. In the middle and southern areas, except in rare cases, the agro-environmental programs were drafted for a surface smaller than 10% of the total cultivable area.

The behavior of the farmers did not match the expectations of the local Administrations (Table 1). In nine Regions, the surface that was engaged in actions of the Regulation exceeded the availability of programs. In some cases it reached limit values, such as in province of Bolzano (204%), in Valle d'Aosta (157%), in Toscana (482%), in Molise (181%), in Basilicata (164%), in Sicily (228%) and in Sardinia (194%). In other six Regions they only reached 40-50 % of the available surface for the agro-environmental engagements. Finally, in two Regions (Abruzzo and Campania) the adoption rate was interested at best 5% of the regional program.

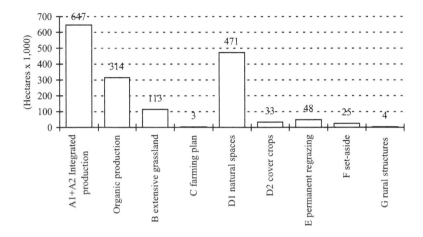

Figure 2 - Adherence to the Regulation 2078/92/EC in Italy: division for engagements

In Italy the agro-environmental programs of the Regulations 2078/92/EC was real-ized in 12.6% of the cultivable area and the adoption reached 88.7% of the initial pro-grams, that is to say 11.6% of the total cultivable area.

The division of adherence forms for every single engagement demonstrated that the main interest has been turned to the introduction of the integrated (action A1 and A2) and organic (action A3 and A4) techniques of cultivation. The success of these measures can be explained by the diffusion, in the recent past, of the integrated pest control in the main agricultural northern regions and by the realization of public and private assistance serv-ices for the management of this innovative technique. It is very significant to notice that the surface addressed to the action D1, which requires the realization of natural areas (hedges, thickets, etc.) on the farm, is less then actions A1+A2 and more then the actions A3+A4 for a total of about 470,000 hectares (Figure 2). It is possible to notice a sensible delay in the adoption of measures in favor of the recreative activity (action G) that in most regions has not been effected. A typical case is offered by Umbria. This region realized, alone, 96% of projects for the public use of the farming areas.

The obtained results are alluring for the last 2 years of adoption.

The effort both by the public administration and the farmers for caring and guardian-ship of territory is surely a patrimony that should not remain isolated. It should be prom-ulgated and it should have an incentive for the future. We must consolidate the matured experience, correcting the eventual distortions. The directives pursued by the European Commission should give an incentive to the interventions on the farms for the introduc-tion of environmental compatible techniques and for farm improvement. This is already made clear by the well-known preliminary proposals of Agenda 2000.

3.   AUCTIONS AND AGRO-ENVIRONMENTAL CONTRACTS

The development of markets for public goods in the countryside is limited by the characteristics of goods involved and by the structure of the markets themselves. Among such characteristics, it is worth mentioning the market structure and the kind of competition, the uncertainty about the quality of products, the uncertainty about the value of the products, and the information asymmetry that can translate into problems of both hidden action or morality (Hanley, 1991).

For this reason, some kind of regulation is considered necessary in order to make the market for public goods in the countryside work. Nevertheless, all policy instruments show some shortcoming. In particular, the flat rate subsidy, widely adopted in Europe and Italy, may be considered the less efficient instrument, while new instruments are emerging in the literature as more efficient alternatives (Fraser, 1993; Gallerani et al. 1999; Richard and Trommeter, 1994, Torquati, 1998; Viaggi, 1996).

Among alternative instruments, very promising tools are green auctions. The interest for auctions depends on the conditions in which they are applied. Compared with a flat rate payment, they yield better results when (Lactaz-Lohmann e Van der Hamsvoort, 1998):

- the information basis is weak;
- the number of potential participants is large;
- the contracts offered are homogeneous;
- the farms are heterogeneous;
- the production of environmental goods or services is separable among farms.

There are basically two ways in which auctions can be used for agro-environmental policies (Lactaz-Lohmann e Van der Hamsvoort, 1998):

- government procurement auctions;
- auctions of certificates.

In the case of government procurement auctions, the commodity being traded is the public good. The farmer submits a bid indicating the amount of money he would ask for implementing the management agreement. The public body is the buyer and the farmer is the seller. In the case of auction certificates, the public body sells the right to obtain pre-determined financial rewards and the farmers make a commitment to provide public goods in exchange.

While the former model is adopted for the Conservation Reserve Program (USA), the latter is used in UK for the Countryside Stewardship Scheme.

Auctions can be classified in four kinds (Rasmusen, 1989):

- english;
- first-price sealed bid;
- second price sealed bid;
- dutch.

The english auction is the most classical one. It is an ascending bidding process in which all bids are known by other possible bidders (it is a kind of "visible" auction) and the higher bid wins. In the case of the dutch auctions the process is still visible, but prices are called from high to low, with the first bidder taking the prize.

In the case of sealed bid auctions, the bidders do not know the offer that other bidders will do. For first price sealed bid, the higher price wins and the bidder pays the price he offered. For second price-sealed bid, the higher price still wins, but the amount paid is the one offered by the second higher bidder.

The kind of auction most common in the literature about public goods in the countryside is the first price sealed bid auction.

The auction is theoretically able to increase the efficiency of the selection process, as it introduces competition between producers so as to act as a cost revelation mechanism. At the same time, it reduces the uncertainty about the value of the commodity being traded, as it is the better informed party who makes the first move (Lactaz-Lohmann e Van der Hamsvoort, 1998; Rasmusen, 1989).

Such advantages translate into the fact that the public body auctioning contracts can involve more land with the same money, or reduce the total cost for the same land involved.

The use of auctions fits well in the general trend towards a "value for money" approach, leading to more efficient ways of implementing agro-environmental schemes. Bidding is also perceived as fair, which is important from a political point of view (Lactaz-Lohmann e Van der Hamsvoort, 1998).

However, the auction revelation mechanism is not perfect, as a certain rent for farms still remains, dependent on how the decision making process is organized.

The main limitations to the instrument of auction, are the following (Lactaz-Lohmann e Van der Hamsvoort, 1998):

-   bidding with a common value element: if a common value element emerges, as it would through learning by farmers over repeated auctions, the result tends to approximate the higher cap for all bidders, hence pushing the result towards the flat aid option. The auction of certificates solves this problem;
-   transaction costs: it is certain that transaction costs exist and it is likely that they are higher for auctions then for other policy instruments. There can be identified ex ante costs (information, making of the contract, etc.) as well as ex post costs (monitoring, etc.). An auction can be administratively more difficult than other instruments, or, at least, the risk due to bad administration can be higher;
-   spatial targeting: since auctions better apply to a wide number of participants, it does not allow a sufficient targeting over the territory.

4.   THE MODEL

The model applied in this paper is based on the approach developed by Lactaz-Lohmann e Van der Hamsvoort (1997), with some adaptation to pluriannual policy measures, such as creation of wetlands.

The model assumes that the farmers hold private information about the profit from agricultural activity on their own farms, both under conventional and conservation technology, denoted as Π_0 e Π_1 respectively. The profit is defined as the net per hectare return, excluding agro-environmental payments.

In the case of wetlands creation, the soil destination will change completely. If, as it often happens, the new soil use does not yield any profit, Π_1 will be a negative number given by the compliance cost, including both investment costs (soil movements, tree plantation, etc.) and maintenance (water regulation, etc.). On the other side, Π_0 will be the full previous profit, which acts in the model as an opportunity cost.

Given the characteristics of the intervention, both Π_1 and Π_0 are to be interpreted as the accumulation of profits and costs over the years, discounted at a certain discount rate r:

$$\Pi_0 = \sum_{i=0}^{n} \Pi_{0i} \cdot \frac{1}{(1+r)^i} \text{ and } \Pi_1 = \sum_{i=0}^{n} \Pi_{1i} \cdot \frac{1}{(1+r)^i} \tag{1a, 1b}$$

If the farmer submits a bid (b) that is accepted, his utility will be:

$$U(\Pi_1 + b)$$

where $U(\cdot)$ represents a Von Neumann-Morgenstern utility function, monotonic, increasing and doubly differentiable (Kreps, 1990). The bid b, in this case, is to be interpreted as the present value of payments to be made over the whole period of n years in which the soil is dedicated to wetlands. If the bid is rejected, the farmer will get:

$$U(\Pi_0)$$

equal to his reservation utility.

It is further assumed that the bidding strategy by the farmer is driven by the maximum acceptable payment β, over which no offer will be accepted. Such payment represents the reserve price and is not known to the farmer. The farmer will submit his bid if the expected utility in case of acceptance is higher than his reservation utility:

$$U(\Pi_1 + b)P(b \leq \beta) + U(\Pi_0)[1 - P(b \leq \beta)] > U(\Pi_0) \tag{2}$$

where P is the probability of acceptance.

It is plausible to assume that each farmer will form some expectation on β, that can be characterized by the density function $f(b)$ and by the distribution function $F(b)$. The probability that an offer is accepted is given by:

$$P(b \leq \beta) = \int_{b}^{\overline{\beta}} f(b)db = 1 - F(b) \tag{3}$$

where $\overline{\beta}$ represents the upper limit of farmer expectations about the maximum acceptable offer.

Substituting 3 in 2 we have:

$$U(\Pi_1 + b)[1 - F(b)] + U(\Pi_0)F(b) > U(\Pi_0) \qquad (4)$$

The decision is determined by the trade off between payoffs and the probability of acceptance. A higher bid increases the net profit, but reduces the probability of acceptance. The farmer's problem is hence that of determining the optimal bid, according to the maximization of his utility, above the reservation utility.

Assuming that the farmers are risk neutral and simply maximize their net payoffs, equation 3 can be transformed in:

$$(\Pi_1 + b - \Pi_0)[1 - F(b)] > 0 \qquad (5)$$

Maximizing equation 4 with respect to b, it is possible to obtain the optimal bid as:

$$b_{opt}^* = \Pi_0 - \Pi_1 + \frac{[1 - F(b)]}{f(b)} \qquad (6)$$

Assuming a rectangular distribution of acceptance probability between $\underline{\beta}$ and $\overline{\beta}$, the following distribution of probability is obtained:

$$f(b) = \left| \begin{array}{lll} 0 & if & b < \underline{\beta} \\ \dfrac{1}{\overline{\beta} - \underline{\beta}} & if & \underline{\beta} \leq b \leq \overline{\beta} \\ 0 & if & b > \overline{\beta} \end{array} \right.$$

$$F(b) = \left| \begin{array}{lll} 0 & if & b < \underline{\beta} \\ \dfrac{b - \underline{\beta}}{\overline{\beta} - \underline{\beta}} & if & \underline{\beta} \leq b \leq \overline{\beta} \\ 0 & if & b > \overline{\beta} \end{array} \right.$$

It can be shown that, in this case, the optimal bid is determined as follows:

$$b_{rn}^* = \max\left\{\frac{\Pi_0 - \Pi_1 + \overline{\beta}}{2}, \underline{\beta}\right\} \tag{7}$$

$$\text{s.t. } b_{rn}^* > \Pi_0 - \Pi_1$$

Hence, the optimal offer for a risk neutral farmer is a growing function of the opportunity cost and of the maximum bid cap. From this formulation some limits of the revelation mechanism are quite evident. For example, even farmers with no participation costs will produce a positive bid, equal to $1/2\,\overline{\beta}$, or, at least $\underline{\beta}$, so behaving as free riders.

5. FIRST RESULTS

As a first empirical testing, the model is applied to the case of wetlands in the Municipality of Mirandola (Modena, Italy), described in section 2. The potential participating area is given by the total arable land cultivated with crops different from vegetables or orchards, as it is assumed that the land used for such cultivation will never be converted into wetlands.

The distribution of compliance costs has been calculated on the basis of available data about average profit foregone, participation costs, and their variance. In order to evaluate these parameters over all of the area, in absence of farm-by-farm estimations, a normal distribution has been assumed. The budget has been assumed at 3.5 Million Euro, corresponding to the amount necessary to expand the area involved up to 5% of total agricultural land of the Commune at the present (Reg. 2078/92/EC) cost of payments. This is quite a large objective compared with present participation.

Table 2 shows the comparison between the best result obtainable with auctions and the best result produced by a flat rate payment. For the auction, this result relates to the case in which $\overline{\beta}$ is assumed equal to 11,052 Euro and $\underline{\beta}$ is assumed equal to 4,107 Euro/ha.

As expected, auctions provide a reduction of rent for farmers and an increase of the area involved given the same budget. Assuming farmers to be risk averse could further improve the result. The change in the area involved is about 3% of the total, while the change in rents is about 44%. The latter number is mostly due to the small amount of rents, that are only about 7% of the total amount paid, which is a very low sum if compared with other experiences in the application of reg. 2078/92/EC (Caggiati et al., 1997).

Such an outcome is clearly a function of the assumptions made. Two of them are particularly important. First of all, the outcome depends on the distribution of expectations of farmers, driven by the level of $\overline{\beta}$ and $\underline{\beta}$, together with the functional form adopted. On the other side, the results of the comparison between a flat rate payment and an auction are a function of the total budget, as increasing the area involved means in-

Table 2 - Comparison between auctions and the results of the flat rate payment

	Flat rate	**Auction RN**
Surface involved	320.63	330.85
Total Payments	3,512,662	3,512,662
Total Compliance costs	3,256,688	3,368,720
Total Rents for farmers	255,974	143,942

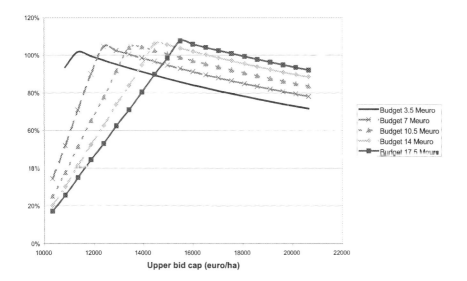

Figure 3. Land involved in the reg. 1257/99/EC with auction as a percentage of flat rate payment

creasing the variability of compliance costs, hence affecting the comparative efficiency of the more or less market-like instruments.

Figure 3 shows the sensitivity of the results to both these variables. The budget is assumed to increase up to 5 times the previous hypothesis. This is in any case well above any possible budget for this kind of measure, but allows a better understanding of the effect of wider areas on the result of the instruments under analysis. $\overline{\beta}$ is assumed to change from 10 to 25 thousand euro, while holding $\underline{\beta}$ constant.

First of all, it is necessary to pay attention to the function corresponding to budget of 3.5 Meuro (main assumption). Note how narrow the space for good results is. In fact expectation of maximum bid cap falls in an area 5-6% wide around the cap, yield the best bid. The participating area will be higher compared with flat rate payments (100%). In all other cases, auction results will be worse, even sharply worse if the expected bid cap is far away from the optimal point.

Unfortunately for the public administration, expectations cannot be fully controlled and the fact that the result is so dependent on such parameters suggests that auctions are not too reliable after all, at least at this territorial scale.

Increasing the budget, hence increasing the area that can be involved, causes a movement towards higher optimal expectation about bid caps and a widening of the range of optimal bid caps around the best one, that, in the extreme case, is about 20% of optimal bid cap size. This means that increasing the size of the territory over which auctions are carried out, also improves the results. This is also true because larger areas and a higher number of farmers reduce opportunities for collusion in setting bids.

In the case of budget equal to 17.5 Meuro, the best results obtainable through auctions allow an increase in area of more than 10% compared with a flat rate payment.

6. DISCUSSION

As the literature suggests, auctions can improve significantly the efficiency of agro-environmental measures. This result is confirmed by some empirical evidence arising from an application in Emilia-Romagna. Nevertheless, results also induce some caution in the interpretation of the improvement in Agro-Environmental Policy obtainable through the use of auctions.

Some reasons are quite straightforward for the results of this paper. If the variability of the area involved is not sufficiently high, the information of the public decision maker about the actual costs of each farm is not too bad, and the expectations of farmers about bid caps are not very good, the result of auctions can be worse than that of a flat rate payment. Even when conditions are met for the auction to be an efficient way of improving the environment, the dependence on farmers' expectation makes the instrument not too reliable.

Wider considerations can reinforce caution. In particular, more issues relating to implementability can be considered, as the administrative culture and organization in Italy is quite different from that of countries in which auctions have been developed. Their use in Italy can be seen as unfair and difficult to manage. Moreover, implementing auctions can require more administrative costs compared to flat rate payments. Thus its feasibility is conditioned to the efficiency of the administrative system, which is not always very satisfying in Italy.

If ever applied, the learning process by farmers will quickly require adaptation. This is particularly true when the costs are comparatively homogeneous as in the tested area.

The use of a sale of certificates would be an interesting alternative, but it would require more work by the farmer in doing the project and likely more work for the public administration in selecting them on the basis of a number of possible criteria.

Even with all these drawbacks, auctions may be able to solve many problems in the fixation of payments where the variability of profits and costs is much higher and when compliance costs are not known. This is a structural situation in agro-environmental policies, as promoted technologies change over time and their effects and response by farmers are not well known in advance. So, for example, a two-stage policy, with temporary auctions then switching to a flat aid, may be an interesting way for some application and for new research.

7. REFERENCES

Caggiati P. – Gallerani V. – Viaggi D. – Zanni G. (1997), *La valutazione delle politiche agro*-ambientali. *Un'applicazione di contabilità ambientale al reg. (CEE) 2078/92 in Emilia-Romagna*, CNR, Bologna.

Fraser I. (1993), *Agri-environmental policy and Discretionary incentive mechanism: the countryside stewardship scheme as a case study*, The Manchester Metropolitan University, Department of Economics and Economic History.

Gallerani V. – Viaggi D. – Zanni G. (1999), Il monitoraggio delle politiche agroambientali con lo strumento della contabilità ambientale, in: *XXXVI Convegno di Studi SIDEA "La competitività dei sistemi agricoli italiani"*, Milano, 9-11 settembre 1999

Hanley N. (ed., 1991), *Farming and the countryside - An economic analysis of external costs and benefits*, CAB international, Wallingford.

Kreps D.M. (1990), *A course in microeconomic theory*, Harvester Wheatsheaf, New York.

Latacz-Lohmann U. – Van der Hamsvoort C. (1997), Auctioning conservation contracts: a theoretical analysis and an application, *American Journal of Agricultural Economics*, **79**: 407-418.

Latacz-Lohmann U. – Van der Hamsvoort C. (1998), Auctions as a means of creating a market for public goods from agriculture, *Journal of Agricultural Economics*, **49**: 334-345.

Rasmusen E. (1989), *Games and information*, Blackwells, Oxford.

Reichelderfer K. – Boggess W.G. (1988), Government decision making and program performance: the case of the conservation reserve program, *American journal of agricultural economics*, **70**: 1-11.

Richard A. – Trommeter M., (1994), La rationalisation des contrats entre pouvoirs publics et agriculteurs: le cas des mesures agri-environnementales, in: Aubert D. (ed.), *Réformer la politique agricole commune: l'apport de la recherche économique*, Actes et Communicationes, 12, INRA, Versailles, pp. 307-323.

Torquati B. (1998), Applicazione e valutazione delle misure agroambientali; un approccio di contrattazione mediante il modello principale-agente, in: *XXXV Convegno SIDEA: L'Agricoltura italiana alle soglie del XXI secolo*, Palermo.

Viaggi D. (1996): *L'utilisation du modèle principal-agent pour l'analyse de l'application du règlement CEE 2078/92*, Final dissertation for the European Master in Agricultural Economics and Agribusiness, Rennes.

A DECISION SUPPORT SYSTEM FOR THE ECONOMIC-ENVIRONMENTAL ASSESSMENT OF CROP PRODUCTION

Guido Maria Bazzani, Paolo Caggiati, and Carlo Pirazzoli [*]

SUMMARY

The "Crop Economic Analyzer Model" (CEAM) is a decision support system (DSS) for the economic-environmental assessment of agricultural crops, designed to answer both public and private needs. The software attempts to reconcile the demand for an eco-compatible agriculture with the farmers' need to obtain an adequate return on their investment.

CEAM operates at the crop level, taking into account all the activities required by the production cycle; this permits evaluators to analytically quantify the utilization of raw materials, labor and machinery and their cost.

The program, which operates as a Microsoft Windows® application, consists of three main modules: problem definition, calculation, and report generation. The user interface is highly friendly. The system supports the creation and storage of personalized archives pertaining to specific farm conditions and analyzed scenarios.

CEAM can be used at the farm level as a decision support tool for technicians and farmers, to estimate ex ante profitability and conformity to regulations, and to evaluate *ex post* compliance with standards, consumption of resources, and economic return on the basis of real data.

It should be emphasized that for some practices, like fertilization, it performs as an expert system, through the internal database on agricultural technologies and regulatory constraints, organized into a comprehensive evaluation process.

At the public level, if adopted by the local authorities for the extension service, it permits the collection of homogeneous data on agricultural practices.

Flexibility coupled with simplicity of use, make CEAM a powerful tool for the guidance and monitoring of agricultural activities.

[*] Guido Maria Bazzani and Paolo Caggiati, CNR Land and Agri-System Management Research Centre (Ge-STA), Bologna. Carlo Pirazzoli, Department of Agricultural Economics and Engineering, University of Bologna.

1. INTRODUCTION

The Crop Economic Analyzer Model (CEAM) is a decision support system (DSS) for the economic-environmental assessment of agricultural crops (Figure 1). The DSS offers support for agricultural activities designed to answer both public and private needs; in particular, it attempts to reconcile the demands of eco-compatible agriculture, according to the most recent agricultural policy guidelines, with the farmers' need to obtain an adequate return on their investment (Jennings and Wattam, 1994).

CEAM is a crop system evaluation model which can be used at the farm level as a decision support tool for technicians and farmers. This puts at their disposal the existing knowledge base with regard to agricultural technology, regulatory constraints and the current costs and revenues, including incentives, while at the same time organizing this information into a comprehensive evaluation process.

The support system described here can be used for different purposes and at different times (Webster, 1990):

at the farm level:

- *ex ante*, to estimate profitability and conformity to regulations;
- during the cultivation phase to estimate returns and conformity to regulations;
- *ex post*, to evaluate compliance with existing standards, consumption of resources, and the economic return on the basis of known data resulting from the decisions effectively made by the farmer;

at the policy-making level:

- to simulate, *ex ante*, the economic impact of new agro-environmental measures;
- to evaluate, *ex post*, the results of the policies adopted on the basis of real data.

2. METHODOLOGICAL ASPECTS

The DSS assists agricultural operators in the pursuit of quality standards which impose restrictions and constraints on the production process, in conformity with the regulations adopted at the local level to implement the agro-environmental policies in force (Castellini, Palmieri, and Pirazzoli, 2000).

The analysis is conducted for each individual activity, and takes into consideration all the farm field operations required by the crop production cycle; this makes it possible to analytically quantify the utilization of raw materials, manpower and machinery (Pirazzoli et al., 1999). The system supports multiple technical and economic evaluations. In particular, the following are highlighted: utilization of resources, crop production parameters, costs and revenues. The first two groups of data are especially useful for conducting a series of technical analyses aimed at verifying compliance with production regulations. The third group permits an assessment of profitability on the basis of the market prices and anticipated subsidies.

A particularly significant feature of the program is its ability to function as an expert system. The definition of certain operations by the user is implemented as a guided procedure based on an internal database, which is compiled beforehand and contains all the

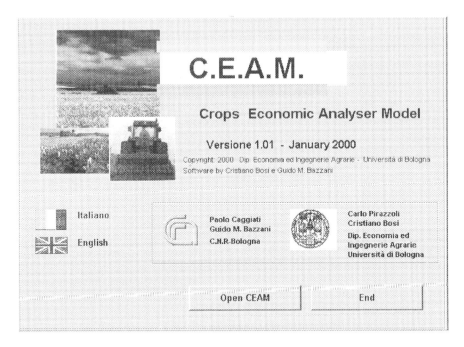

Figure 1. CEAM - Opening screen

information necessary for identifying agricultural methods compatible with the production regulations in force. One of the first operations to be supported was the definition of the fertilizing plan; its environmental impact made this a priority. Operationally, the guided choice of the plan takes place through an interaction between parameters entered by the operator, which characterize the situation in question, and parameters contained in the internal database, on the basis of "if–then" and "do case" constructs. The exogenous parameters used in this case include: soil type defined on the basis of pH, texture, presence of organic matter, preceding crop; and anticipated yields, related to the seeding/transplant method. The interaction of these parameters with the internal database enables CEAM to output the recommended fertilization level, and to verify that the choices made comply with the regulations in force. Similar procedures are planned for other operations such as pest control, weed control and irrigation, which are of particular importance in eco-compatible agriculture.

Concerning the economic analysis of the crop, the program operates on three different levels:

- "direct cost" takes into consideration only those costs directly attributable to the crop and effectively sustained, such as fuel, fertilizers, herbicides, pesticides, water, rental costs and outside labour.

- "full cost" also takes into consideration those shared costs which are only partially attributable to the crop, such as the cost of maintenance, insurance and depreciation of machines used for more than one crop.
- "total cost" also takes into account theoretical costs, that is to say the remuneration of resources owned by the farmer, which do not entail a cash outlay but nonetheless represent a cost which can be quantified using the opportunity cost criterion.

By subtracting the three above mentioned costs from the crop revenue, it is possible to calculate three profitability estimates — defined respectively as the gross margin, net income proprietary and net income entity — which describe the ability of the crop to remunerate the production resources. This breakdown of the economic analysis permits a total evaluation of the necessary investment and its profitability, and provides useful information for farm decision-making.

3. THE SOFTWARE

The program operates as a Microsoft Windows® application. System requirements are a Pentium 100 Mhz processor or better, and at least 4 MB of RAM. Once installed, the program can be started by clicking its icon on the program bar, or in the various other ways provided for by the operating system.

The program appears in a single window (MDI) which contains all the forms of the application. The main menu adapts to the user's choices, making it possible to progressively configure the operations based on the previously entered parameters. The operator is thus assisted by the adaptive presentation of the various forms and options.

CEAM consists of three main modules: data entry (problem definition), calculation (model elaboration), and report generation (results) (Figure 2). The user interface is provided through a set of predefined forms for accessing an internal database on the regulations in force and the recommended agricultural methods for different plant-growth conditions. The system also supports the creation and storage of personalized archives, in Microsoft Access® file format, pertaining to specific farm conditions, and manpower and machinery availability. It is also possible to save the scenarios processed by operators (Figure 3) for the evaluation of alternative agricultural methods. The program output is provided both on-screen and saved as a Microsoft Excel® spreadsheet for subsequent use by the operators.

4. FUTURE PROSPECTS

The three principal characteristics of the program are: greatly facilitated data entry, thanks to guided forms which provide access to the databases and validate the choices made; flexibility, which permits a highly realistic representation of very diverse farm situations; and speed of simulation, which, together with the generation of a clear yet comprehensive program output, makes CEAM an extremely useful tool for repeated simulation of different situations. The above mentioned elements make CEAM

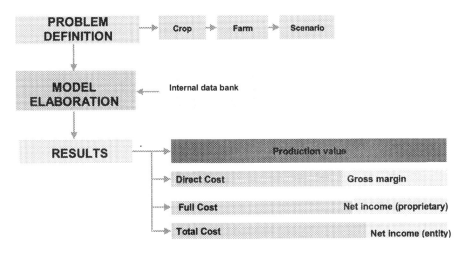

Figure 2. CEAM - Flow-chart

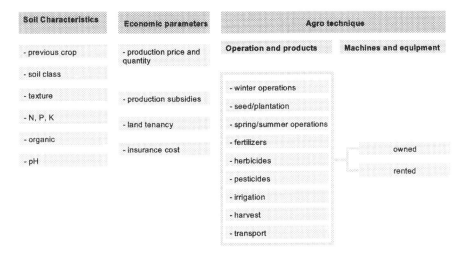

Figure 3. C.E.A.M. – Scenario definition

particularly suitable for evaluating the risks and uncertainties associated with crops and agricultural methods.

The need to set up and maintain up-to-date databases in a dynamic and evolving context requires constant effort on the part of public operators. The modular structure of the program supports the gradual extension of the DSS to new crops and production areas. This flexibility, coupled with its simplicity of use, make this — in our view — a powerful tool which can be adopted on a large scale, becoming a reference methodology for guiding and monitoring agricultural activities (Caggiati, 1997).

5. REFERENCES

Caggiati P. (1997), *L'evoluzione degli strumenti informatici per la gestione delle imprese agricole italiane*, Editura Universitatii A. I. Cuza, Iasi (Romania), pp. 26-38.

Castellini A. – Palmieri A. – Pirazzoli C. (2000), *Primi risultati sulla redditività del pomodoro da industria ottenuto con sistemi produttivi a ridotto impatto ambientale*, V Giornate Scientifiche SOI 2000, Sirmione, march 29.

Jennings D. – Wattam S. (1994), *Decision Making. An Integrated Approach*, Pitman Publishing, London.

Pirazzoli C. et al. (1999), *La peschicoltura nell'Unione Europea: comparazione economica tra i principali sistemi produttivi*, CSO, Ferrara.

Webster J.P.G. (1990), Reflections on the Economics of Decision Support Systems, in: *Proceedings of 3rd International Congress for Computer Technology*, May 27-30, Frankfurt, pp. 307-317.

TRAFFIC NOISE AND HOUSING VALUES
A Hedonic Approach

Giuliano Marella and Paolo Rosato [*]

SUMMARY

Noise is one of the most serious forms of pollution and one of the main causes of the deterioration of the quality of life in urban areas. This paper analyzes the effects of traffic noise on house prices in the historic center of Ferrara. The analysis was done by analyzing the distribution of house prices and noise pollution, using multiple regression. This study also demonstrates that the discomfort produced by noise is highly variable and that it increases more than proportionally to the noise produced. There is however a clear and significant negative relationship between the discomfort produced by noise and housing values. Average housing devaluation is 1,900 ITL per percentage point of people disturbed by noise per square meter of surface area. Noise pollution, therefore, clearly affects property values, which in turn legitimizes the recent protective norms for urban environment.

1. INTRODUCTION

Noise, an acoustic phenomenon which creates discomfort, has become one of the most serious forms of pollution in recent years. Among all the sources of noise, transport activities involve the greatest number of people. In the countries belonging to the Organization for Economic Co-Operation and Development (OECD), around 110 million people are exposed to a level of noise above the threshold considered acceptable within the norms of different countries (65 dB(A)), principally for traffic (Venezia, 1997). In Italy a

---

[*] Giuliano Marella, Department of Mechanics and Management Innovation (DIMEG), University of Padua. Paolo Rosato, Department of Civil Engineering, University of Trieste. The paper is the result of the collaboration between the authors. Paolo Rosato has written section 3.2 and Giuliano Marella has written sections 2 and 3.1. Both authors were involved in writing sections 1 and 4. The authors are grateful to Danilo Cazzaro for the precious suggestions on the relationships between noise and annoyance and to Elena Chiarelli for data collection.

Economic Studies on Food, Agriculture, and the Environment
Edited by Canavari *et al.*, Kluwer Academic/Plenum Publishers, 2002

361

survey conducted for 20 years in 39 Italians cities shows an average level of noise of 71 dB(A) during the day and 63.5 dB(A) during the night. [1]

The definition of acoustic pollution is very elusive because it depends not only on the characteristics of noise, but also on the type of anthropic reaction. In fact, acoustic damage caused by a noise level above 80 dB can be analyzed with audiometric exams and is widely documented in the literature. The extra auditory damage (annoyance, stress, depression, insomnia, less production in the job) caused by a lower level of noise is not well known and studied.

Nevertheless, there are several national and international studies that have done an economic evaluation of the social impact caused by urban traffic noise. These assessments were made with different methodologies, referring to the property depreciation (Furlan, 1998) or the willingness to pay of people to reduce the damaged caused by noise pollution (Colombino and Locatelli Biey, 1996).

This paper estimates the effect of noise pollution on residential property value. The work is divided into two sections. The first illustrates the technical problem of assessing noise pollution indicators. The second contains a synthetic description of the sample and the hedonic model for the economic assessment of noise impact. The subject of the study was the historic center of the city of Ferrara in the eastern Emilia-Romagna region (Italy).

2.   NOISE AND HOUSING VALUE

The definition and implementation of a hedonic model to assess the influences of traffic on property values requires that particular attention be paid to the choice of independent variables that represent noise. We can use two alternative approaches to the choice of variables: the first is linked to the technical measure of noise, the second is linked to the noise perception of the population. In the first case, the measurement of noise has been well defined by physicists, and the data collections promoted by public administrations for the fulfillment of regulations, permit us to map the noise pollution in some urban centers. Nevertheless the approach linked to noise perception by the population is more suitable for the assessment. In fact, the establishment of property prices is not influenced by the physical measure of the noise, but is influenced by the perception of noise by the people who are involved in the property market, both on the supply and demand sides. Therefore, it could be convenient to adopt as an explanatory variable in the model the percentage of people who declare to be disturbed by noise. This measure can be determined without direct data collection, starting from an equivalent noise level through a proper function.

As far as the technical measure is concerned, the acoustic sensation of a pure sound can be well determined from the pressure level of sound (measured in decibels -dB-), and its frequency. Through a normal Fletcher-Munson audiogram, we don't consider the pure sounds in measuring noise pollution but noise level. In this case it is not enough to know the intensity and the frequency of noise to obtain a corresponding sensation index. To assess the sensation derived from a noise it is necessary to use an instrument for measuring the level of sound pressure: the sound-level meter. It can weigh the noise signal using

[1] Data collections in the main Italian cities were carried out by Lega per l'Ambiente and Ferrovie dello Stato during the event «Treno Verde» in 1991.

acoustic filters according to three weighting curves (A, B or C). In urban noise studies, the measure of sonorous pressure levels is carried out according to the weighting curve A, which represents the most common approach to handle a sonorous signal and to correlate it with the reaction auditory apparatus (and, as a consequence, with the sonorous sensation).

Since urban noise is changeable in time, it's necessary to use a measure of energetic level linked to the variable of time to assess the noise damages and annoyance. The most widely used parameter is called continuous equivalent noise level (LEQ), measured in dB(A). If we have a variable sonorous signal, referring to a particular period of observation, LEQ is the sonorous pressure level of a constant signal, which corresponds to variable one, in the same period.

Continuous equivalent noise level was assumed as a parameter of reference from Italian regulation for acoustic pollution. The Prime Minister's Decree (d.p.c.m.) dated November 14, 1997, "Determination of limit values for sonorous emission" classifies the municipality in six different classes based on different uses of the land. Moreover, it determines the maximum values for the sonorous emissions in each class, using the measure of LEQ expressed in dB(A). This regulation distinguishes between daytime LEQ – measured between 6:00 a.m. and 10:00 p.m. – and night – measured as the rest of the day. The maximum values during the night are normally more restrictive than those in the day.

In the case of the urban center of Ferrara, the availability of data for acoustic pollution depends on a methodical data collection promoted by the Regional Environmental Protection Agency. Data collection began in 1991, after the emanation of the d.p.c.m. dated March 1, 1991, entitled "Maximum limits of noise in houses and environment". The data were collected during the day, with a sound-level meter, and were organized to individualize the LEQ in dB(A).

After the data collection we defined a function, which links the LEQ with the percentage of people disturbed by the noise level. Many social acoustic investigations were carried out in different countries beginning in the seventies. These investigations correlate the acoustic indexes with the human reaction in terms of annoyance level (Bertoni et al., 1988). Different families of functions related the LEQ measured with the sound-level meter with the percentage of people who declare they were disturbed by the noise level. Many functions present in the international literature can not be applied to the case of Ferrara because they are linked to sources of noise different from urban traffic (trains and aircraft), and because the noise level was measured with units different from daytime LEQ, as "day night level" (DNL).

The curve elaborated from the database of the city of Modena was very useful (Bertoni et al., 1994). It is the result of a survey of a wide sample of people (900 residents from the urban area). Modena has a real estate and town-planning design quite similar to Ferrara, with similar sources of noise. The function relating the percentage of people disturbed by noise level (Annoyance), with the independent variable represented by the acoustic daytime noise level, is measured in LEQ and expressed in dB(A). The relation that links the two variables is:

$$\text{Annoyance (\%)} = 0.13 \text{ dB(A)}^2 - 12.87 \text{ dB(A)} + 323.30$$

As far as the use of this formula in the hedonic model is concerned (Figure 1), the day time LEQ is more useful than the nighttime measure one for the interpretation of

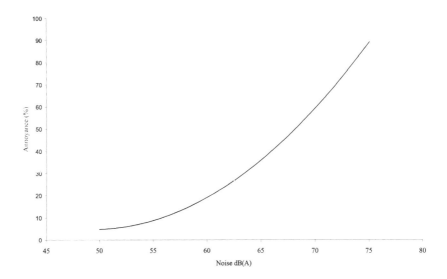

Figure 1. The relationship between noise and annoyance

property values. In fact, the establishment of the price is usually made during daytime visits to the property. Moreover, the data collections of noise pollution in Ferrara and in other cities are made during the day.

3.   THE CASE STUDY: FERRARA HISTORICAL CENTER

Noise economic impact has been evaluated in a part of the historical center of Ferrara. The area analyzed makes up a continuous and homogeneous urban structure and is characterized by a condominium composed of a numbers of units. Within this rather loosely knit urban fabric there are numerous architectural–historical elements that add value to the real estate capital and allow Ferrara to be defined as a city on a "human scale".

A survey was conducted in 1998 through a random sampling of direct interviews with real estate professionals. The 88 transactions examined refer to properties well distributed in the area examined. A chart was compiled for each property describing the characteristics that could have a potential influence on its value.

3.1.   The data

The most important qualitative and quantitative sample characteristics are reported in Tables 1 and 2. The sample was subjected to statistical analyses to define its characteristics and to ascertain the possibility of using multiple regression analyses. The average surface area of the residences is 134 sq. m with a fairly wide dispersion around the mean

Table 1.   Quantitative sample characteristics

	Min	Max	Average	D. S.	Skewness	Kurtosis
Surface (sq. m)	38	340	134.27	54.80	1.279	2.506
Price (ITL/sq. m)	1,869,231	2,721,600	2,230,860	203,468	.436	-.644
Noise (LEQ)	52.00	74.00	62.7727	5.1231	.086	-.415
Annoyance (%)	5.58	82.80	31.0423	18.4517	.914	.301

Table 2.   Qualitative sample characteristics

	State	Frequency	Percentage
Building with good fixtures	0	69	78.4
	1	19	21.6
	Total	88	100.0
Apartment with good fixtures	0	51	58.0
	1	37	42.0
	Total	88	100.0
Well-preserved apartment	0	47	53.4
	1	41	46.6
	Total	88	100.0
Independent heating system	0	21	23.9
	1	67	76.1
	Total	88	100.0
Air conditioning system	0	67	76.1
	1	21	23.9
	Total	88	100.0
Floor level without elevator	0	53	60.2
	1	20	22.7
	2	12	13.6
	3	2	2.3
	4	1	1.1
	Total	88	100.0

(D.S. 54.8). The distribution is slightly asymmetric (Skew.: 1.279) with a concentration in the extreme values a little superior to the normal distribution (Kur.: 2.506).

The sample of property prices have little variation. The average price per sq. m is 2.2 million ITL with a standard deviation equal to 0.2. The distribution is similar to the normal one, with a value that is acceptable in the skewness index (0.4) and in that of kurtosis (-0.6).

The average noise (LEQ) in the areas surrounding the residences is 63 dB(A) with a modest dispersion around the mean (D.S. 5.1). The distribution is fairly normal. Likewise, the percentage of people disturbed by noise shows a normal distribution around the mean of 31%.

The state of housing preservation is good given that 47% of the stock is well maintained. Indeed, 42% of the housing reveals high quality fixtures. The percentage with good quality fixtures decreases for the building containing the apartment (22%).

Regarding technical equipment, 76% of the apartments have an independent heating system. The percentage is less when considering air-conditioning systems (24%). Generally, buildings are equipped with an elevator and the percentage of apartments located above the first floor without an elevator is low (17%).

3.2.   The model

The assessment of the impact of noise on property value was determined through multiple regression. Widely used in economic analyses, this technique is particularly efficient in investigating and quantifying the dependency relationships between the value of the goods and their characteristics (Rosen, 1974). Numerous authors have illustrated its potential in studying real estate markets. In the Italian literature, the first contributions involved the study of residential property values (Simonotti, 1988; Simonotti, 1991; Curto and Simonotti, 1994; Del Giudice, 1994; Micelli, 1998).

Multiple regression analysis permits the building of a linear model such as:

$$Y = \beta_0 + \beta_1 X_1 + \beta_2 X_2 + ...\beta_n X_n + e$$

where:

Y   = property value;
X_n = variables influencing value;
β_0 = constant: net of the effect induced by the variable X_n;
β_n = coefficients that measure how the value varies with respect to the average variation of the variable X_n;
e = stochastic variable representing error.

The model is defined by identifying the variables (X_n) and the coefficients (β_n). [2]

The statistical properties of the model are illustrated in Table 3; the coefficients of the linear model are reported in Table 4.

The identified model is interesting given that it includes all the variables currently thought to influence value in the local real estate market. Moreover, with respect to the number of variables present in the model, the coefficient of determination R^2 takes on an acceptable value (0.81). Furthermore, all of the coefficients of the independent variables show an high level of significance and present a relative value that is consistent with the opinion of market professionals.

The model reported in Table 4 shows the incidence of every single variable on the determination of price per square meter of housing. The reference point is established by the constant (2.1 million per sq. m) that estimates the property value of an apartment in a condominium, situated in a quiet environment (noise equal to 52 dB), with mediocre preservation and economical fixtures. In addition, it is equipped with an elevator, if located above the first floor, but without an air conditioning system.

The impact of noise on property value is clearly negative. Average housing devaluation is 1,900 ITL per percentage point of people disturbed by noise per square meter of

---

[2] The coefficients were estimated through the technique of weighted least squares (WLS). The weighting variable was the apartment surface.

Table 3.	Characteristics of the regression model
Parameter	Value
Multiple R	.911
R^2	.830
R^2 Adjusted	.816
Standard Error	87,419

Table 4. The MRA model

	B	Std. error	Beta	t	Sig.
Constant	2,064,156.305	29,299.787		70.450	.000
Annoyance (%)	-1,905.299	581.515	-.173	-3.276	.002
Building with good fixtures	90,076.777	27,316.104	.183	3.298	.001
Apartment with good fixtures	260,223.213	32,436.401	.635	8.023	.000
Apartment with good preservation	72,418.350	28,068.251	.179	2.580	.012
Air conditioning system	129,038.031	24,579.496	.272	5.250	.000
Indipendent heating system	62,639.740	24,329.643	.132	2.575	.012
Floor level without elevator	-24,830.990	11,072.496	-.109	-2.243	.028

Dependent variable: Price per sq. m

surface area. This means that going from a quiet environment [50-55 dB(A)] to one in the proximity of a busy road [70-75 dB(A)], housing values decrease by around 6%. Traffic pollution, summarized by the percentage of people disturbed by traffic noise, takes on a significant role in the formation of property values. Figure 2 shows the trend of the value per sq. m of an apartment, with economical fixtures and preservation, with respect to the noise level. The noise effect is always negative but the impact is contained between 50 and 60 dB(A) before increasing decisively over 65 dB(A).

As far as intrinsic characteristics are concerned, the quality of the fixtures is highly appreciable. The role of the state of preservation appears to be more contained, because buying often implies maintenance work. Finally, the model shows the significant role of technical equipment (elevators, air conditioning and heating systems).

4. CONCLUDING REMARKS

The paper has analyzed the impact of traffic noise on housing value in the historic center of Ferrara. The analysis was done by comparing the distribution of house prices and noise pollution, using multiple regression. The study has shown that noise can be considered a proxy of many forms of pollution produced by motor traffic and that it is nearly impossible to distinguish the effect of single components. This study also demonstrates that the discomfort produced by noise is highly variable and that it increases more than proportionally to the noise produced. There is however a clear and significant negative relationship between the discomfort produced by noise and housing values. Average housing devaluation is 1,900 ITL per percentage point of people disturbed by noise per square meter of surface area. This means that going from a quiet environment to one in the proximity of a busy road, housing values decrease by around 6%. Noise pollution, therefore, clearly affects property values, which in turn legitimizes the recent protective norms for urban environment.

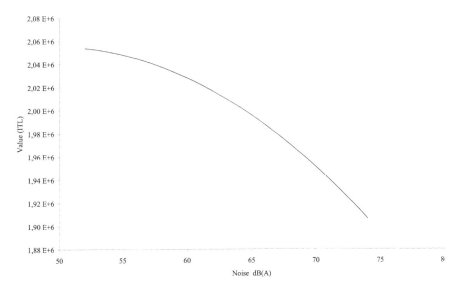

Figure 2.   The relationship between noise and housing value

The preceding evaluation, from an appraisal and planning perspective, should certainly be verified through more extensive investigation given that it derives from analyses done on a rather homogeneous sample of apartment types. Probably, modifications of building typology could lead to a substantial reduction of the role of noise (i.e. soundproofing). Nonetheless is possible to affirm that the real estate market is certainly sensitive to traffic pollution and that this sensitivity will be more accentuated over time. In other words, it reaffirms the concept that a negative externality, while maintaining a prevalently public aspect, becomes a partially private "bad" if, as in the case of a residential building, it produces a depreciation of the asset or an additional cost to prevent the damages.

5.   REFERENCES

Bertoni D. – Franchini A. – Magnoni M. (1988), *Il rumore urbano e l'organizzazione del territorio*, Pitagora, Bologna.
Bertoni D. et al. (1994), *Gli effetti del rumore dei sistemi di trasporto sulla popolazione*, Pitagora, Bologna.
Colombino U. – Locatelli Biey M. (1996), Valutazione contingente della disponibilità a pagare per una riduzione dei rumore, *Genio Rurale*, **1**: 11-17.
Curto R. (1993), *Qualità edilizia, qualità ambientale e mercato immobiliare:un'applicazione della Multiple Regression Analysis (MRA) al caso della città storica*, in: XIV Conferenza Italiana di Scienze Regionali, Bologna, 4-5 ottobre.
Curto R. (1994), *La quantificazione e costruzione di variabili qualitative stratificate nella Multiple Regression Analysis (MRA) applicata ai mercati immobiliari*, in: Atti del II° Simposio Italo Spagnolo Ce.S.E.T.- A.E.V.A., Centro Studi di Estimo e di Economia Territoriale, pp. 195-222.

Curto R. – Simonotti M. (1994), Una stima dei prezzi impliciti in un segmento del mercato immobiliare di Torino, *Genio Rurale*, **3**: 66-73.

Del Giudice V. (1994), Un modello di stima del peso dei caratteri immobiliari nella formazione del prezzo degli immobili, *Genio Rurale*, **5**: 21-27.

Furlan S. (1998), *External Costs in Urban Areas*, Proceedings of World Meeting of Environmental Economists, Venice, June 25-27.

Micelli E. (1998), Qualità urbana e valori immobiliari, *Genio Rurale*, **1**: 54-60.

Venezia E. (1997), Internalizzazione delle esternalità ambientali, in: *Esternalità e trasporti, Atti della IV Riunione Scientifica Annuale*, Torbianelli A.V., ed., Società Italiana degli Economisti dei Trasporti-ISTIEE, Trieste, pp. 373-400.

Rosen S. (1974), Hedonic Prices and Implicit Market::Product Differentiation in Pure Competition, *Journal of Political Economy*, **82**: 34-55.

Simonotti M. (1988), L'analisi di regressione nelle valutazioni immobiliari, *Studi di Economia e Diritto*, **3**: 369-401.

Simonotti M. (1991), Un'applicazione della multiple regression analysis nella stima di appartamenti, *Genio Rurale*, **2**: 9-15.